AF692439

Smart Wearable Technology: Thermal Management and Energy Applications

Smart Wearable Technology: Thermal Management and Energy Applications

Guest Editors

Zhanxiao Kang
Qing Chen

MDPI

Basel • Beijing • Wuhan • Barcelona • Belgrade • Novi Sad • Cluj • Manchester

Guest Editors
Zhanxiao Kang
School of Fashion and Textiles
The Hong Kong Polytechnic
University
Hong Kong

Qing Chen
Shanghai International
Fashion Innovation Center
Donghua University
Shanghai
China

Editorial Office
MDPI AG
Grosspeteranlage 5
4052 Basel, Switzerland

This is a reprint of the Special Issue, published open access by the journal *Processes* (ISSN 2227-9717), freely accessible at: www.mdpi.com/journal/processes/special_issues/1SKF7MT5FP.

For citation purposes, cite each article independently as indicated on the article page online and as indicated below:

Lastname, A.A.; Lastname, B.B. Article Title. *Journal Name* **Year**, *Volume Number*, Page Range.

ISBN 978-3-7258-3818-9 (Hbk)
ISBN 978-3-7258-3817-2 (PDF)
https://doi.org/10.3390/books978-3-7258-3817-2

© 2025 by the authors. Articles in this book are Open Access and distributed under the Creative Commons Attribution (CC BY) license. The book as a whole is distributed by MDPI under the terms and conditions of the Creative Commons Attribution-NonCommercial-NoDerivs (CC BY-NC-ND) license (https://creativecommons.org/licenses/by-nc-nd/4.0/).

Contents

About the Editors

Zhanxiao Kang

Dr. Zhanxiao KANG is currently a Research Assistant Professor at Hong Kong Polytechnic University. His general interests include heat and mass transfer, heat pipes, thermal diodes, computational fluid dynamics, and wearable technology for human body thermal management. Dr. Kang is actively engaged in projects related to flexible heat pipes, high-performance thermal rectifiers, sweat management, evaporative heat dissipation, and wearable air conditioners. He has published about 50 peer-reviewed papers in prestigious academic journals. Furthermore, Dr. Kang also serves as a reviewer for over 40 academic journals, including *Nature Communications*, *Advanced Materials*, *Advanced Science*, etc.

Qing Chen

Dr. Qing CHEN is an associate professor from Donghua University. Shen has been engaged in the design and development of knitted products for many years, mainly researching the design and development of functional and intelligent knitted fabrics and clothing. She has hosted one National Natural Science Foundation project, one Shanghai Natural Science Foundation project, and one Jiangsu Provincial Natural Science Foundation project, and participated in over 10 enterprise industry university research projects. She has published over 50 papers and 8 authorized invention patents, and received one second-place prize and one third-place prize from the China Business Federation Science and Technology Progress Award, as well as three second-place prizes and one third-place prize from the China Textile Industry Federation Science and Technology Award.

Preface

The rapid advancement of smart wearable technology is transforming how we interact with our environment, offering innovative solutions to enhance human comfort, safety, and performance. *Smart Wearable Technology: Thermal Management and Energy Applications* brings together cutting-edge research at the intersection of material science, energy efficiency, and intelligent design. This book highlights breakthroughs in wearable systems that address thermal regulation, energy sustainability, and real-time monitoring—critical areas for healthcare, industrial work, and everyday life.

As global challenges like climate change and energy demands grow, the need for adaptive, efficient wearable technology becomes urgent. This volume explores advanced textiles and devices that regulate body temperature, from cooling garments for heat stress to self-heating fabrics for extreme cold. It also examines energy-harvesting solutions, such as solar-integrated textiles and piezoelectric systems, which reduce reliance on traditional power sources.

A key focus is the integration of smart technologies like AI and IoT for real-time physiological monitoring and responsive fabric systems. Research on respiratory tracking, activity recognition, and defect detection in conductive textiles demonstrates the potential of data-driven wearables. Practical applications, such as cooling systems for healthcare PPE, showcase how these innovations improve safety and comfort in demanding environments.

This book serves as a valuable resource for researchers, engineers, and industry professionals, offering insights into both current advancements and future directions. We thank the contributors for their pioneering work and invite readers to explore the transformative potential of smart wearables.

Zhanxiao Kang and Qing Chen
Guest Editors

processes

MDPI

Review

Personal Wearable Thermal and Moisture Management Clothing: A Review on Its Recent Trends and Performance Evaluation Methods

Junming Zhou [1], Jinming Zhao [1], Xiaolei Guo [1], Yuxing Hu [1], Xiaofeng Niu [1,*] and Faming Wang [2]

1 College of Urban Construction, Nanjing Tech University, Nanjing 211816, China; junmingzhou@njtech.edu.cn (J.Z.); 202261224064@njtech.edu.cn (J.Z.); 202361224073@njtech.edu.cn (X.G.); 202161224021@njtech.edu.cn (Y.H.)
2 Division Animal and Human Health Engineering, Department of Biosystems (BIOSYST), KU Leuven, Kasteelpark Arenberg 30, BE-3001 Leuven, Belgium; faming.wang@kuleuven.be
* Correspondence: newton@njtech.edu.cn

Abstract: Personal wearable systems designed to manage temperature and moisture are gaining popularity due to their potential to enhance human thermal comfort, safety, and energy efficiency, particularly in light of climate change and energy shortages. This article presents the mechanisms of thermal and moisture management, recent advances in wearable systems for human thermal and moisture management, and methods for their performance evaluation. It evaluates the pros and cons of various systems. The study finds that most wearable systems for thermal and moisture management are being examined as individual topics. However, human heat and moisture management have noteworthy interactions and impacts on human thermal comfort. There are certain limitations in the methods used for evaluating personal heat and moisture management in wearable systems. This review suggests future research directions for wearable systems to advance this field and overcome these limitations.

Keywords: high-temperature and high-humidity environments; wearable clothing; thermal and moisture management; performance evaluation indicators

Citation: Zhou, J.; Zhao, J.; Guo, X.; Hu, Y.; Niu, X.; Wang, F. Personal Wearable Thermal and Moisture Management Clothing: A Review on Its Recent Trends and Performance Evaluation Methods. *Processes* **2023**, *11*, 3063. https://doi.org/10.3390/pr11113063

Academic Editors: Zhanxiao Kang, Qing Chen and Blaž Likozar

Received: 6 October 2023
Revised: 19 October 2023
Accepted: 24 October 2023
Published: 25 October 2023

Copyright: © 2023 by the authors. Licensee MDPI, Basel, Switzerland. This article is an open access article distributed under the terms and conditions of the Creative Commons Attribution (CC BY) license (https://creativecommons.org/licenses/by/4.0/).

1. Introduction

Personnel working in high-temperature and high-humidity environments, such as outdoor medical personnel, workers in mining, metallurgy, and other industries who need to wear heat-protective clothing, as well as firefighters, are not equipped with air conditioning and other temperature-control devices due to the limitations of the workplace. In addition, these personnel are also required to carry out high-intensity labor, which is prone to generating excessive heat and then forming a high-temperature and high-humidity microenvironment, resulting in uncompensated heat stress [1]. Physiological studies indicate that high-temperature environments can decrease human concentration and execution, while also triggering a sense of anxiety, which can significantly impact work efficiency. Additionally, if body temperature exceeds 39 °C, individuals are at risk of a heatstroke, and at a temperature of 40.6 °C, it can be life-threatening, as demonstrated in Figure 1 [2]. Therefore, maintaining a constant human body temperature not only ensures work efficiency but is also an important prerequisite for good health. In the face of the growing demand for personal cooling/dehumidification, the concept of personal thermal and moisture management wearable systems has been proposed [3]. Personal thermal and moisture management wearable systems are one of the common solutions to protect the body from heat stress, especially for those who work in hot and humid areas, which is considered to be an effective personal protection measure [4].

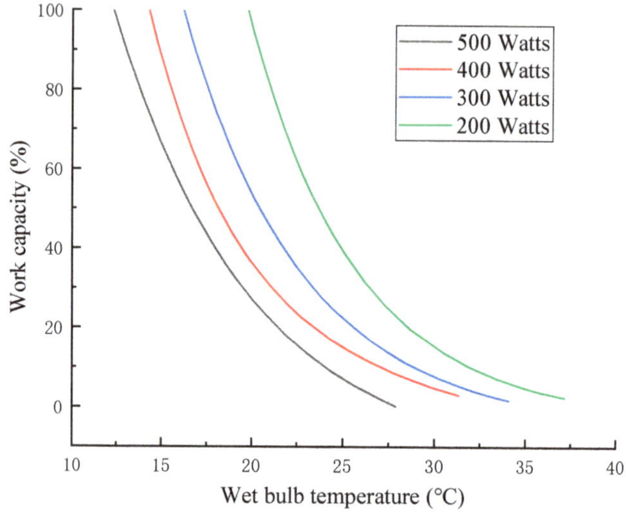

Figure 1. Association between work capacity and wet bulb temperature for different work intensities.

The human body maintains its thermal equilibrium through the transfer of heat and sweat, which can occur via four primary pathways: thermal conduction, thermal convection, thermal radiation, and evaporation, as illustrated in Figure 2 [5]. Many personal thermal and moisture management wearable systems have been designed for human temperature and humidity regulation [6]. These devices can improve thermal comfort for the human body and provide better temperature and humidity control in hot and humid environments. These mechanisms play a crucial role in regulating body temperature and ensuring optimal physiological functioning. The development of these wearable systems represents a significant advancement in the field of personal thermal management [7]. By optimizing heat and moisture transfer, these devices can help individuals maintain a comfortable body temperature and mitigate the risk of heat-related ailments. Additionally, they have the potential to enhance overall well-being and performance in various settings, including sports, occupational environments, and daily activities [8,9].

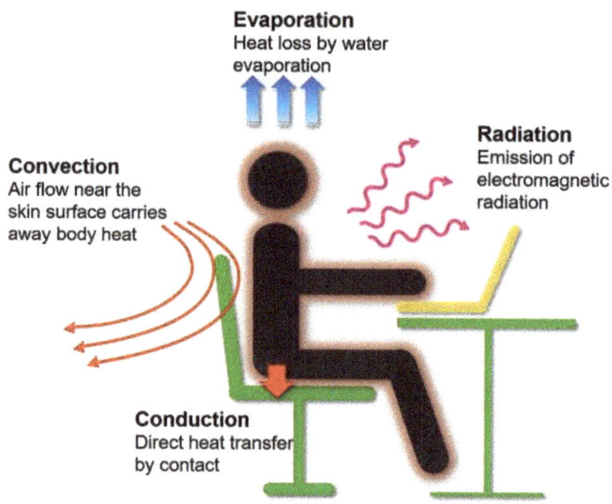

Figure 2. Personal thermal and moisture management [5].

In this paper, we explore the design, functionality, and performance of these personal thermal and moisture management wearable systems. We examine their potential benefits and limitations, discuss the underlying principles of heat and moisture transfer, and present recent advancements in this rapidly evolving field. Through a comprehensive analysis, we aim to shed light on the significance of these wearable technologies and their potential applications in improving human thermal comfort and well-being.

2. Thermal and Moisture Management Mechanism

The human–clothing–environment heat and moisture transfer mechanism includes both heat and moisture transfer. When talking about human heat and moisture management, clothing plays an important role in this process.

2.1. Thermal Transfer Mechanism

Heat transfer from the human body surface to the environment is a complex and variable process involving the interaction of several factors. Among them, the air layer between the skin and the garment, the thermal conductivity properties of the garment material, and the heat transfer from the air layer between the garment and the environment are the key elements. The modes of heat transfer include conduction, convection, radiation, and evaporative heat dissipation, which interact and work together to influence the heat exchange process between the body and the environment.

Clothing plays a crucial role in the heat exchange between the human body and the environment, and it can provide thermal insulation and heat preservation. In order to better understand the heat transfer performance of garments, researchers have abstracted and idealized the heat transfer process of garments, thus establishing garment heat transfer models, as shown in Figure 3. However, these models do not fully reflect the situation of real human beings wearing garments, because the heat exchange between the human body and the environment is very complex in real situations.

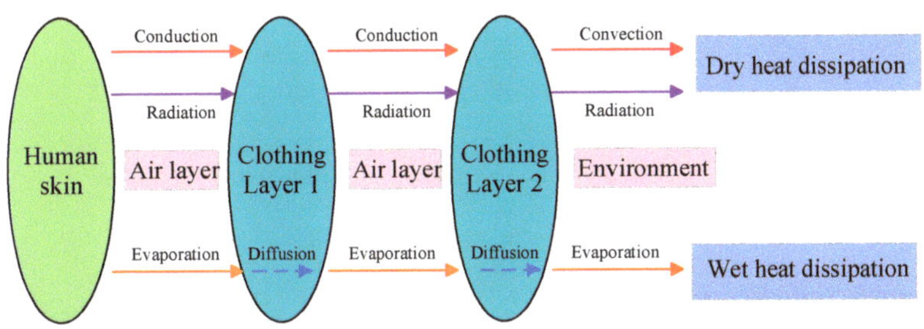

Figure 3. Thermal transfer mechanism of clothing.

When the human body wears looser clothing, is in a state of motion, or is in an environment with high airflow velocities, there is not only conduction and radiation but also convective heat dissipation between the human body and the clothing and the air layer between the clothing and the clothing. In addition, convective heat dissipation also occurs between the air layer and the environment, and when the garment material is fluffy, radiative heat dissipation also occurs within the garment. Therefore, in order to evaluate the heat transfer performance of garments more comprehensively, researchers have proposed the concept of garment thermal resistance [10]. Garment thermal resistance integrates several factors, including the heat transfer performance of the garment material, the heat transfer performance of the air layer between the body and the garment, the fit of the garment to the body, and the flow of the air layer within the garment. By measuring the size of the thermal resistance of a garment, we were able to evaluate the heat transfer performance of the garment more comprehensively [11]. It is worth noting that current cooling suits can put the human body in a comfortable temperature range, but there still is a risk of condensation, putting the human skin in a wet environment; thus, humidity control is particularly important.

2.2. Moisture Transfer Mechanism

Water exists on the surface of human skin in two forms: gaseous and liquid. When water spreads through a garment into the environment, the process is known as the moisture permeability of the garment. Moisture permeability is the diffusion of water from the surface of the human body into the environment through both the openings of the garment and the fabric of the garment. Figure 4 shows a schematic diagram of the moisture permeation of a garment, which includes the pathways of moisture transfer.

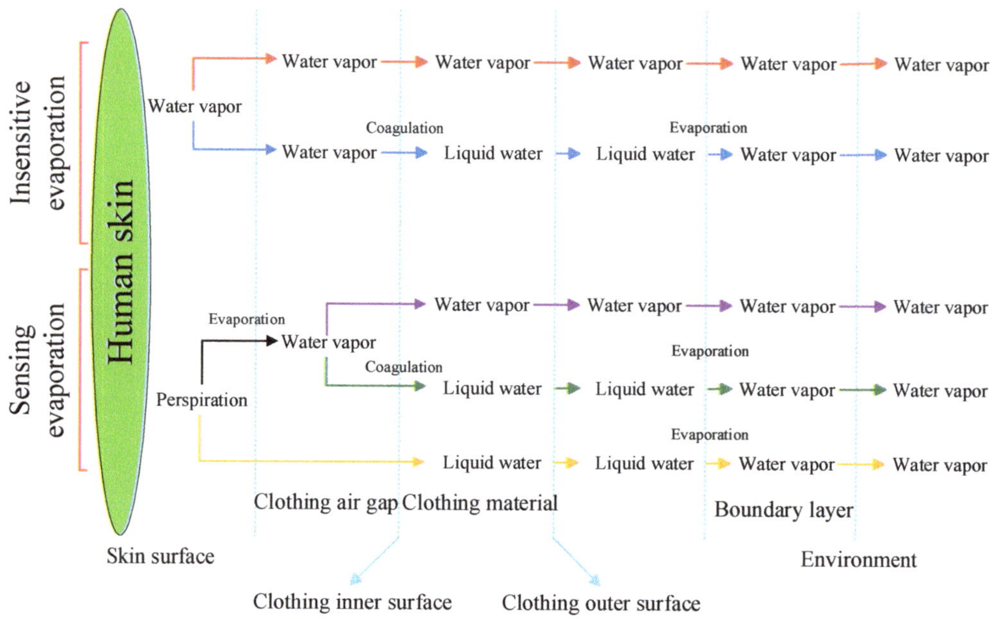

Figure 4. Moisture transfer mechanism of clothing.

In order to assess the moisture permeability of garments, commonly used indicators are the garment moisture resistance and the moisture permeability index [12]. The garment moisture resistance is the degree to which a garment impedes the diffusion of water, i.e., the ability of the garment material to block water. The moisture permeability index is a measure of the moisture permeability of a garment, which takes into account the moisture resistance of the garment and the rate of diffusion of water vapor. The higher the moisture permeability index, the better the moisture permeability of the garment and the more effective the transfer of moisture from the surface of the human body to the environment.

By studying the moisture-wicking properties of garments, we can design and select garments that are more appropriate for different environmental conditions and activity levels. In addition, garments with good moisture-wicking properties can help the human body stay dry and comfortable inside special garments that are airtight (e.g., epidemic-proof garments, firefighting garments), reducing the retention of perspiration on the surface of the skin and thus providing a better wearing experience [13]. This is of great importance for firefighters, outdoor workers, and people in everyday life. Therefore, studying the moisture permeability and wicking properties of garments is a key step in improving the comfort and functionality of garments.

3. Classification and Characteristics of Thermal Management Wearable Clothing

Thermal management wearable system refers to the use of gases, liquids, phase change materials, and other cooling media for temperature control of the microenvironment under the clothing, to mitigate the heat stress of the human body in a hot environment [14]. There are various classifications of thermal management wearable systems based on different characteristics. First, according to the classification based on the cooling coverage, they can be divided into full-body-cooling garments [15] and localized-cooling garments [16]. Secondly, according to the classification of whether energy is consumed in the cooling process, it can be divided into active-cooling garments [17] and passive-cooling garments [18]. In addition, according to the classification based on the cooling medium used for cooling, it can be classified into evaporative-cooling clothing [19], air-cooling clothing [20], liquid-cooling clothing [21], phase change-cooling clothing [14], radiant-cooling clothing [18], and hybrid-cooling clothing [22]. Finally, the heat transfer pathways can be categorized as thermal convection, thermal conduction, and thermal radiation [6]. In the following part of this paper, we introduce and characterize the thermal management wearable system according to heat transfer pathways.

3.1. Thermal Convection Wearable Clothing

Thermal convection wearable clothing is a type of clothing specifically designed to regulate the body's temperature. It is based on the principle that heat generated by the body is carried away by means of convection heat transfer to reduce body temperature or maintain a comfortable temperature. Gas-cooled garments and liquid-cooled garments are two common types of thermal-convection wearable clothing.

(1) Air-cooling clothing

Air-cooling clothing typically employs low-temperature compressed air or ambient air as a cooling medium. This is achieved by circulating the air through pipelines or a sandwich of clothing, which then comes into contact with the surface of the human body. The resulting convection heat transfer and evaporation then help to dissipate heat away from the body [20]. Air-cooling clothing can be classified into two types: duct-type air-cooling clothing and fan-type air-cooling clothing, as depicted in Figure 5. Sayed et al. proposed an air/CO_2-cooling garment, which utilized a cylinder to store high-pressure gas, and the compressed gas expanded as it passed through the orifice, thereby cooling the human microenvironment, as shown in Figure 5a [23]. To assess the cooling efficiency of the clothing, a total of 19 male participants were part of an experimental procedure. Compressed air was provided to the air-cooling clothing to facilitate an efficient airflow to the body. The temperature of the air supplied was temperature-dependent for workplace comfort. Nevertheless, duct-type air-cooling clothing has drawbacks, including a high weight, noise, and a low reliability, hindering portable design implementation.

To meet the needs of certain industries that prioritize both comfort and portability, researchers have devised cooling clothing with tiny fans that can fit each individual perfectly [24]. Wang et al. developed an air-cooled suit of dry ice combined with ventilation that successfully stopped the mascot actor's core temperature from rising during the performance, as illustrated in Figure 5b [25]. Twelve healthy male volunteers took part in the research. There were no reports of heat-related illnesses, as well as pulmonary, esophageal, and cardiovascular diseases. The mean age, weight, height, body surface area, and body mass index (mean $\pm$ SD) were 21.3 $\pm$ 1.9 years, 65.3 $\pm$ 6.5 kg, 1.75 $\pm$ 0.04 m, 1.79 $\pm$ 0.09 m^2, and 21.3 $\pm$ 1.9 kg/m^2, respectively. The primary aim of cooling garments is to mitigate heat stress in the human body by fostering the flow of ambient air throughout the body for enhanced evaporative cooling and convective heat transfer, as noted in Ref. [21]. Nevertheless, in environments featuring high temperature and high humidity, where sweat accumulates on the body surface, air-cooling clothing's efficacy is significantly restricted, leading to discomfort and potential metabolic imbalances, as discussed in Ref. [26]. Therefore, it may be necessary to select alternative cooling equipment for environments with high temperature and humidity levels.

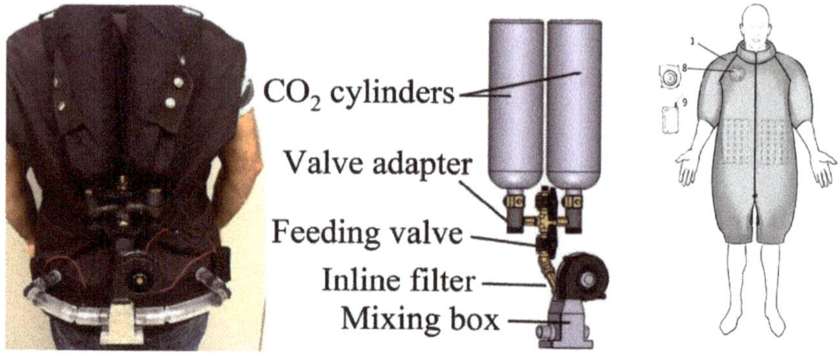

(a) (b)

Figure 5. (**a**) Duct-type air-cooling clothing [23]; (**b**) Fan-type hybrid dry ice and air-cooling clothing [25].

(2) Liquid-cooling clothing

The fundamental principle of liquid-cooling clothing is for the cryogenic liquid generated by the cold-source equipment to transfer heat with the human skin through the cooling tubing inside the clothing. This process lowers the body's surface temperature and alleviates the heat strain caused by high temperatures [14]. Unfortunately, there are numerous flaws in the current design of liquid-cooled clothing. First, many of the current liquid-cooling clothing's cooling units are complicated, with some employing compressors to cool the refrigerant, adding excess weight to the system and being inconvenient for personnel to carry [27]. Second, the majority of liquid-cooling clothing is tailored for specific environments, making them unsuitable for general workplaces [28]. Third, scholars have utilized metal materials to construct piped systems for liquid-cooling clothing [29]. Although metals offer excellent thermal conductivity, they tend to be rigid and uncomfortable to wear. These suits are categorized as thermoelectric and ice liquid-cooled depending on the type of cold source used [30].

The thermoelectric liquid-cooling clothing, illustrated in Figure 6a, comprises four main components: a cooling device, a cooling piping system, a micropump, and a basic garment [17]. The circulating cooling water in the thermoelectric suit absorbs body-generated heat through the piping, effectively lowering the body's surface temperature [31]. In this process, the temperature of the hot side rises, and this temperature rise negatively impacts the cooling performance of the semiconductor cooling plate. Based on their experiments, Zhang et al. [17] found that the cooling effect improved with the increasing flow rate under an optimal operating voltage. The cooling suit's average maximum temperature difference was 5.5–39.2 °C. Thermoelectric liquid-cooling clothing has certain limitations, including a low coefficient of performance (COP), a reduced efficiency in high-capacity applications, and wide temperature ranges [32].

The ice liquid-cooling clothing, illustrated in Figure 6b, is typically worn with a fanny pack or backpack containing an ice bag. The coolant exchanges heat with the ice storage container, and a micropump propels the coolant through a circulating pipeline within the clothing to exchange heat with the human body [30]. This sort of liquid-cooled attire diminishes the device's intricacy, reduces cost, and enhances portability. As a result, it is more favored in the market and possesses greater practical application value. Nevertheless, liquid-cooling clothing necessitates the timely replacement of its cooling source. For a liquid-cooling suit, it is necessary to configure the type of cooling medium, inlet temperature, and flow rate according to the actual situation. Table 1 shows the type of medium, medium temperature, medium flow rate, and cooling efficiency for liquid-cooling clothing.

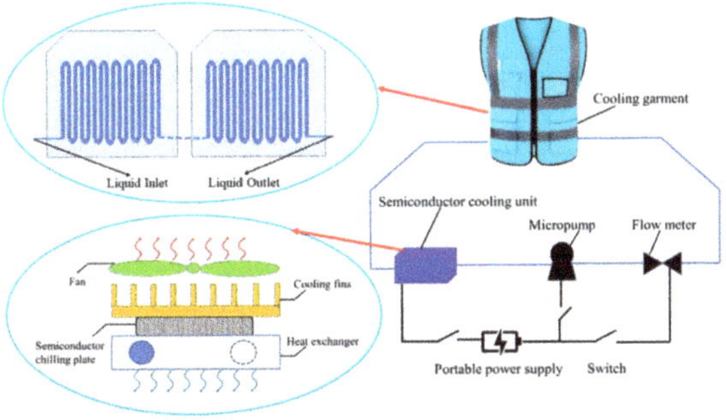

(a)

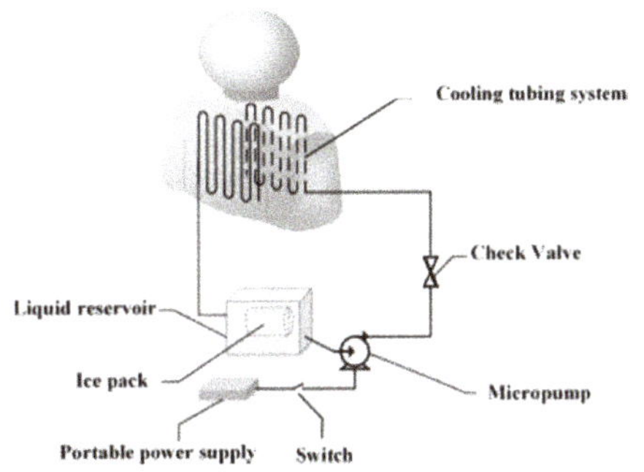

(b)

Figure 6. (**a**) Thermoelectric liquid-cooling clothing [17]; (**b**) ice liquid-cooling clothing [30].

Table 1. Cooling medium parameters and cooling efficiency.

References	Environmental Temperature	Environmental Humidity/%	Liquid Temperature/°C	Liquid Flow Rate/L·min^{-1}	Cooling Efficiency/W
[21]	30 °C	40	-	3.8	300
[17]	39.2 °C	60	16	0.9	160.43
[33]	30 °C	-	15.7	-	340.4
[34]	45 °C	-	-	0.54	243.2
[25]	34.0 ± 0.5 °C	78 ± 5	5.3	0.5	169.2
[35]	35 °C	60	22	-	210

3.2. Thermal Transfer Wearable Clothing

Thermal transfer wearable clothing is a specialized garment that removes heat generated by the body through heat transfer, regulating body temperature. A prevalent type of heat transfer wearable clothing is the phase change-cooling suit. This cooling utilizes phase-change materials (PCMs) that absorb a considerable amount of latent heat at the phase-change point, resulting in cooling the human body. PCMs are a significant energy storage option [36,37], categorized into three groups: organic materials, inorganic materials, and mixed cocrystals of the two [38]. Specific features are presented in Table 2 [39]. The phase-change cooling suits usually have pockets on the front and back that are specifically designed for the PCMs. Before usage, the materials should be stored in a cold environment to preserve their coldness, and they are later placed in the pockets. Once the PCM absorbs heat and undergoes a phase change, it is taken out and placed back in a cold environment to store cold for the next usage [40]. The cooling effectiveness of the phase-change cooling suit is determined by the specific phase-change material used. Different materials extract varying degrees of heat during the phase-change process, resulting in differences in cooling capacity. The cooling duration of the suit typically lasts between 2 and 5 h [41]. Refer to Figure 7 for a schematic illustration of the phase-change cooling suit's structure [3].

Table 2. Comparison of different phase-change materials.

Description	Organic PCMs		Inorganic PCMs		Eutectic Material
Materials	Fatty acids	Paraffinic	Metals	Hydrated salt	Mixtures
Melting point	30~70 °C	40~70 °C	20~100 °C	−20~100 °C	−20~100 °C
Latent heat	150~250 kJ/kg	150~250 kJ/kg	100~500 kJ/kg	200~400 kJ/kg	100~500 kJ/kg
Advantages	Good stability, good reusability, well-defined phase-transition point [42]	Good stability, good reusability, adjustable melting point [43]	Good stability, good thermal conductivity, nontoxic and harmless [44]	High latent heat, good reusability, inexpensive [45]	High thermal conductivity, high latent heat, good stability, adjustable melting point [46]
Drawbacks	Low melting point, low phase-transition temperature range, easily contaminated, toxic, and slightly corrosive	Corrosive to plastic containers. High volume change, volatile, flammable	High melting point, high weight, small phase-transition temperature range, more expensive	Corrosive and slightly toxic. Poor stability, susceptible to moisture, easily crystallized	Heavy weight and high cost

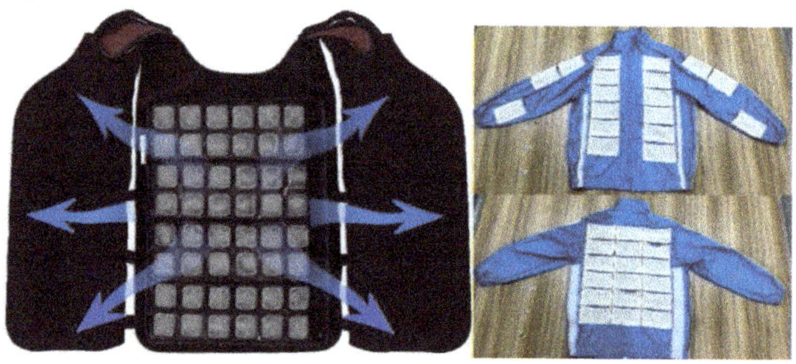

Figure 7. Phase-change cooling clothing [3].

Phase-change cooling clothing has a simple structural design, does not require the installation of refrigeration equipment like air-cooled suits and liquid-cooled suits, is easy to put on and take off, and has a good cooling effect. However, the production cost of phase-change materials is high, the weight of phase-change materials increases the burden of the cooling suit, and the phase-change materials installed in the pockets of the clothing lead to phase-change cooling suits with poor breathability, preventing the evaporation of sweat. To improve the cooling performance of phase-change cooling clothing, the common method is to select phase-change materials with high thermal properties and temperatures close to the skin temperature and reasonably distribute phase-change materials according to the temperatures of different parts of the human body. In addition, adding an insulating layer outside the phase-change material to reduce the heat absorption of the phase-change material from the external environment, and adding metal nanoparticles, graphene, or other thermal fillers to the phase-change material to enhance the thermal conductivity of the phase-change material is also one of the methods to improve the cooling performance of phase-change cooling suits [14]. Tesar and Kordik [47,48] conducted experimental tests on PCM-based clothing for soldiers in hot desert climates. The moderate melting-point temperature of n-ecosane (35.7 °C) made it highly advantageous as it allowed for the maintenance of a temperature of 35 °C. Butts et al. [49] utilized PCM to cool 20 male participants in their clothing and observed a significant reduction in their thermal, psychological, and perceptual strains when compared to males not exposed to PCM under test conditions of 34.2 °C and 54.7% RH.

3.3. Thermal Radiation Wearable Clothing

The utilization of radiative cooling in thermal protective clothing exhibits potential for energy conservation owing to its passive nature. Human skin displays emissivity values greater than 0.95 [50,51] and can be regarded as being almost entirely comprised of blackbody radiation. This elevated emissivity characterizes the human skin as a special infrared emitter. In a typical indoor setting, up to 50% of the total heat loss arises from radiation [52]. Nevertheless, traditional clothing design does not make provisions for the radiation component. Radiant cooling clothing is a developing approach to cooling the body. It achieves this by enhancing the body's radiative heat loss to the environment, as well as reflecting heat radiation from sunlight through the spectral selectivity of the fabric's microfiber structure [53]. The use of passive cooling technology in radiant cooling suits means that they do not require an external energy supply, bringing the benefits of simplicity and energy savings. Since the solar spectrum is mainly between 0.3 and 4.0 μm [53], it is promising and feasible to enhance solar reflection and dissipate human thermal radiation. Zhao et al. developed a layered nanofiber textile with improved thermal insulation and radiant heat management for effective personal heat management in harsh temperatures, achieving a temperature reduction of 7.2 °C over white cotton in hot environments, as shown in Figure 8 [54].

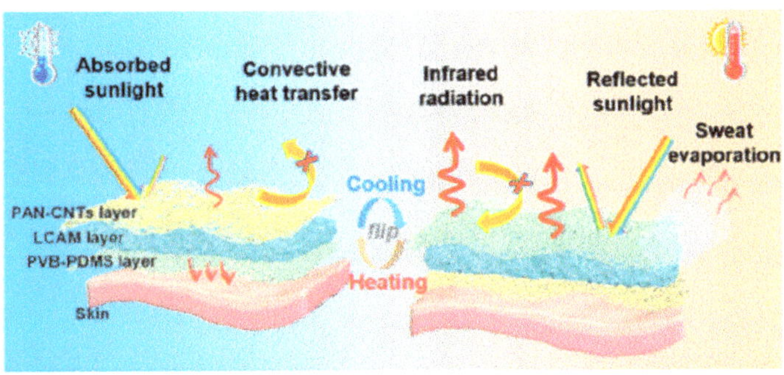

Figure 8. Thermal-radiation cooling clothing [54].

In recent years, numerous research teams have focused on creating textiles that possess exceptional radiative cooling capabilities. Cai et al. [53] were the first to present a spectrally selective, radiatively cooled textile suited for outdoor cooling, with an over 90% reflectance of solar irradiance and a satisfactory transmission of thermal radiation to the human body. This was achieved through avoiding overheating of the simulated skin, ranging from 5–13 °C when compared with cotton fabrics. Irfan et al. [55] developed a nanotextile using nanoparticle-doped polymeric materials and electrostatic spinning technology. Results showed 91% solar reflectance, 81% mid-infrared transmission, and a cooling performance of 9 °C compared to cotton textiles.

These findings suggest that radiatively cooled clothing has great potential for effective cooling and various applications. In addition, a thorough examination of radiant thermal cooling and heating, which includes reflective surfaces, has been conducted [56]. This review offers significant insights into radiant-cooling technologies. Radiant-cooling technology has a critical role in personal thermal management, particularly in the context of clean energy radiation that does not contribute to carbon emissions. However, this technique is limited by weather conditions and cannot be fully applied in all circumstances [57]. Lin et al. [58] conducted a comprehensive review study addressing atmospheric conditions in various regions, including application challenges. Nonetheless, many issues still need to be addressed for passive personal cooling clothing, such as material stability, durability, and comfort. Thus, there are currently no established products available for practical use. Therefore, extensive research and development are required to enhance the effectiveness and dependability of novel radiative cooling materials and manufacturing techniques. Table 3 presents an outline of thermoregulation textiles that rely on radiant heat transfer.

Table 3. Summary of thermoregulation textiles based on thermal radiation.

References	Types	Materials	Results
[51]	Radiation-cooled, infrared, transparent, visible, opaque fabric	Parallel-aligned polyethylene fibers with a low infrared absorbance	The infrared transmittance of the fiber is 0.972
[59]	Radiation-cooled fabrics are opaque, transparent to mid-infrared	Nanoporous polyethylene	It reduces skin temperature by up to 2.7 °C.
[60]	Radiation-cooled fabrics are opaque, transparent to mid-infrared	Homogeneous and continuous nano-polyethylene	It reduces skin temperature by up to 2.3 °C.
[61]	Cooling and heating of fabrics by mid-infrared emitted radiation	Polyethylene nanolayer	Maximum core temperature reduction up to 36.59 °C
[62]	Mid-infrared solar reflective emission cooling textiles	Integrated solar reflectors and thermal emitters	5 °C cooling in direct sunlight
[63]	Mid-infrared solar reflective emission radiation heating and cooling of textiles	Photonic structure	Temperature difference of 20 °C for radiant textiles
[64]	Low and medium infrared-emitting radiant fabrics	Silver nanowire composite coated on cotton	Average reflectivity is 66% higher than conventional fabrics

3.4. *Hybrid-Cooling Wearable Clothing*

Hybrid-cooling wearable clothing refers to garments that employ a combination of two or three heat-transfer mechanisms [65].

Hybrid-cooling wearable clothing is specifically engineered to optimize the benefits of various cooling techniques to achieve a superior cooling efficiency [22]. Figure 9 demonstrates a hybrid-cooling wearable clothing that combines microfans and PCM to improve cooling efficiency while addressing the drawbacks of air-cooling and phase change-cooling clothing [66]. Wang et al. [25] designed a suit based on dry ice and ventilated fans, utilizing the sublimation of dry ice to reduce the wearer's load and reinforce air circulation to effectively alleviate thermal discomfort in mascot actors. The experiment demonstrated the effectiveness of the hybrid-cooling suit in preventing the mascot actor's core body temperature from exceeding 38.0 °C and consistently maintaining an average skin temperature below 36.5 °C. Ni et al. [67] proposed a hybrid-cooling garment that combined PCM and a ventilation fan, resulting in the improved regulation of heart rate and skin temperature. One of the benefits of this design was that the PCM, without additional cooling, absorbed a significant amount of potential [37,68–70]. Kang et al. [71] provided a similar hybrid design that combined PCM and a ventilation fan, which was also proven effective in thermal management. The design of hybrid-cooling suits can be customized to different usage scenarios and necessities. For instance, in high-temperature environments, liquid and air cooling can be combined to take advantage of the effective cooling properties of liquid and the rapid cooling features of air, resulting in faster and more efficient cooling. In addition, hybrid-cooling wearable clothing could incorporate additional features, such as protection and breathability, to better meet user needs.

3.5. *Characterization of Thermal Management Wearable Clothing*

Personal cooling clothing effectively lowers the body's surface temperature in high-temperature settings, reduces discomfort and fatigue caused by heat, improves work efficiency, and protects individuals' health. The suit finds practical uses in various settings such as industrial production, fieldwork, and others that require cooling. The performance characteristics of various cooling suits are compared in the introduction above. Please refer to Table 4 for a summary of the results.

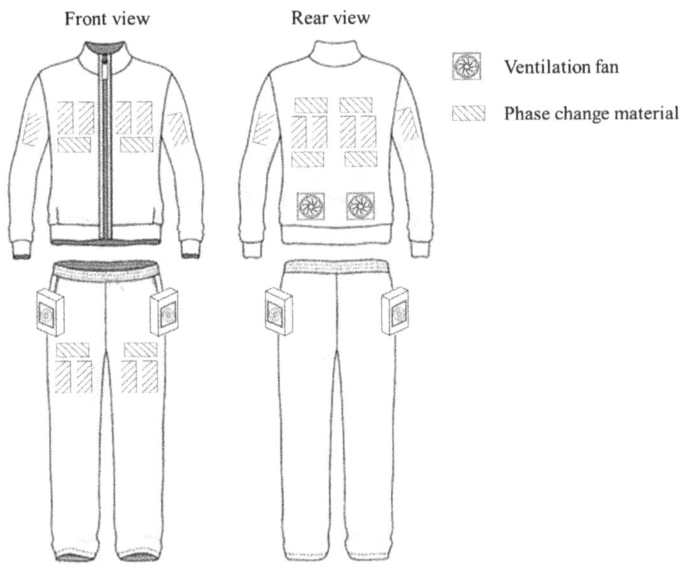

Figure 9. Hybrid-cooling wearable clothing [66].

Table 4. Comparison of performance characteristics of several cooling clothing [72].

Types	Cooling Method	Total Weight	Cooling Temperature	Mean Cooling Power (6 h)	Notice
WCDS	Thermoelectric and blowers	1.2 kg	26.5~28.8 °C	51.7 W	-
Air-cooling clothing	Ventilating fans	0.7 kg	32.5 °C	9 W	-
Liquid-cooling clothing	Ice water and water circulation tubing system	4.3 kg	11.1 °C	37.1 W	Condensation water
Ice-cooling clothing	Ice gel packs	1.9 kg	4.2 °C	25.6 W	Condensation water

Note: -, not available.

In summary, although these cooling methods can effectively regulate the temperature of the environment around the human body to a more suitable temperature if the temperature of the cooling medium is lower than the dew point of the air, it will lead to the emergence of a condensation phenomenon. Condensation is the process by which water vapor on the surface or inside a cooling suit condenses into water droplets when it comes in contact with a cooler surface [73].

The occurrence of condensation is a challenge to human comfort. When condensation occurs, garments around the human body become wet, which not only causes discomfort but also leads to increased heat loss, further affecting the body's perceived temperature. Damp garments can also cause discomfort or health problems, such as a feeling of coldness and moisture that harbors bacteria [74]. As a result, cooling alone to reduce body temperature is no longer sufficient to meet the body's comfort needs.

There are a number of factors that need to be considered to address the problems associated with the risk of condensation. Firstly, the design of cooling suits should take full account of factors such as ambient temperature and humidity to determine the appropriate cooling water temperature and avoid the occurrence of condensation caused by cold water that is too low. Secondly, the material and structure of the cooling suit should have good moisture permeability and fast drying properties to quickly remove moisture and improve wear comfort [75].

4. Classification and Characteristics of Moisture Management Wearable Clothing

The quantity of sweat that does not evaporate from the skin's surface per hour is typically between 30 and 80 g when the surrounding temperature is not high. Under such circumstances, the sweat can easily spread into the surrounding environment through the gaps among clothing fibers. Yet, during physical exertion or hotter weather, sweat production significantly increases resulting in a larger quantity of sweat accumulating on the skin's surface [72]. At present, it is no longer sufficient for clothing to simply be breathable and permeable to moisture in order to effectively dissipate sweat. Rather, clothing must be able to absorb significant amounts of sweat before releasing the resulting moisture into the surrounding air through evaporation. In cases where the surrounding air is dry or windy, sweat can evaporate quickly. However, if the humidity of the surroundings exceeds 60%, individuals may feel increasingly uncomfortable. Therefore, being in a highly humid environment or sweating during exercise and causing your clothing to become wet can influence and alter the thermal and humidity comfort of your attire.

Personal moisture management wearable clothing is a type of clothing with a unique design that is capable of effective sweat regulation in addition to body temperature regulation. Moisture transport in textiles has a significant impact on the physiological comfort of the human body. When the human body produces sweat, a series of wetting, penetration, transport, and evaporation processes occur with the textile. At the same time, the temperature, humidity, and wind speed of the environment will also have an impact on the exchange of moisture and heat, which further affects the comfort of the human body.

4.1. Moisture Management Materials

Textile regulation of humidity is the process of transferring perspiration through the textile and usually involves two aspects depending on the different aggregation states of humidity, i.e., the regulation of sweat vapor and liquid sweat [76]: (1) the movement of moisture in the gaseous components of the material, e.g., the transfer of water molecules within fibers or in the space between fibers; (2) the interaction of humidity with the solid components of the material, e.g., repulsion or adsorption of water molecules with hydrophobic or hydrophilic groups for humidity transfer.

(1) Water vapor regulation

Gaseous perspiration wearable clothing is an important sweat transfer method in wet management wearable systems. Evaporative-cooling garments are used to remove heat from the body by absorbing heat through the evaporation of moisture as shown in Figure 10 [19]. Evaporative-cooling garments can be made from a variety of highly absorbent polymers that are more effective than conventional textiles [77–80]. When the human body perspires, the sweat penetrates the inner fabric layers of the moisture management garment. These inner fabric layers are usually made of special materials that have a large surface area and good breathability. Through the action of this material, sweat can evaporate quickly, removing heat from the body's surface and expelling moisture from the body. In this way, individual moisture-regulating clothing can help the human body stay dry and comfortable, avoiding excessive sweat accumulation on the skin surface, which causes discomfort and a feeling of wetness and coldness [6]. At this point, water vapor moves through the air space between fibers by diffusion or through their internal molecular structure when the vapor is under a vapor concentration gradient. Current sweat vapor-conditioning methods generally involve the design of materials that facilitate the transfer of sweat vapor from the body to the environment, which is achieved through pore size design, i.e., materials with a high water vapor permeability [81].

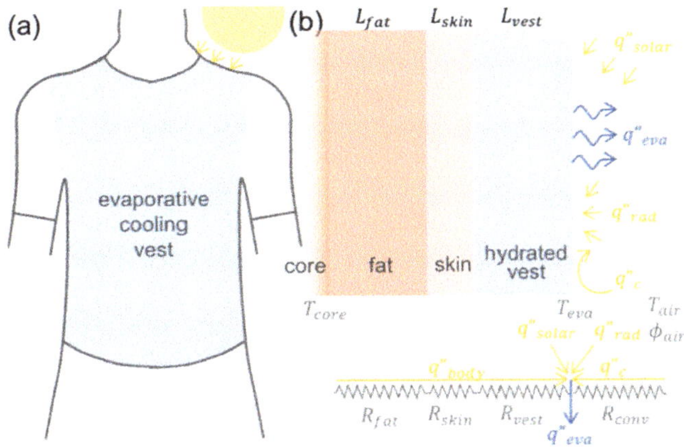

Figure 10. (**a**) Schematic diagram of evaporative cooling clothing; (**b**) heat and mass transfer processes involved in evaporative cooling [19].

The evaporative moisture-regulating clothing material has strong water absorption properties and good breathability and comfort, which can effectively help sweat to be discharged, thus enhancing the effect of evaporative cooling [79]. It is worth noting that the effect of evaporative-cooling clothing is greatly affected by environmental factors, and in high-temperature and high-humidity environments, it is difficult to evaporate the moisture inside the clothing, so the effect of evaporative cooling will be limited. On the contrary, in an environment with a low relative humidity, the effect of evaporative cooling will be better because the moisture is easier to evaporate, thus removing body heat [82]. Therefore, the advantages of evaporative-cooling garments are that they are light to wear, easy to use, simple to manufacture, and suitable for industrial mass production; the disadvantages are that the liquid adhering to the inner surface of the garment seriously affects the comfort of the human body, inhibits the evaporation of sweat, and is more effective in environments with a low relative humidity.

(2) Liquid vapor regulation

As an intermediate layer between the skin and the environment, textiles play a critical role in achieving personal comfort and safety by managing the thermal and moisture conditions of the localized body. When liquid sweat is in direct contact with the moisture-regulating clothing material, liquid sweat is transferred through the material in both in-plane and trans-plane directions [83]. Many studies have focused on sweat-directed transplanar transfer, i.e., the transfer of liquid sweat from the skin to the outer surface of the material to keep the skin dry. Many researchers have achieved a directional liquid transfer of materials through a fabric structure design by mimicking plant structures [84]. Miao et al. investigated sandwich-structured textiles with hierarchical nanofiber networks and Janus wettability, as shown in Figure 11 [85]. The human body temperature covered by this sandwich-structured textile decreased by about 4.2 °C with rapid sweat evaporation compared to commercial cotton textiles. The driving force for the movement of liquid sweat within the material was capillary forces, which were influenced by intermolecular forces between the liquid and the surrounding fiber surfaces, as well as the size and configuration of the space within the textile. Although water transfer in materials is accompanied by energy transfer, existing material strategies have paid less attention to the energy transfer of liquid sweat. Most material designs for liquid sweat conditioning do not quantify or consider the thermal energy transfer involved [86]. As shown in Figure 11, sweat can affect cooling efficiency in two ways: (I) by increasing the evaporative distance from the skin through cross-plane transfer and (II) by increasing the evaporative surface area through in-plane transfer.

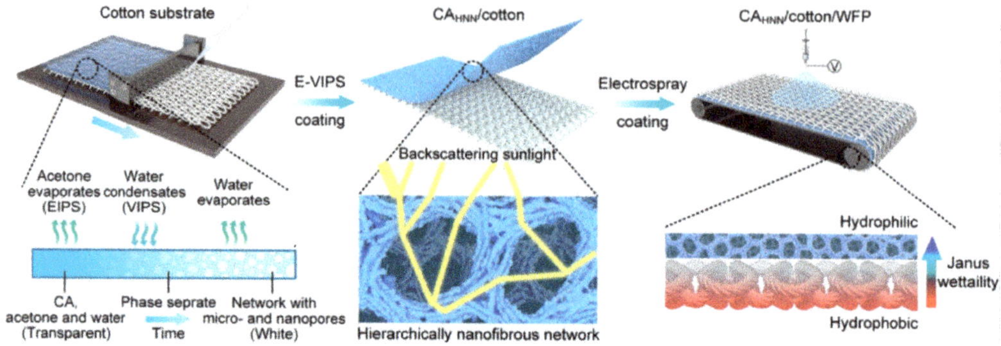

Figure 11. Schematic representation of textiles with hierarchical nanofiber networks and Jannus wettability [85].

4.2. Moisture Management Clothing

The effectiveness of thermoregulatory clothing is limited in hot and humid environments because less moisture is transferred to the pairs within the garment, with the attendant risk of sweating and condensation within the thermoregulatory clothing. Wang and Hu [87] showed that the average thermal sensation in humans under hot conditions was related to the sensation of sweating. Keeping the skin and microclimate air dry can be achieved by using solid desiccant packs. Solid desiccants such as silica gel can reduce the moisture content of the microclimate air between the fabric layers within the cooling undershirt through an adsorption process. However, the released adsorbed heat can increase the temperature of the microclimate air if the thermoregulation suit is not sufficient to absorb the heat generated by the desiccant and lost from the human torso [88]. It is of interest to determine whether a thermoregulation device in combination with a desiccant pack would improve the cooling performance of the undershirt.

The purpose of the PCM composite desiccant packs proposed by Mariam Itani et al. [89] was to provide cooling to the human body working in hot conditions and to adsorb water vapor in the microclimate air, thus increasing the evaporation of sweat from the skin layer and improving the cooling of the human body. Experimental modeling and validation were conducted to find the state of the microclimate air and its moisture content under the influence of using PCM desiccant packs. The idea was to use a solid desiccant in addition to the PCM package to keep the skin and the macroclimatic air layer dry to prevent condensation and enhance sweat evaporation from the skin, as in Figure 12a [89]. A wearable cooling and dehumidification system was designed by Lou et al., as shown in Figure 12b [72]. A lightweight air-ventilated undershirt was worn underneath a protective suit, and an air cooling and moisture-removal (ACMR) chamber was attached to the lower back area of the knitted undershirt. In a simulated hospital environment, the optimal cooling effect corresponded to a 3.5 °C drop in ambient temperature and a 6% reduction in relative humidity. The properties of sweat-expulsion clothing and desiccant-adsorption wearable clothing are demonstrated in Table 5. Specific properties may vary depending on product design and material selection. When selecting an appropriate moisture management garment, it is recommended that a comprehensive evaluation be performed based on individual needs and environmental conditions.

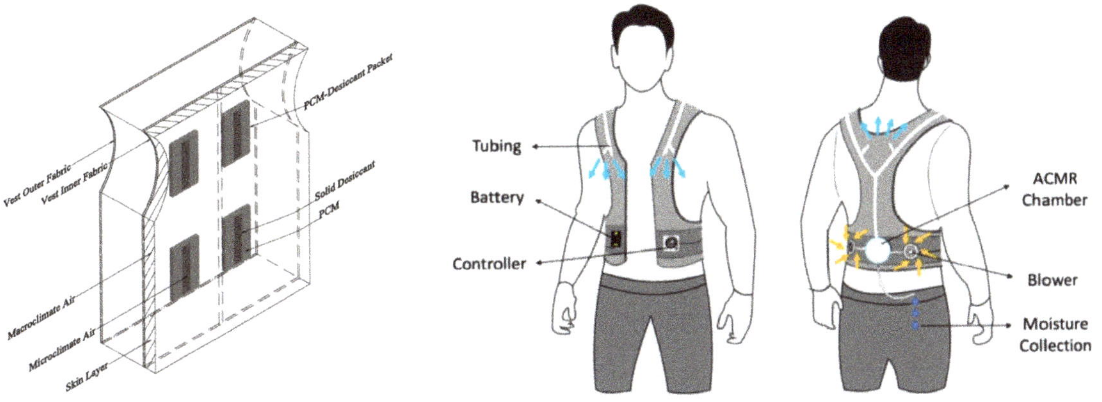

(**a**) (**b**)

Figure 12. Schematic diagram of (**a**) PCM composite desiccant packages [89] and (**b**) ACMR for clothing [72].

Table 5. Comparison of sweat expulsion clothing and desiccant adsorption wearable clothing.

Types	Principle	Dehumidification Effect	Drawback
Moisture management materials	Surface evaporation	140–160 $g/m^2/h$ [74]	Depending on ambient humidity
Desiccant-dehumidification clothing	The adsorption capacity of the desiccant	Decrease from 21.23 g/kg to 19.74 g/kg [89]	Dehumidification process is exothermic
Condensation-dehumidification clothing	Cold-surface condensation	26.3 g/h [72]	High energy consumption, condensation

In summary, there is a risk of temperature increase in dehumidified clothes during the process of humidity regulation, especially when desiccant-dehumidified clothes use a desiccant to absorb water vapor around the human body. However, this method of dehumidification can effectively regulate the humidity of the environment around the human body to keep it appropriate. However, at the same time, the absorption of water vapor releases latent heat, which causes the temperature of the desiccant to increase [89]. This phenomenon of temperature increase has a certain impact on human comfort. As the temperature of the desiccant increases, the temperature of the garment around the human body increases accordingly, reducing comfort. As a result, dehumidification alone can no longer satisfy the human body's need for comfort. Despite the rapid development of materials, there are both challenges and opportunities. Can new textiles maintain all aspects of their performance after many normal washings? Consideration should also be given to whether the materials are sufficiently readily available and cost-effective for large-scale industrial production.

A number of factors need to be considered to address the problems associated with rising temperatures. First, the design of dehumidifying garments should take full account of the thermal conductivity and heat dissipation properties of the desiccant to minimize the extent of a temperature rise. Secondly, the material and structure of the dehumidifying garment should have good air permeability and heat dissipation to quickly dissipate heat and maintain a suitable temperature on the surface of the garment. In addition, the use of intelligent control systems can monitor and regulate the dehumidification process to ensure a balance between dehumidification performance and comfort.

5. Performance Evaluation Indicators of Personal Wearable Thermal and Moisture Management Clothing

At present, there are many methods for evaluating the thermal and moisture comfort of garments, mainly analyzing and evaluating them from physical [90], physiological [91], and psychological [65] perspectives, and the evaluation methods can be divided into subjective and objective evaluations. An objective evaluation uses physical indicators and their fabric performance indicators and human physiological indicators, and a subjective evaluation is based on the psychological indicators of the human body [92–94].

5.1. Objective Evaluation Methods

The objective evaluation method is an evaluation method that uses objectively changing data that can be specifically measured and the results derived by instruments [94]. This evaluation method mainly evaluates the thermal and moisture comfort of garments from the following aspects: (1) using the thermal and humidity properties of textile materials as well as the thermal and humidity properties of the entire garment to evaluate the thermal and humidity comfort of the garment [95]; (2) using the garment microenvironment as the basis for the study of thermal and humidity comfort, and reflecting the influence of fabrics on the human body's sense of comfort by measuring the changes in temperature and humidity of the climatic region between the fabric and the skin, and proposing evaluation indices [96]; (3) analyzing human physiological data, clothing physiology points out some important physiological indexes, including body core temperature [97], average skin temperature [90], average body temperature, metabolic heat production, heat balance difference, heat loss, perspiration, heart rate [98], and blood pressure [99,100].

Typically, a two-step method is used for active sweat measurement. The measurement process of this method is as follows: first, the dry heat dissipation is measured while wearing a thousand dummy garments, and then the total heat dissipation (including the evaporative heat dissipation portion) is measured for a wet dummy attire. By calculating the wet heat dissipation minus the dry heat dissipation, the result of evaporative heat dissipation can be obtained [101]. In contrast, the "Walter" sweat warmer dummy for passive sweating uses a different measurement method [102]. In the measurement process, the total moisture resistance is calculated by considering the skin as a layer of the garment, and then the moisture resistance of the skin is subtracted from the total moisture resistance to obtain the moisture resistance of the garment and the surface air layer. Different from the two-step method, the evaporated sweat and wet-state heat dissipation of the warm body dummy are measured first in the measurement process to get the wet resistance and thousand-state heat dissipation, and then the thermal resistance is calculated based on the dry-state heat dissipation. As an instrument for testing the overall thermal and humidity performance of garments, the warm body dummy can simulate the mass transfer and heat transfer between the human body and the environment. Its measurements are objective, accurate, and repeatable [103]. To avoid ethical issues, inter- and intrasubject variability, and high costs, sweat dummies are often used to simulate sweat-induced evaporative cooling. Table 6 shows the different sweating simulation methods dummies used [104].

Table 6. Dummies with different methods of sweating simulation.

References	The Dummy	Characterization
[105]	Tore	Prewetted tight fabric skin applied to a dry thermal manikin
[102]	Walter	Water-filled manikin with a waterproof but vapor permeable surface
[104]	Coppelius	Manikin with an inner skin spreading water superficially and an outer vapor-permeable skin
[106]	Newton	Manikin with a supply of water to a fabric skin by means of sweating outlets distributed over the manikin's surface
[104]	ADAM	Manikin with a porous metal surface with superficial sweating

5.2. Subjective Evaluation Methods

The subjective evaluation method is a method of measuring the thermal comfort of garments which utilizes the psychological thermal comfort feelings of subjects in a specific environment for the assessment and is also known as the psychological method. As a supplement and validation of the objective evaluation method, the steps of this method are to predesign a questionnaire form and allow subjects to rate the comfort indicators of the garment according to their personal psychological feelings during the garment-wearing experiment. The scales and the meaning of subjective votes are shown in Table 7. The determination of the indicators and the division of the scoring scale are the key links in the subjective evaluation method, which still needs to be further improved.

Table 7. Subjective rating scales.

Scales	Thermal Sensation	Thermal Comfort	Thermal Satisfaction	Thermal Preference	Sweat Feeling	Cold Stimuli Sensation
+4	Very hot	Extremely uncomfortable	Very unsatisfied			Very strong cold stimuli sensation
+3	Hot	Very uncomfortable	Unsatisfied	Much warmer	The very strong feeling of sweating	Strong cold stimuli sensation
+2	Warm	Uncomfortable	Slightly unsatisfied	Warmer	The strong feeling of sweating	Medium cold stimuli sensation
+1	Slightly warm	Slightly uncomfortable	Satisfied	A little warmer	Slight feeling of sweating	A little cold stimuli sensation
0	Neutral	Comfortable	Very satisfied	No change	No feeling of sweating	No cold stimuli sensation
−1	Slightly cool			A little cooler		
−2	Cool			Cooler		
−3	Cold			Much cooler		
−4	Very cold					

The sweating dummy is also a tool for assessing the comfort of garments, effectively avoiding moral and physiological factors in human experimentation, and is not subject to psychiatric factors. It has a good reproducibility. However, the sweating dummy is still a device that simulates the shape and physiological characteristics of the human body, and its assessment results do not fully and realistically reflect the actual subjective feelings of humans.

5.3. Thermal and Moisture Regulation Model

The thermal and moisture regulation model is a method for evaluating the thermal and humidity comfort of garments, which simulates the processes of human body temperature regulation and thermal and moisture transfer of garments by establishing a thermal and moisture transfer model. The thermal and moisture regulation model has been widely used at home and abroad to evaluate the thermal and humidity comfort of clothing, which mainly includes the human body temperature regulation model, the thermal and moisture transfer model of clothing, and the human body–clothing–environment system model.

5.3.1. Human Thermoregulation Model

The human thermoregulation model consists of equations describing heat transfer and regulatory responses in the body. Over the last half-century, scholars have conducted substantial research and developed numerous human thermoregulation models. These models differ in the delineation of body nodes, and different methods of delineating body nodes correspond to different forms of equations for describing the regulatory responses of the body [107,108].

(1) The two-node model

The simplest two-node model, in which the human body is represented by a segment with two nodes in the core and skin layers, has gained popularity in engineering applications due to its simplicity, ease of implementation in computer tools, and faster computation time [109].

Gagge et al. laid the foundation for a two-node model of the whole body. The model was based on a direct, one-dimensional, transient heat transfer from the core to the skin layer to the environment. However, the thermoregulatory systems of sweating, skin blood flow, and shivering were based on empirical relationships developed for the average young adult. The accuracy of empirical models, especially the model constants, is closely related to experimental conditions, parameter ranges, quality of measurements, and physiological differences in subjects [110]. Doherty et al. [111] found that the model underpredicted skin wettability by 0.16 °C, and core temperature by 0.31 °C, but overpredicted skin temperature by 0.48 °C. At the same time, Ooka et al. [112] found that the model predicted a skin temperature 0.75 °C higher and a core temperature 0.32 °C lower. The model deviated from the experimental results. In addition to this, Takada et al. [107] found that the model underestimated skin temperature by up to 1.8 °C and core temperature by up to 0.5 °C, under cold-exposure conditions. Therefore, further improvement of the thermoregulatory system of the two-node model is needed, and a corresponding improvement scheme should be proposed. The flow diagram of a two-node model of the thermoregulatory system is shown in Figure 13 [113]. In this system dynamics model, the core and skin nodes of the Gagge model are modeled as two stocks that store the body's energy, which is then converted into the temperatures of two nodes, T_{skin} and T_{core}.

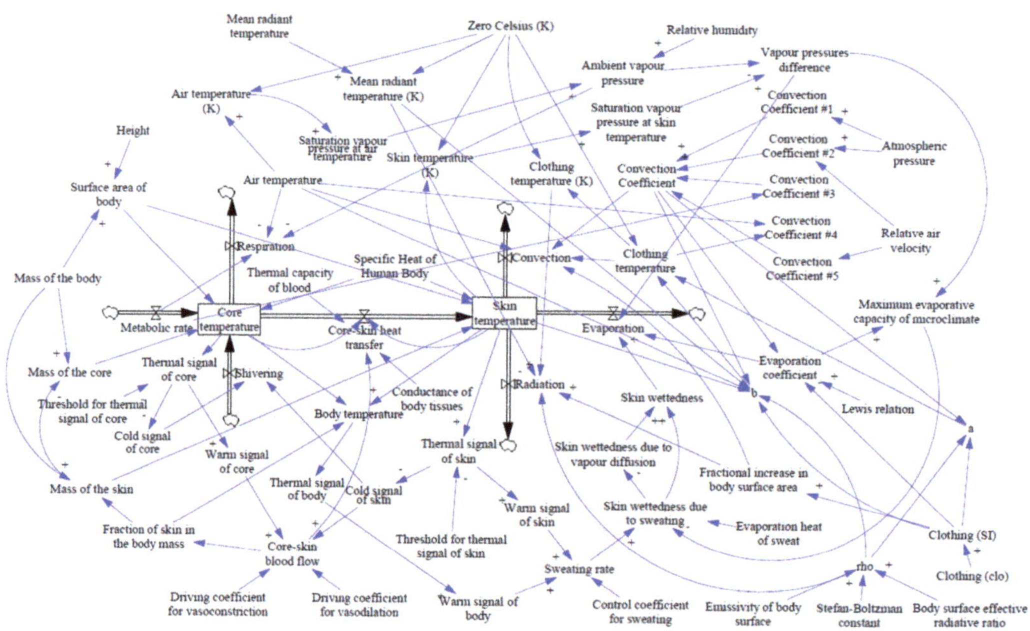

Figure 13. Flow diagram of a two–node model of the thermoregulatory system [113].

To overcome this problem, recent studies have proposed improved two-node models to more accurately assess the effects of personal cooling systems. These improvements include the consideration of local insulation and heat transfer in clothing, as well as modeling the heat exchange characteristics of cooling devices [114].

(2) Multinode model

The multinode body heat model is an extended and more complex version of the two-node model, as shown in Figure 14 [115]. Various models have been proposed to predict the state of heat regulation in the human body. Perhaps the most influential is the 25-node model proposed by Stolwijk [116], consisting of 24 body "nodes" (six body segments with four compartments each) and a central blood "node". Each node has a certain amount of metabolic heat generation, convective heat exchange with the central blood chamber, and convective heat exchange with neighboring nodes. Subsequently, some scholars have also improved this human thermal regulation model to make it suitable for application in a wider range of scenarios, such as cold environments, space environments, and so on. Researchers have made various enhancements to the model to accurately forecast the patterns of local responses in different parts of the human body. Munir et al. [117] reassessed the dynamic characteristics of the Stolwijk model and examined several modifications to the model.

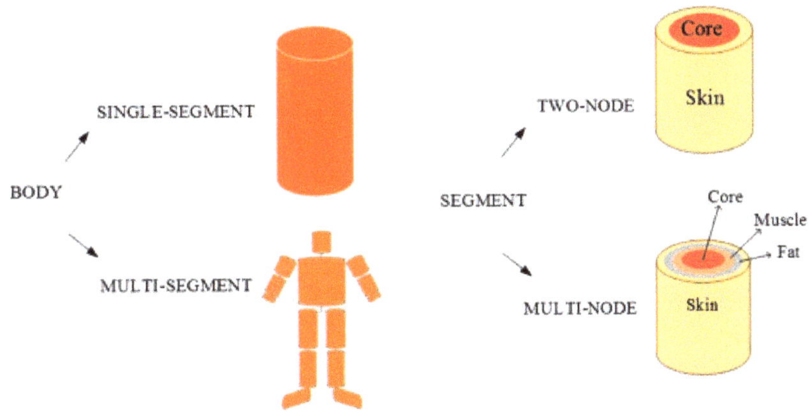

Figure 14. Comparison of the model segmentation, and the two-node and multinode models [115].

Multinodal models use partial differential equations to describe the heat transfer problem in the human body. The differential equations are usually discretized using the finite difference method, finite volume method, or finite element method to obtain the linear algebraic equations at each node. The advantage of the multinode model is that it can be divided into sections and nodes as needed, has better flexibility and accuracy, and can be applied to dynamic, nonstationary environments. The disadvantage is that the accuracy is not as good as the multinodal model in environments with large temperature gradient changes. The characteristics of the specific model can be seen in Table 8.

Table 8. The characteristic of the specific models [114,118].

Model Name	Model Classification	Model Characteristics
Human thermoregulation model	Two-node model Multinode model	When there are large temperature differences in the human body, more accurate simulation results can be obtained by using multinode and multiunit models.
Heat and moisture transfer modeling of clothing	Steady-state model Transient model	Simulate the heat and moisture transfer process of garments to derive the heat and moisture properties of garments.
Human body–clothing–environment system model	-	The human body–clothing environment as a whole combines a model of human thermoregulation and a model of heat and moisture transfer from clothing.

5.3.2. Evaluation of Thermal and Moisture Management Performance

There is no generalized performance evaluation method that can be used to assess the performance of wearable systems in managing thermal and moisture. The performance assessment of wearable systems involves two main aspects: (i) wearer comfort and safety; (ii) functionality of the wearable system. Ismail et al. [119] investigated the effect of uniform crosswinds on ventilation, heat, and moisture transfer to a clothed human body by simplifying the human body as a vertical cylinder with a uniform annular air layer. However, the existing evaluation methods are not sufficiently developed, especially for the performance evaluation of wearable systems under specific hot and humid conditions. Due to the wide variety of wearable systems and the different functions and technologies involved, specific assessment methods need to be developed for different types of wearable systems. These evaluation methods should be able to comprehensively consider the impact of thermal and moisture on the comfort and safety of wearable systems and accurately assess the performance of thermal and moisture regulation and energy conversion. Existing standard evaluation methods for thermal and moisture management of wearable systems are shown in Table 9 [75].

Table 9. Existing standard evaluation methods for thermal and moisture management of wearable systems.

Thermal/Moisture Transfer	Thermal/Moisture Transfer	Indexes	Test Equipment
Human body → environment	Heat transfer	Thermal resistance	Hot plate; thermal torso; thermal manikin
	Moisture transfer	Liquid-water transfer indexes	Moisture management tester
	Coupled heat and moisture transfer	Evaporative resistance	Sweating hot plate; sweating torso; sweating thermal manikin
		Water vapor transmission	Test dish, balance
Environment → human body	Heat transfer	Absorbed thermal energy	Thermal protective performance; tester, radiant protective performance tester

From the perspective of thermal and moisture management, the performance evaluation of wearable systems can include the following aspects: (i) wearable comfort and safety under specific thermal and moisture conditions, taking into account the thermal and moisture regulation performance as well as the thermal and moisture conditions of the human body and the environment; (ii) wearable functionality under specific thermal and moisture conditions, including thermal and moisture energy conversion and physiological sensing. In order to evaluate the performance of wearable systems under such hot and humid conditions, there is a need to provide standardized and repeatable evaluation methods to support the development of wearable systems.

Our research group is currently developing a cooling and dehumidifying garment that uses membrane separation to reduce skin temperature and absorb surface sweat, thereby creating a cool and comfortable microenvironment within the garment. Therefore, future research should further explore technologies that simultaneously regulate temperature and humidity to improve thermal and moisture comfort for the human body and provide better protection.

6. Conclusions

The presented analysis provided a comprehensive overview of wearable systems for personal thermal and moisture management, discussing their functions, latest developments, and performance evaluation approaches. The findings contribute to the understanding of the challenges and considerations involved in regulating temperature and humidity within wearable systems. In the light of this study, the main findings derived can be stated as follows:

(1) Although current cooling methods are effective in regulating the surrounding temperature, condensation can result if the temperature of the cooling medium is below the dew point of the air.
(2) Moisture removal from textiles is strongly influenced by the humidity of the surrounding environment. In addition, when a desiccant is used to absorb water vapor from around the body, the desiccant absorbs the water vapor, releasing latent heat and leading to an increase in body temperature.
(3) Temperature and humidity regulation is a complex coupled process, and there are no generalized performance evaluation methods available to assess the performance of wearable systems in managing temperature and moisture.
(4) Addressing the issues involved requires considering a combination of cooling and dehumidification factors, such as the use of intelligent control systems, which play a vital role in maintaining the delicate balance between effective dehumidification performance and optimal comfort.

Funding: The research work presented in this article was financially supported by the Postgraduate Research & Practice Innovation Program of Jiangsu Province (No. SJCX23_0493).

Data Availability Statement: Further inquiries can be directed to the corresponding authors.

Conflicts of Interest: The authors declare no conflict of interest.

References

1. Song, G.; Wang, F. *Firefighters' Clothing and Equipment: Performance, Protection, and Comfort*; CRC Press: Boca Raton, FL, USA, 2019; pp. 217–268.
2. Kjellstrom, T.; Kovats, R.S.; Lloyd, S.J.; Holt, T.; Tol, R.S.J. The Direct Impact of Climate Change on Regional Labor Productivity. *Arch. Environ. Occup. Health* **2009**, *64*, 217–227. [CrossRef] [PubMed]
3. Wang, H.; Xu, Z.; Ge, B.; Li, J. Experimental Study on a Phase Change Cooling Garment to Improve Thermal Comfort of Factory Workers. *Build. Environ.* **2023**, *227*, 109819. [CrossRef]
4. Wu, G.; Liu, H.; Wu, S.; Liu, Z.; Mi, L.; Gao, L. A Study on the Capacity of a Ventilation Cooling Vest with Pressurized Air in Hot and Humid Environments. *Int. J. Ind. Ergon.* **2021**, *83*, 103106. [CrossRef]
5. Peng, Y.; Cui, Y. Advanced Textiles for Personal Thermal Management and Energy. *Joule* **2020**, *4*, 724–742. [CrossRef]
6. Sajjad, U.; Hamid, K.; Tauseef-ur-Rehman; Sultan, M.; Abbas, N.; Ali, H.M.; Imran, M.; Muneeshwaran, M.; Chang, J.-Y.; Wang, C.C. Personal Thermal Management—A Review on Strategies, Progress, and Prospects. *Int. Commun. Heat Mass Transf.* **2022**, *130*, 105739. [CrossRef]
7. Su, Y.; Tian, M.; Li, J.; Zhang, X.; Zhao, P. Numerical Study of Heat and Moisture Transfer in Thermal Protective Clothing against a Coupled Thermal Hazardous Environment. *Int. J. Heat Mass Transf.* **2022**, *194*, 122989. [CrossRef]
8. Hu, B.; Shi, X.-L.; Zou, J.; Chen, Z.-G. Thermoelectrics for Medical Applications: Progress, Challenges, and Perspectives. *Chem. Eng. J.* **2022**, *437*, 135268. [CrossRef]
9. Zhang, Q.; Wang, S.; Wang, X.; Jiang, Y.; Li, J.; Xu, W.; Zhu, B.; Zhu, J. Recent Progress in Daytime Radiative Cooling: Advanced Material Designs and Applications. *Small Methods* **2022**, *6*, 2101379. [CrossRef]
10. Wu, Q.; Liu, J.; Zhang, L.; Zhang, J.; Jiang, L. Effect of Temperature and Clothing Thermal Resistance on Human Sweat at Low Activity Levels. *Build. Environ.* **2020**, *183*, 107117. [CrossRef]
11. Yang, J.; Wang, F.; Song, G.; Li, R.; Raj, U. Effects of clothing size and air ventilation rate on cooling performance of air ventlaiton clothing in a warm codnition. *Int. J. Occup. Saf. Ergon.* **2022**, *28*, 354–363. [CrossRef]
12. Kim, H.A. Moisture Vapor Permeability and Thermal Wear Comfort of Ecofriendly Fiber-Embedded Woven Fabrics for High-Performance Clothing. *Materials* **2021**, *14*, 6205. [CrossRef] [PubMed]
13. Gao, H.; Deaton, A.S.; Fang, X.; Watson, K.; DenHartog, E.A.; Barker, R. Effects of Environmental Temperature and Humidity on Evaporative Heat Loss through Firefighter Suit Materials Made with Semi-Permeable and Microporous Moisture Barriers. *Text. Res. J.* **2022**, *92*, 219–231. [CrossRef]
14. Wang, F.; Pang, D.; Liu, X.; Liu, M.; Du, W.; Zhang, Y.; Cheng, X. Progress in Application of Phase-Change Materials to Cooling Clothing. *J. Energy Storage* **2023**, *60*, 106606. [CrossRef]
15. Ke, Y.; Wang, F.; Xu, P.; Yang, B. On the Use of a Novel Nanoporous Polyethylene (nanoPE) Passive Cooling Material for Personal Thermal Comfort Management under Uniform Indoor Environments. *Build. Environ.* **2018**, *145*, 85–95. [CrossRef]
16. Liu, B.; Wang, H.; Lin, H.; Su, Y.; Wei, G.; Xu, Z. Effect of Phase Change Cooling Vest on Related Thermal Regulation Factors in Moderately Hot Environments. *Build. Environ.* **2023**, *242*, 110566. [CrossRef]
17. Zhang, M.; Li, Z.; Wang, Q.; Xu, Y.; Hu, P.; Zhang, X. Performance investigation of a portable liquid cooling garment using thermoelectric cooling. *Appl. Therm. Eng.* **2022**, *214*, 118830. [CrossRef]

18. Ma, Z.; Zhao, D.; Wang, F.; Yang, R. A Novel Thermal Comfort and Energy Saving Evaluation Model for Radiative Cooling and Heating Textiles. *Energy Build.* **2022**, *258*, 111842. [CrossRef]
19. Rykaczewski, K. Rational Design of Sun and Wind Shaded Evaporative Cooling Vests for Enhanced Personal Cooling in Hot and Dry Climates. *Appl. Therm. Eng.* **2020**, *171*, 115122. [CrossRef]
20. Choudhary, B.; Udayraj; Wang, F.; Ke, Y.; Yang, J. Development and experimental validation of a 3D numerical mdel based on CFD of the human torso wearing air ventilation clothing. *Int. J. Heat Mass Transf.* **2020**, *147*, 118973.
21. Wang, F.; Song, W. An investigation of thermophysiological responses of human while using four personal cooling strategies during heatwaves. *J. Therm. Biol.* **2017**, *70*, 37–44. [CrossRef]
22. Wan, X.; Wang, F.; Udayraj. Numerical Analysis of Cooling Effect of Hybrid Cooling Clothing Incorporated with Phase Change Material (PCM) Packs and Air Ventilation Fans. *Int. J. Heat Mass Transf.* **2018**, *126*, 636–648. [CrossRef]
23. Al Sayed, C.; Vinches, L.; Dupuy, O.; Douzi, W.; Dugue, B.; Hallé, S. Air/CO_2 Cooling Garment: Description and Benefits of Use for Subjects Exposed to a Hot and Humid Climate during Physical Activities. *Int. J. Min. Sci. Technol.* **2019**, *29*, 899–903. [CrossRef]
24. Zhao, M.; Gao, C.; Wang, F.; Kuklane, K.; Holmér, I.; Li, J. A Study on Local Cooling of Garments with Ventilation Fans and Openings Placed at Different Torso Sites. *Int. J. Ind. Ergon.* **2013**, *43*, 232–237. [CrossRef]
25. Wang, F.; Chow, C.S.W.; Zheng, Q.; Ke, Y.; Yang, B.; Zheng, X.; Noor, N.; Zhang, Q.; Zhang, C.; Zhou, H. On the Use of Personal Cooling Suits to Mitigate Heat Strain of Mascot Actors in a Hot and Humid Environment. *Energy Build.* **2019**, *205*, 109561. [CrossRef]
26. Peng, Y.; Li, W.; Liu, B.; Jin, W.; Schaadt, J.; Tang, J.; Zhou, G.; Wang, G.; Zhou, J.; Zhang, C.; et al. Integrated Cooling (i-Cool) Textile of Heat Conduction and Sweat Transportation for Personal Perspiration Management. *Nat. Commun.* **2021**, *12*, 6122. [CrossRef]
27. Ernst, T.C.; Garimella, S. Demonstration of a Wearable Cooling System for Elevated Ambient Temperature Duty Personnel. *Appl. Therm. Eng.* **2013**, *60*, 316–324. [CrossRef]
28. Yang, J.; Zhang, Y.; Huang, Y.; Chen, W. Effects of Liquid Cooling Garment on Physiological and Psychological Strain of Firefighter in Hot and Warm Environments. *J. Therm. Biol.* **2023**, *112*, 103487. [CrossRef]
29. Bartkowiak, G.; Dabrowska, A.; Marszalek, A. Assessment of an Active Liquid Cooling Garment Intended for Use in a Hot Environment. *Appl. Ergon.* **2016**, *58*, 182–189. [CrossRef]
30. Guo, T.; Shang, B.; Duan, B.; Luo, X. Design and Testing of a Liquid Cooled Garment for Hot Environments. *J. Therm. Biol.* **2015**, *49–50*, 47–54. [CrossRef]
31. Chen, W.Y.; Shi, X.L.; Zou, J.; Chen, Z.G. Thermoelectric Coolers for On-Chip Thermal Management: Materials, Design, and Optimization. *Mater. Sci. Eng. R Rep.* **2022**, *151*, 100700. [CrossRef]
32. Jeong, E.S. A New Approach to Optimize Thermoelectric Cooling Modules. *Cryogenics* **2014**, *59*, 38–43. [CrossRef]
33. Xu, Y.; Li, Z.; Wang, J.; Zhang, M.; Jia, M.; Wang, Q. Man-Portable Cooling Garment with Cold Liquid Circulation Based on Thermoelectric Refrigeration. *Appl. Therm. Eng.* **2022**, *200*, 117730. [CrossRef]
34. Shu, W.; Wang, J.; Zhang, X.; Luo, X. A Statistical Study to Evaluate the Performance of Liquid Cooling Garments Considering Thermal Comfort. In Proceedings of the ASME 2019 International Technical Conference and Exhibition on Packaging and Integration of Electronic and Photonic Microsystems, Anaheim, CA, USA, 7–9 October 2019; Volume 12. [CrossRef]
35. Zhang, M.; Li, Z.; Wang, Q.; Yuan, T.; Xu, Y. Research on Refrigerant Optimization and Characteristic Parameters Based on Thermoelectric Refrigeration Cooling Garment. *Appl. Therm. Eng.* **2022**, *212*, 118606. [CrossRef]
36. Rehman, T.-; Ali, H.M. Experimental Study on the Thermal Behavior of RT-35HC Paraffin within Copper and Iron-Nickel Open Cell Foams: Energy Storage for Thermal Management of Electronics. *Int. J. Heat Mass Transf.* **2020**, *146*, 118852. [CrossRef]
37. Tauseef-ur-Rehman; Ali, H.M.; Janjua, M.M.; Sajjad, U.; Yan, W.M. A Critical Review on Heat Transfer Augmentation of Phase Change Materials Embedded with Porous Materials/Foams. *Int. J. Heat Mass Transf.* **2019**, *135*, 649–673. [CrossRef]
38. Nazir, H.; Batool, M.; Osorio, F.J.B.; Isaza-Ruiz, M.; Xu, X.; Vignarooban, K.; Phelan, P.; Inamuddin; Kannan, A.M. Recent Developments in Phase Change Materials for Energy Storage Applications: A Review. *Int. J. Heat Mass Transf.* **2019**, *129*, 491–523. [CrossRef]
39. Khan, Z.; Khan, Z.; Ghafoor, A. A Review of Performance Enhancement of PCM Based Latent Heat Storage System within the Context of Materials, Thermal Stability and Compatibility. *Energy Convers. Manag.* **2016**, *115*, 132–158. [CrossRef]
40. Butts, C.L.; Torretta, M.L.; Smith, C.R.; Petway, A.J.; McDermott, B.P. Effects of a Phase Change Cooling Garment during Exercise in the Heat. *Eur. J. Sport Sci.* **2017**, *17*, 1065–1073. [CrossRef] [PubMed]
41. Qiao, Y.; Cao, T.; Muehlbauer, J.; Hwang, Y.; Radermacher, R. Experimental Study of a Personal Cooling System Integrated with Phase Change Material. *Appl. Therm. Eng.* **2020**, *170*, 115026. [CrossRef]
42. Zhang, R.; Chen, D.; Chen, L.; Cao, X.; Li, X.; Qu, Y. Preparation and Thermal Properties Analysis of Fatty Acids/1-Hexadecanol Binary Eutectic Phase Change Materials Reinforced with TiO_2 Particles. *J Energy Storage* **2022**, *51*, 104546. [CrossRef]
43. Chen, Q.; Wang, H.; Gao, H.; Wang, X.; Ma, B. Effects of Porous Silicon Carbide Supports Prepared from Pyrolyzed Precursors on the Thermal Conductivity and Energy Storage Properties of Paraffin-Based Composite Phase Change Materials. *J. Energy Storage* **2022**, *56*, 106046. [CrossRef]
44. Chen, K.; Guo, L.; Wang, H. A Review on Thermal Application of Metal Foam. *Sci. China Technol. Sci.* **2020**, *63*, 2469–2490. [CrossRef]

45. Ma, F.; Liu, L.; Ma, L.; Zhang, Q.; Li, J.; Jing, M.; Tan, W. Enhanced Thermal Energy Storage Performance of Hydrous Salt Phase Change Material via Defective Graphene. *J. Energy Storage* **2022**, *48*, 104064. [CrossRef]
46. Ayaz, H.; Chinnasamy, V.; Jeon, Y.; Cho, H. Thermo-Physical Studies and Corrosion Analysis of Caprylic Acid–Cetyl Alcohol Binary Mixture as Novel Phase Change Material for Refrigeration Systems. *Energy Rep.* **2022**, *8*, 7143–7153. [CrossRef]
47. Tesař, V.; Kordík, J. Energy Storage in Macro-Capsules for Thermal Comfort Garments. *J. Energy Storage* **2019**, *25*, 100842. [CrossRef]
48. Tesař, V.; Kordík, J. Melting N-Eicosane in Scaled-up Model of Capsule for Use in Hot Climate Thermal-Comfort Garments. *EPJ Web Conf.* **2019**, *213*, 02086. [CrossRef]
49. Butts, C.L.; Smith, C.R.; Ganio, M.S.; McDermott, B.P. Physiological and Perceptual Effects of a Cooling Garment during Simulated Industrial Work in the Heat. *Appl. Ergon.* **2017**, *59*, 442–448. [CrossRef]
50. Sanchez-Marin, F.J.; Calixto-Carrera, S.; Villaseñor-Mora, C. Novel Approach to Assess the Emissivity of the Human Skin. *JBO* **2009**, *14*, 024006. [CrossRef]
51. Tong, J.K.; Huang, X.; Boriskina, S.V.; Loomis, J.; Xu, Y.; Chen, G. Infrared-Transparent Visible-Opaque Fabrics for Wearable Personal Thermal Management. *ACS Photonics* **2015**, *2*, 769–778. [CrossRef]
52. Hardy, J.D.; DuBois, E.F. Regulation of Heat Loss from the Human Body. *Proc. Natl. Acad. Sci. USA* **1937**, *23*, 624–631. [CrossRef]
53. Cai, L.; Song, A.Y.; Li, W.; Hsu, P.-C.; Lin, D.; Catrysse, P.B.; Liu, Y.; Peng, Y.; Chen, J.; Wang, H.; et al. Spectrally Selective Nanocomposite Textile for Outdoor Personal Cooling. *Adv. Mater.* **2018**, *30*, 1802152. [CrossRef] [PubMed]
54. Gu, B.; Xu, Q.; Wang, H.; Pan, H.; Zhao, D. A Hierarchically Nanofibrous Self-Cleaning Textile for Efficient Personal Thermal Management in Severe Hot and Cold Environments. *ACS Nano* **2023**, *17*, 18308–18317. [CrossRef] [PubMed]
55. Iqbal, M.I.; Lin, K.; Sun, F.; Chen, S.; Pan, A.; Lee, H.H.; Kan, C.-W.; Lin, C.S.K.; Tso, C.Y. Radiative Cooling Nanofabric for Personal Thermal Management. *ACS Appl. Mater. Interfaces* **2022**, *14*, 23577–23587. [CrossRef] [PubMed]
56. Zhu, F.; Feng, Q. Recent Advances in Textile Materials for Personal Radiative Thermal Management in Indoor and Outdoor Environments. *Int. J. Therm. Sci.* **2021**, *165*, 106899. [CrossRef]
57. Li, Z.; Chen, Q.; Song, Y.; Zhu, B.; Zhu, J. Fundamentals, Materials, and Applications for Daytime Radiative Cooling. *Adv. Mater. Technol.* **2020**, *5*, 1901007. [CrossRef]
58. Lin, K.T.; Han, J.; Li, K.; Guo, C.; Lin, H.; Jia, B. Radiative Cooling: Fundamental Physics, Atmospheric Influences, Materials and Structural Engineering, Applications and Beyond. *Nano Energy* **2021**, *80*, 105517. [CrossRef]
59. Hsu, P.C.; Song, A.Y.; Catrysse, P.B.; Liu, C.; Peng, Y.; Xie, J.; Fan, S.; Cui, Y. Radiative Human Body Cooling by Nanoporous Polyethylene Textile. *Science* **2016**, *353*, 1019–1023. [CrossRef]
60. Peng, Y.; Chen, J.; Song, A.Y.; Catrysse, P.B.; Hsu, P.-C.; Cai, L.; Liu, B.; Zhu, Y.; Zhou, G.; Wu, D.S.; et al. Nanoporous Polyethylene Microfibres for Large-Scale Radiative Cooling Fabric. *Nat. Sustain.* **2018**, *1*, 105–112. [CrossRef]
61. Hsu, P.C.; Liu, C.; Song, A.Y.; Zhang, Z.; Peng, Y.; Xie, J.; Liu, K.; Wu, C.L.; Catrysse, P.B.; Cai, L.; et al. A Dual-Mode Textile for Human Body Radiative Heating and Cooling. *Sci. Adv.* **2017**, *3*, e1700895. [CrossRef]
62. Raman, A.P.; Anoma, M.A.; Zhu, L.; Rephaeli, E.; Fan, S. Passive Radiative Cooling below Ambient Air Temperature under Direct Sunlight. *Nature* **2014**, *515*, 540–544. [CrossRef]
63. Li, W.; Shi, Y.; Chen, Z.; Fan, S. Photonic Thermal Management of Coloured Objects. *Nat. Commun.* **2018**, *9*, 4240. [CrossRef] [PubMed]
64. Yu, Z.; Gao, Y.; Di, X.; Luo, H. Cotton Modified with Silver-Nanowires/Polydopamine for a Wearable Thermal Management Device. *RSC Adv.* **2016**, *6*, 67771–67777. [CrossRef]
65. Wei, W.; Wu, B.; Guo, Y.; Hu, Y.; Liao, Y.; Wu, C.; Zhang, Q.; Li, Y.; Chen, J.; Hou, C.; et al. A Multimodal Cooling Garment for Personal Thermal Comfort Management. *Appl. Energy* **2023**, *352*, 121973. [CrossRef]
66. Xu, P.; Kang, Z.; Wang, F.; Udayraj. A Numerical Analysis of the Cooling Performance of a Hybrid Personal Cooling System (HPCS): Effects of Ambient Temperature and Relative Humidity. *Int. J. Environ. Res. Public Health* **2020**, *17*, 4995. [CrossRef]
67. Ni, X.; Yao, T.; Zhang, Y.; Zhao, Y.; Hu, Q.; Chan, A.P.C. Experimental Study on the Efficacy of a Novel Personal Cooling Vest Incorporated with Phase Change Materials and Fans. *Materials* **2020**, *13*, 1801. [CrossRef]
68. Sajawal, M.; Rehman, T.; Ali, H.M.; Sajjad, U.; Raza, A.; Bhatti, M.S. Experimental Thermal Performance Analysis of Finned Tube-Phase Change Material Based Double Pass Solar Air Heater. *Case Stud. Therm. Eng.* **2019**, *15*, 100543. [CrossRef]
69. Dhumane, R.; Qiao, Y.; Ling, J.; Muehlbauer, J.; Aute, V.; Hwang, Y.; Radermacher, R. Improving System Performance of a Personal Conditioning System Integrated with Thermal Storage. *Appl. Therm. Eng.* **2019**, *147*, 40–51. [CrossRef]
70. Gharbi, S.; Harmand, S.; Jabrallah, S.B. Experimental Study of the Cooling Performance of Phase Change Material with Discrete Heat Sources—Continuous and Intermittent Regimes. *Appl. Therm. Eng.* **2017**, *111*, 103–111. [CrossRef]
71. Kang, Z.; Udayraj; Wan, X.; Wang, F. A New Hybrid Personal Cooling System (HPCS) Incorporating Insulation Pads for Thermal Comfort Management: Experimental Validation and Parametric Study. *Build. Environ.* **2018**, *145*, 276–289. [CrossRef]
72. Lou, L.; Zhou, Y.; Yan, Y.; Hong, Y.; Fan, J. Wearable Cooling and Dehumidifying System for Personal Protective Equipment (PPE). *Energy Build.* **2022**, *276*, 112510. [CrossRef]
73. Ge, F.; Wang, C. Exergy Analysis of Dehumidification Systems: A Comparison between the Condensing Dehumidification and the Desiccant Wheel Dehumidification. *Energy Convers. Manag.* **2020**, *224*, 113343. [CrossRef]
74. Lou, L.; Chen, K.; Fan, J. Advanced Materials for Personal Thermal and Moisture Management of Health Care Workers Wearing PPE. *Mater. Sci. Eng. R Rep.* **2021**, *146*, 100639. [CrossRef]

75. Wang, F.; Zhou, X.; Wang, S. Development processes and property measurements of moisture absorption and quick dry fabrics. *Fibres Text. East. Eur.* **2009**, *17*, 46–49.
76. Zeng, S.; Pian, S.; Su, M.; Wang, Z.; Wu, M.; Liu, X.; Chen, M.; Xiang, Y.; Wu, J.; Zhang, M.; et al. Hierarchical-Morphology Metafabric for Scalable Passive Daytime Radiative Cooling. *Science* **2021**, *373*, 692–696. [CrossRef] [PubMed]
77. Rother, M.; Barmettler, J.; Reichmuth, A.; Araujo, J.V.; Rytka, C.; Glaied, O.; Pieles, U.; Bruns, N. Self-Sealing and Puncture Resistant Breathable Membranes for Water-Evaporation Applications. *Adv. Mater.* **2015**, *27*, 6620–6624. [CrossRef]
78. Nayak, R.; Kanesalingam, S.; Houshyar, S.; Wang, L.; Padhye, R.; Vijayan, A. Evaluation of Thermal, Moisture Management and Sensorial Comfort Properties of Superabsorbent Polyacrylate Fabrics for the Next-to-Skin Layer in Firefighters' Protective Clothing. *Text. Res. J.* **2018**, *88*, 1077–1088. [CrossRef]
79. Bhuiyan, M.A.R.; Wang, L.; Shanks, R.A.; Ding, J. Polyurethane–Superabsorbent Polymer-Coated Cotton Fabric for Thermophysiological Wear Comfort. *J. Mater. Sci.* **2019**, *54*, 9267–9281. [CrossRef]
80. Yang, Y.; Rana, D.; Lan, C.Q.; Matsuura, T. Development of Membrane-Based Desiccant Fiber for Vacuum Desiccant Cooling. *ACS Appl. Mater. Interfaces* **2016**, *8*, 15778–15787. [CrossRef]
81. Li, Z.; Zhu, M.; Shen, J.; Qiu, Q.; Yu, J.; Ding, B. All-Fiber Structured Electronic Skin with High Elasticity and Breathability. *Adv. Funct. Mater.* **2020**, *30*, 1908411. [CrossRef]
82. Guan, M.; Li, J. Garment Size Effect of Thermal Protective Clothing on Global and Local Evaporative Cooling of Walking Manikin in a Hot Environment. *Int. J. Biometeorol.* **2020**, *64*, 485–499. [CrossRef]
83. Wang, F.; Annaheim, S.; Morrissey, M.; Rossi, R.M. Real evaporative cooling efficiency of one-layer tight-fitting sportswear in a hot environment. *Scand. J. Med. Sci. Sports* **2013**, *24*, e129–e139. [CrossRef] [PubMed]
84. Chen, Q.; Fan, J.T.; Sarkar, M.K. Biomimetics of Branching Structure in Warp Knitted Fabrics to Improve Water Transport Properties for Comfort. *Text. Res. J.* **2012**, *82*, 1131–1142. [CrossRef]
85. Miao, D.; Cheng, N.; Wang, X.; Yu, J.; Ding, B. Sandwich-Structured Textiles with Hierarchically Nanofibrous Network and Janus Wettability for Outdoor Personal Thermal and Moisture Management. *Chem. Eng. J.* **2022**, *450*, 138012. [CrossRef]
86. Guan, M.; Annaheim, S.; Li, J.; Camenzind, M.; Psikuta, A.; Rossi, R.M. Apparent Evaporative Cooling Efficiency in Clothing with Continuous Perspiration: A Sweating Manikin Study. *Int. J. Therm. Sci.* **2019**, *137*, 446–455. [CrossRef]
87. Wang, H.; Hu, S. Experimental Study on Thermal Sensation of People in Moderate Activities. *Build. Environ.* **2016**, *100*, 127–134. [CrossRef]
88. Zhao, M.; Gao, C.; Wang, F.; Kuklane, K.; Holmér, I.; Li, J. The Torso Cooling of Vests Incorporated with Phase Change Materials: A Sweat Evaporation Perspective. *Text. Res. J.* **2013**, *83*, 418–425. [CrossRef]
89. Itani, M.; Ghaddar, N.; Ghali, K. Innovative PCM-Desiccant Packet to Provide Dry Microclimate and Improve Performance of Cooling Vest in Hot Environment. *Energy Convers. Manag.* **2017**, *140*, 218–227. [CrossRef]
90. Luo, M.; Wang, Z.; Zhang, H.; Arens, E.; Filingeri, D.; Jin, L.; Ghahramani, A.; Chen, W.; He, Y.; Si, B. High-Density Thermal Sensitivity Maps of the Human Body. *Build. Environ.* **2020**, *167*, 106435. [CrossRef]
91. Liu, W.; Lian, Z.; Deng, Q.; Liu, Y. Evaluation of Calculation Methods of Mean Skin Temperature for Use in Thermal Comfort Study. *Build. Environ.* **2011**, *46*, 478–488. [CrossRef]
92. Chen, Y.; Su, Y.; Liu, G.; Zhao, P.; Tian, M.; Li, J. Design and Performance Evaluation of a Firefighting Protective Suit with an Incorporated Liquid-Cooled System. *Int. J. Occup. Saf. Ergon.* **2023**, *26*, 1–10. [CrossRef]
93. Hou, J.; Yang, Z.; Xu, P.; Huang, G. Design and Performance Evaluation of Novel Personal Cooling Garment. *Appl. Therm. Eng.* **2019**, *154*, 131–139. [CrossRef]
94. Golbabaei, F.; Heydari, A.; Moradi, G.; Dehghan, H.; Moradi, A.; Habibi, P. The Effect of Cooling Vests on Physiological and Perceptual Responses: A Systematic Review. *Int. J. Occup. Saf. Ergon.* **2020**, *28*, 223–255. [CrossRef] [PubMed]
95. Lapka, P.; Furmanski, P.; Wisniewski, T. Assessment of thermal performance of protective garments: The advanced numerical model. *Int. J. Numer. Methods Heat Fluid Flow* **2017**, *27*, 1078–1097. [CrossRef]
96. Rathour, R.; Das, A.; Alagirusamy, R. Study on the Influence of Constructional Parameters on Performance of Outer Layer of Thermal Protective Clothing. *J. Text. Inst.* **2023**, *114*, 1336–1346. [CrossRef]
97. Pearson, S.J.; Highlands, B.; Jones, R.; Matthews, M.J. Comparisons of Core Temperature between a Telemetric Pill and Heart Rate Estimated Core Temperature in Firefighters. *Saf. Health Work* **2022**, *13*, 99–103. [CrossRef] [PubMed]
98. Moran, D.S. Stress Evaluation by the Physiological Strain Index (PSI). *J. Physiol. Pharmacol.* **2000**, *11*, 403–423. [CrossRef]
99. Udayraj; Talukdar, P.; Das, A.; Alagirusamy, R. Heat and Mass Transfer through Thermal Protective Clothing—A Review. *Int. J. Therm. Sci.* **2016**, *106*, 32–56. [CrossRef]
100. Udayraj; Wang, F. A Three-Dimensional Conjugate Heat Transfer Model for Thermal Protective Clothing. *Int. J. Therm. Sci.* **2018**, *130*, 28–46. [CrossRef]
101. Liu, G.; Liang, S.; Hu, S. Calculation Method of Mean Skin Temperature Weighted by Temperature Sensitivity of Various Parts of Human Body. *J. Therm. Biol.* **2021**, *100*, 102995. [CrossRef]
102. Fan, J.; Chen, Y.S. Measurement of Clothing Thermal Insulation and Moisture Vapour Resistance Using a Novel Perspiring Fabric Thermal Manikin. *Meas. Sci. Technol.* **2002**, *13*, 1115. [CrossRef]
103. Wang, F. Measurements of clothing evaporative resistance using a sweating thermal manikin: An overview. *Ind. Health* **2017**, *55*, 473–484. [CrossRef] [PubMed]

104. Wang, F.; Havenith, G.; Mayor, T.S.; Kuklane, K.; Leonard, J.; Młynarcyk, M.; Hodder, S.; Wong, C.; Kishino, J.; Dai, X. Clothing real evaporative resistance determined by means of a sweating thermal manikin: A new round-robin study. In Proceedings of the 10th Internatinal Meeting on Thermal Manikin and Modelling (10i3m), Tampere, Finland, 7–9 September 2014.
105. Wang, F.; Kuklane, K.; Gao, C.; Holmér, I. Development and validity of a universal empirical equation to predict skin surface temperature on thermal manikins. *J. Therm. Biol.* **2010**, *35*, 197–203. [CrossRef]
106. Wang, F.; Lai, D.; Shi, W.; Fu, M. Effects of fabric thickness and material on apparent 'wet' conductive thermal resistance of knitted fabric 'skin' on sweating manikins. *J. Therm. Biol.* **2017**, *70*, 69–76. [CrossRef] [PubMed]
107. Takada, S.; Kobayashi, H.; Matsushita, T. Thermal Model of Human Body Fitted with Individual Characteristics of Body Temperature Regulation. *Build. Environ.* **2009**, *44*, 463–470. [CrossRef]
108. Li, F. A 3D Finite Element Thermal Model for Clothed Human Body. *JFBI* **2013**, *6*, 149–160. [CrossRef]
109. Enescu, D. Models and Indicators to Assess Thermal Sensation Under Steady-State and Transient Conditions. *Energies* **2019**, *12*, 841. [CrossRef]
110. Takada, S.; Sakiyama, T.; Matsushita, T. Validity of the Two-Node Model for Predicting Steady-State Skin Temperature. *Build. Environ.* **2011**, *46*, 597–604. [CrossRef]
111. Doherty, T.; Arens, E.A. *Evaluation of the Physiological Bases of Thermal Comfort Models*; American Society of Heating, Refrigerating and Air-Conditioning Engineers: Peachtree Corners, GA, USA, 1988.
112. Ooka, R.; Minami, Y.; Sakoi, T.; Tsuzuki, K.; Rijal, H.B. Improvement of Sweating Model in 2-Node Model and Its Application to Thermal Safety for Hot Environments. *Build. Environ.* **2010**, *45*, 1565–1573. [CrossRef]
113. Melnikov, V.; Krzhizhanovskaya, V.V.; Lees, M.H.; Sloot, P.M.A. System Dynamics of Human Body Thermal Regulation in Outdoor Environments. *Build. Environ.* **2018**, *143*, 760–769. [CrossRef]
114. Joshi, A.; Wang, F.; Kang, Z.; Yang, B.; Zhao, D. A three-dimensional thermoregulatory model for predicting human thermophysiological responses in various thermal environments. *Build. Environ.* **2022**, *207*, 108506. [CrossRef]
115. Katić, K.; Li, R.; Zeiler, W. Thermophysiological Models and Their Applications: A Review. *Build. Environ.* **2016**, *106*, 286–300. [CrossRef]
116. Stolwijk, J.A. Mathematical models of thermal regulation. *Ann. N. Y. Acad. Sci.* **1980**, *335*, 98–106. [CrossRef]
117. Munir, A.; Takada, S.; Matsushita, T. Re-Evaluation of Stolwijk's 25-Node Human Thermal Model under Thermal-Transient Conditions: Prediction of Skin Temperature in Low-Activity Conditions. *Build. Environ.* **2009**, *44*, 1777–1787. [CrossRef]
118. Kang, Z.; Wang, F.; Udayraj. An advanced three-dimensional thermoregulatory model of the human body: Development and validation. *Int. Commun. Heat Mass Transf.* **2019**, *107*, 34–43. [CrossRef]
119. Ismail, N.; Ghaddar, N.; Ghali, K. Predicting Segmental and Overall Ventilation of Ensembles Using an Integrated Bioheat and Clothed Cylinder Ventilation Models. *Text. Res. J.* **2014**, *84*, 2198–2213. [CrossRef]

Disclaimer/Publisher's Note: The statements, opinions and data contained in all publications are solely those of the individual author(s) and contributor(s) and not of MDPI and/or the editor(s). MDPI and/or the editor(s) disclaim responsibility for any injury to people or property resulting from any ideas, methods, instructions or products referred to in the content.

processes

MDPI

Review

Textiles for Very Cold Environments

Tomasz Blachowicz [1,*], Maciej Malczyk [1], Ilda Kola [2], Guido Ehrmann [3], Eva Schwenzfeier-Hellkamp [4] and Andrea Ehrmann [4]

1 Institute of Physics—Center for Science and Education, Silesian University of Technology, 44-100 Gliwice, Poland
2 Department of Textile and Fashion, Polytechnic University of Tirana, 1019 Tirana, Albania; ikola@fim.edu.al
3 Virtual Institute of Applied Research on Advanced Materials (VIARAM)
4 Institute for Technical Energy Systems (ITES), Bielefeld University of Applied Sciences and Arts, 33619 Bielefeld, Germany; eva.schwenzfeier-hellkamp@hsbi.de (E.S.-H.); andrea.ehrmann@hsbi.de (A.E.)
* Correspondence: tomasz.blachowicz@polsl.pl

Abstract: Textiles are often used to protect people from cold environments. While most garments are designed for temperatures not far below 0 °C, very cold regions on the earth near the poles or on mountains necessitate special clothing. The same is true for homeless people who have few possibilities to warm up or workers in cooling chambers and other cold environments. Passive insulating clothing, however, can only retain body heat. Active heating, on the other hand, necessitates energy, e.g., by batteries, which are usually relatively heavy and have to be recharged regularly. This review gives an overview of energy-self-sufficient textile solutions for cold environments, including energy harvesting by textile-based or textile-integrated solar cells; piezoelectric sensors in shoes and other possibilities; energy storage in supercapacitors or batteries; and heating by electric energy or phase-change materials.

Keywords: personal protective equipment; energy harvesting; energy storage; heating; flexible solar cells; temperature sensors

Citation: Blachowicz, T.; Malczyk, M.; Kola, I.; Ehrmann, G.; Schwenzfeier-Hellkamp, E.; Ehrmann, A. Textiles for Very Cold Environments. *Processes* **2024**, *12*, 927. https://doi.org/10.3390/pr12050927

Academic Editors: Zhanxiao Kang and Qing Chen

Received: 11 April 2024
Revised: 28 April 2024
Accepted: 29 April 2024
Published: 1 May 2024

Copyright: © 2024 by the authors. Licensee MDPI, Basel, Switzerland. This article is an open access article distributed under the terms and conditions of the Creative Commons Attribution (CC BY) license (https://creativecommons.org/licenses/by/4.0/).

1. Introduction

Keeping people warm in a cold environment is one of the main purposes of clothes. While thousands of years ago, natural fibers from animals and plants, as well as fur parts, were woven into warming garments [1], nowadays, chemical fibers are being investigated more and more regarding their applicability in warming clothes [2–4]. Besides passive warming by reflecting thermal radiation back towards the human body [5,6], creating air layers as insulation to reduce heat convection [7,8] or increasing the ratio of absorbed thermal radiation from the sun [9,10], there are nowadays many attempts to provide active heating.

Amongst the physical effects that can be used for heating up garments, there are phase-change materials (PCMs) [11–13], as well as different possibilities for converting electric energy into heat [14–16]. While the first are energy-self-sufficient, they have the disadvantage that they can only work in a material-dependent temperature range and cannot be controlled by the user. Electric heating of garments, also called Joule heating [17], on the other hand, needs the possibility of storing electric energy in batteries or supercapacitors to make it available when it is needed and ideally to harvest energy by the garment to make it independent from external power supplies.

This review gives an overview of recent research regarding energy-self-sufficient textiles for active heating to protect people from very cold environments, such as workers in cold storage houses, astronauts, people living near the poles or homeless people. The next sections review different possibilities for actively heating textiles by PCMs or electrically to harvest and store energy in textiles, and then the main part discusses recent research on energy-self-sufficient heating textiles and their potential applications.

2. Active Heating with Textiles

Active heating always requires energy, usually either in the form of electric energy or as latent heat stored in PCMs or subcooled liquids, whereupon the latter are usually not applied in textile fabrics but in simple heat pads activated by buckling a metal sheet. Here we give an overview of different ways of active heating with textile fabrics.

2.1. Phase-Change Materials

Phase-change materials can be integrated into textiles in the form of fibers, microcapsules or nanocapsules [18–20]. PCMs store a large amount of latent heat when molten and release it during solidifying [21]. This process, as shown in Figure 1, makes them especially interesting for textiles used in environments with fluctuating temperatures, e.g., for astronauts' space suits [22]. During melting or crystallization, the temperature of the PCM remains approx. constant. During an increase in temperature, the PCM can absorb heat that is stored in the liquefied phase, while decreasing external temperatures will lead to the release of this stored heat energy by the solidification of the PCM [23].

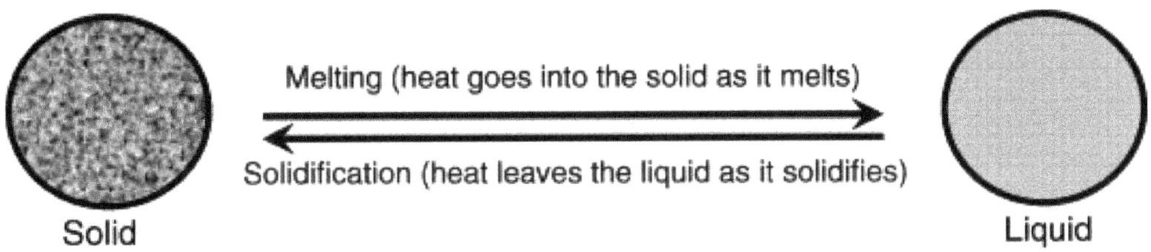

Figure 1. Schematic representation of phase-change process. Reprinted from [22], Copyright 2008, with permission from Elsevier.

Different materials have not only different temperature ranges for storing and releasing heat but also different energy storage properties. While water at 0 °C or even 100 °C can also be regarded as a PCM, several materials provide much higher energy storage capacity combined with high thermal conductivity in a suitable temperature range [24]. For paraffin waxes, Farid et al. reported melting temperatures of 44–64 °C and latent heats of 167–210 kJ/kg [25]. A better suitable temperature range of 30–65 °C was found for fatty acids, which could store latent heat of 153–182 kJ/kg [26]. Glauber salt has a melting temperature of 32 °C, a high latent heat of 254 kJ/kg and is an inexpensive material but shows problems with phase segregation, similar to most other hydrated salts [27].

Amongst the potential problems with using PCMs for active heating/cooling, the usually very low thermal conductivity should be mentioned [28]. To increase their thermal conductivity, metallic or graphitic nano- or microstructures can be embedded in the PCM, whereupon too much additional conductive material would reduce the specific energy storage capacity [29,30].

Depending on the environmental conditions, this range of energies stored as latent heat defines how long a PCM can maintain its melting/solidifying temperature, naturally scaling with the mass of PCM included in a garment. However, the calculation is less simple than may be expected. For a PCM inside a spherical capsule of approx. 100 mm diameter, Tan et al. compared experimental observations during melting with a computational approach [31]. They found thermally stable structures at the top of the capsule, while unstable structures were found at the bottom of the capsule, as depicted in Figure 2a, leading to clear differences in the calculated and measured temperatures, especially in the lower half of the capsule, as visible in Figure 2b [31]. Besides, calculations with different numerical simulations may lead to slightly varying results [32]. In addition, even small changes in the melting temperature near the environmental temperature will lead to large changes in the additional necessary energy to keep a defined temperature [33]. Besides the melting temperature, thermal conductivity was found to have the largest influence on the cooling energy demand [34]. Due to these potential sources of relatively large errors, this review will concentrate on the measured effects of PCMs in garments in cold environments reported in different studies instead of extrapolating the measured findings to other environmental situations.

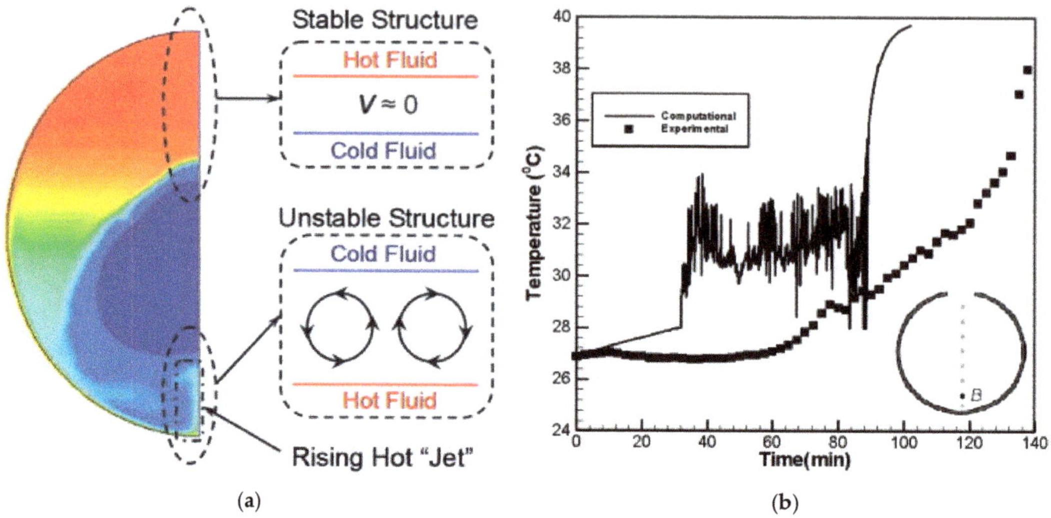

Figure 2. (**a**) Thermally stable and unstable layer structures along the symmetry axis; (**b**) comparison of the computed and measured temperatures at point B at a distance of 0.369 diameters below the center. Blue areas depict the cold, unmolten solid phase, while red areas are hot, molten liquid. Reprinted from [31], Copyright 2009, with permission from Elsevier.

2.2. *Joule Heating*

The most often used possibility for converting electrical energy into heat is based on the so-called Joule heating, also called resistance heating or electro-thermal conversion [35]. Theoretically, the heating power P can simply be calculated by the Joule–Lenz law as $P = RI^2 = U^2/R$, with the resistance R and the current I flowing through it. The thermal energy Q_W is equal to the electrical energy E_{el} and can be calculated as the integral over P, or in the case of time-independent power, it becomes $Q_W = E_{el} = Pt$ with the duration t of heating [17]. This means that conductive materials are necessary for electro-thermal conversion, such as intrinsically conductive polymers or polymers filled with metallic nanoparticles, graphene, graphite, carbon nanotubes, carbon black or, nowadays, MXenes [17]. For use in garments, all of them have specific challenges, such as relatively low conductivity and missing long-term stability of conducting polymers, health risks posed by metallic nanoparticles and MXenes, or low washing resistance of graphene and other materials [17]. Nevertheless, Joule heating is a very often studied method for producing clothing with active heating. It should be mentioned that while the resistance of the heating circuit in a textile fabric is essential for the heating power, this value is often not given or only partly given as linear resistance of a yarn or sheet resistance of a coating whose thickness may be different in different areas of a garment. This makes calculations of the heating power and necessary energy often complicated or even impossible.

2.3. *Peltier Element Heating*

Peltier elements are based on the Peltier effect, or thermoelectric effect, which converts electrical energy into a temperature gradient inside a semiconducting material [36]. Several factors influence the figure of merit, which defines the performance of organic, inorganic or hybrid Peltier materials, which should have low thermal conductivity, high electric conductivity and a high Seebeck coefficient [37]. Due to the interconnections between these parameters, optimization is not straightforward but necessitates complicated methods such as strengthening phonon scattering [38], band engineering [39] or nanostructuring materials [40].

It must be mentioned that, unlike Joule heating textiles, Peltier elements do not heat up through the whole material but have one warmer and one cooler side. Schmidl et al., e.g., prepared Al-doped ZnO-coated polyester spacer fabrics and measured a temperature difference between both sides of the spacer fabric of up to 12 K, which was converted upon changing the polarity of the electrical contacts [41]. This fact must be taken into account when an active heating garment based on Peltier elements is created. On the other hand, Peltier elements in clothing offer heating and cooling depending on the polarity of the applied voltage, making them highly suitable for varying weather conditions [42].

3. Energy-Harvesting Textiles

While PCMs store thermal energy received from the environment in the form of latent heat, electrically heated textiles need to harvest and store energy that can be used for active heating. Mainly, such devices are based on triboelectric nanogenerators (TENGs), solar cells and thermoelectric devices [43], but there are also other mechanisms integrated into different smart textiles to harvest body heat energy, biochemical energy, etc. [44]. This section gives an overview of potential methods of harvesting energy by clothing.

3.1. Triboelectric Nanogenerator

When a person is moving, this mechanical energy can principally be harvested by a triboelectric (nano)generator (TE(N)G) [45]. Movements of the human body, such as walking or arm movements, offer several ten watts of kinetic power [46]. A TENG can be used to transfer movements into contact electrification and electrostatic induction, in this way harvesting energy from human motion [47,48]. Generally, a TENG consists of two materials with different electron affinity, resulting in electrostatic induction when these materials are coupled and moved relative to each other during body movements [49]. This induced potential difference can be reduced by a current through a load connecting the electrodes attached to the triboelectric materials, thus resulting in an AC power output for a periodic movement [49]. Depending on the positioning and movement of triboelectric materials and electrodes, TENGs can be separated into vertical contact-separation mode, lateral sliding mode, single-electrode mode and freestanding triboelectric-layer modes, as depicted in Figure 3 [49]. Most textile-based TENGs work in the vertical contact separation mode due to the ease of embedding them into shoe insoles or other textiles that are steadily pressed or stretched [49].

The first textile-based realizations of TENGs were based on coated yarns, e.g., one with carbon nanotubes and the other one with polytetrafluoroethylene (PTFE), to convert motion or vibration energy into electric energy by the electrostatic effect, leading to relatively small output power densities of ~1 mW/m^2 [50]. A similar value of 1.8 mW/m^2 was reported for alternating polyimide (PI)/polyurethane (PU) strips on a sleeve and alternating polydimethylsiloxane (PDMS)/aluminum strips on the torso for a movement frequency of 1.5 Hz [51]. Slightly higher power densities of 27 mW/m^2 were reported for a TENG that combined shear-thickening fluid and magneto-sensitive films, in this way adding impact-resistant properties [52]. A much higher power density of 953 mW/m^2 was found for a single-thread TENG, produced using a stainless-steel core with a silicone rubber shell and placed on a textile in a sine-like shape [53]. An approx. doubled power density of 2 W/m^2 was found for a PEDOT:PSS-coated textile and PTFE under foot stepping with a frequency of 2 Hz [54]. Still higher values of 3.2 W/m^2 were found for a freestanding triboelectric-layer TENG prepared from Ni-coated electrodes and a parylene triboelectric layer [55]. Even 8.9 W/m^2 was reached by a TENG in which the palm skin was used as a freestanding triboelectric layer and a silicone rubber- and Ni-coated polyester woven fabric as an electrode [56]. Developing textile-based TENGs further, an extremely high power density of 336 W/m^2 was reported for an AlNP-coated top textile and a nanostructured PDMS bottom textile [57].

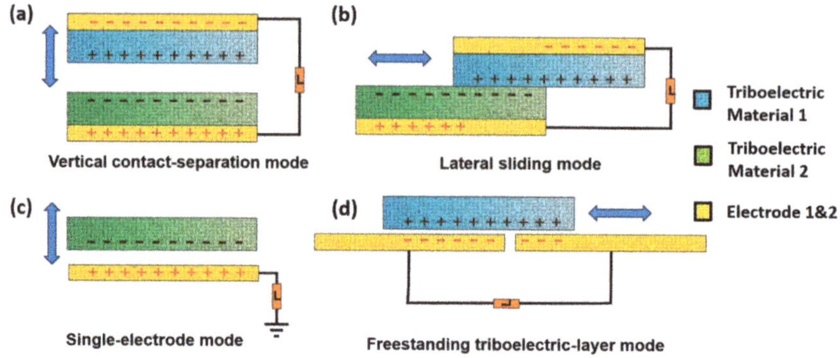

Figure 3. Fundamental modes of TENGs: (**a**) Vertical contact-separation mode; (**b**) lateral sliding mode; (**c**) single-electrode mode; (**d**) freestanding triboelectric-layer mode. Reprinted from [49], Copyright 2019, with permission from Elsevier.

It should be mentioned that these areal power densities are usually peak instantaneous values, while many applications—such as active heating—necessitate a DC power [49]. Besides, the comparison above is based on different areas, varying frequencies and different forces and loads, making a rating of recently investigated textile TENGs complicated. In addition, wash-and-wear resistance, as well as general durability, are in many cases not given, and the areal power density does not take into account the thickness or mass of the respective textile. However, power densities in the range of some ten milliwatts per square meter up to a few watts per square meter are typical for recent textile TENGs and can be used as a base for calculations of harvestable electric energies.

3.2. Solar Energy

Solar energy is the primary source of energy on Earth; thus, harvesting is a natural approach. Different photovoltaic (PV) materials are able to convert the photons of sunlight into electrical energy using the photovoltaic effect [58]. The photovoltaic conversion efficiency can be calculated as $\eta = P_{out}/P_{in}$, with the output electrical power P_{out} and the input solar radiation power P_{in} [58]. While conventional PV systems are rigid, nowadays flexible PV cells can also be produced, e.g., based on dye-sensitized solar cells (DSSCs) [59,60].

To produce photovoltaic textiles, usually flexible, organic PV films are mounted on a fabric, or textile fibers/yarns or textile layers are prepared to have photovoltaic properties [60,61]. Typical efficiencies of such textile-based solar cells are around 1% for optimum production methods, dyes—in the case of DSSCs—and conductivities of the electrodes [62–64] but can be much smaller if non-toxic materials are used for a planned application near the human body [65–67]. Assuming a conversion efficiency of 1% and a perfect orientation of the textile towards the sun, the average solar radiation flux density of approx. 165 W/m^2 [68] would enable energy harvesting of 16.5 W/m^2 with textile-based solar cells.

Generally, such textile-based solar cells are problematic to tailor, which is only scarcely addressed [69]. Wash is even more complicated and necessitates encapsulation, e.g., by lamination and thus modifying the original textile haptics, or by using very small, encapsulated modules, which only slightly increase the bending rigidity of a textile fabric [70–72]. However, these challenges, combined with the well-known problem that efficiencies are reduced for larger cells [73], typically reduce the aforementioned optimum solar energy harvesting.

3.3. Thermoelectric Devices

While Peltier elements can be used to create a temperature difference between both sides of the element by applying a voltage (cf. Section 2.3), the Seebeck effect can, vice versa, be used to generate a thermoelectric potential energy due to a temperature difference between both electrodes on the respective thermoelectric element [74]. Generally, a thermoelectric element consists of a pair of *p*- and *n*-doped semiconductors between a hot and a cold side in which the heat flow from the hot to the cold surface is accompanied by an electric current, as depicted in Figure 4 [75]. The generated voltage V is proportional to the temperature difference ΔT according to $V = \alpha \Delta T$ with the Seebeck coefficient α [75]. The power generation also depends on the cross-sectional area of the thermoelectric device and the length of their legs, besides the aforementioned material properties, such as the Seebeck coefficient and electrical and thermal conductivity [75].

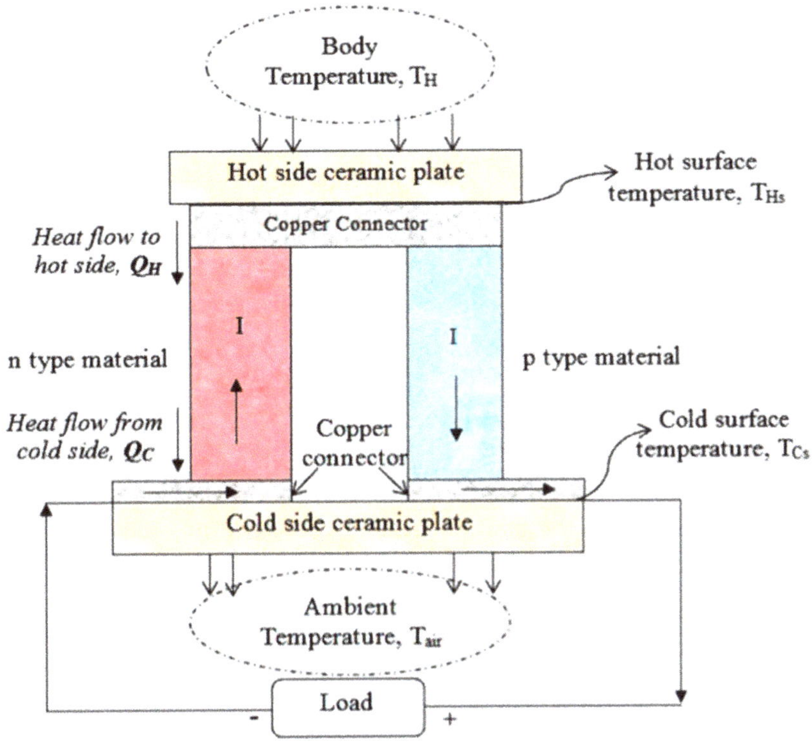

Figure 4. Single thermoelectric pair consisting of *n*-type and *p*-type materials. Heat flows from hot side to top side ($Q_H \rightarrow Q_C$), and electrical current (I) flows from *n*-type to *p*-type material due to a temperature gradient ($\Delta T = T_{Hs} - T_{Cs}$). Reprinted from [75], Copyright 2017, with permission from Elsevier.

With a wearable thermoelectric generator, calculated for a skin temperature of 34 °C, ambient temperature of 22 °C and a heat flow of 20 mW/cm^2, the human body could maximally generate 180 µW/cm^2, i.e., 1.8 W/m^2, with power generation around 1–37 mW for different body parts [76,77]. For larger temperature differences than 12 K, the generated energy would be accordingly higher. While this value is approx. one order of magnitude lower than the previously mentioned theoretical value for textile-based solar cells, it is similar to the range of values reported for TENGs, as discussed before.

Several researchers have investigated textile-based thermoelectric generators and optimized them. Lund et al. used a combination of PEDOT:PSS-coated silk threads and silver-plated polyamide threads to produce a thermoelectric generator by hand-sewing through nine layers of felted wool, connected the legs on the textile surface by coating with a silver-containing paste, and measured the power generation for the hot plate at 35 °C and the cold plate at −30 °C to −5 °C, with results as depicted in Figure 5 [78]. With a device area of $(5.3\ cm)^2$ and an optimized design, they reached a maximum of 1.2 μW for ΔT = 65 K [78], i.e., approx. 427 $\mu W/m^2$ if the whole surface area is taken into account, not only the embroidered parts.

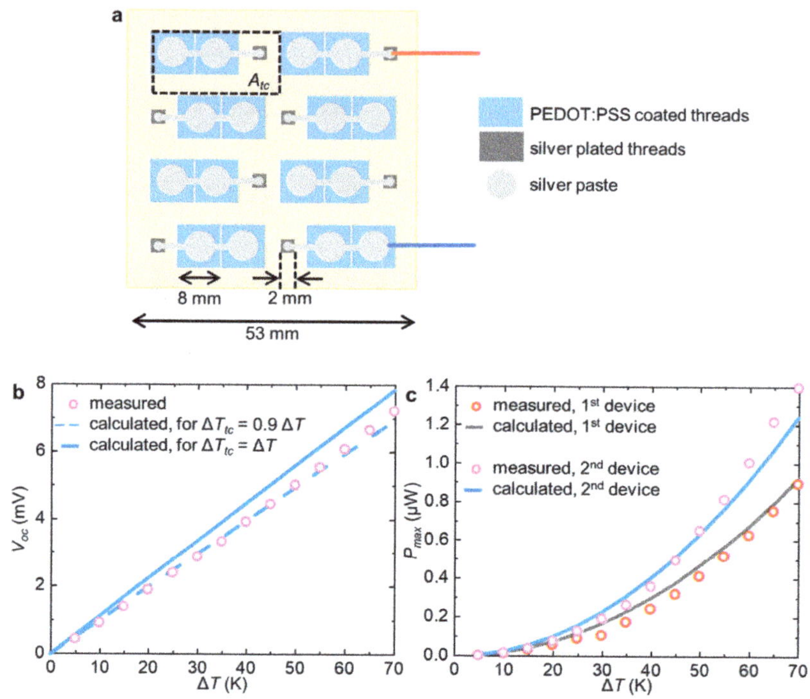

Figure 5. (**a**) Schematic of optimized textile thermopile with optimized leg areas, red and blue lines are connectors to the voltage measuring equipment. (**b**) Measured open-circuit voltage V_{oc} (circles) as a function of the temperature gradient ΔT between the hot plate and the cooler, and calculated data for V_{oc} as a function of ΔT, where $\Delta T_{tc} = \Delta T$ (solid line) and $\Delta T_{tc} = 0.9{\cdot}\Delta T$ (dashed line). (**c**) Measured generated power (circles, red for the 1st thermopile, purple for the 2nd thermopile) as a function of ΔT and calculated generated power (solid lines, grey for the 1st thermopile, blue for the 2nd thermopile) assuming $\Delta T_{tc} = 0.9{\cdot}\Delta T$ and electrical contact resistance of 1.2 Ω (1st thermopile) or 1.1 Ω (2nd thermopile) per thermocouple. Reprinted from [78], originally published under a CC-BY license.

With 864 connected legs coated with PEDOT:PSS and poly([Na(NiETT)], respectively, on an area of $(25\ cm)^2$, a temperature difference of 3 K resulted in a power of 13 μW, whereupon the authors calculated that densely filling 8% of the whole body surface area with the legs of a thermoelectric generator would result in 1 mW power output [79]. Combining PEDOT:PSS and a p-type semiconductor with n-type Ag_2Te, a power output of 6 mW/m^2 was reached by a temperature difference of 20 K [80], while Ag_2Se with Ag connection resulted in a power density of 2.3 W/m^2 for ΔT = 30 K [81].

For a one-dimensional thermoelectric generator with PEDOT:PSS and carbon nanotube/polyethylenimine hydrogel fibers, a value of 4.8 W/m^2 was found for a temperature difference of 60 K [82]. On the other hand, a three-dimensional woven textile produced from a yarn with alternating n- and p-type segments reached a power output of approx. 0.65 W/m^2 for a typical temperature difference of 55 K [83]. In many cases, however, only a maximum power is given for a textile-based thermoelectric generator [84], making it more complicated to estimate the maximum thermoelectric power that could be generated by garments.

3.4. Other Energy Harvesting Methods

Among the other methods of energy harvesting by garments, piezoelectric textiles must be mentioned, which are regularly investigated. As depicted in Figure 6, piezoelectric materials perform a charge separation when a pressure is applied or released, leading to a voltage [85]. Amongst the possible materials that show the piezoelectric effect, such as ZnO, $BaTiO_3$ or $GaPO_4$, especially polyvinylidene fluoride (PVDF) has often been investigated due to its good mechanical properties, flexibility and chemical and thermal stability [85–87]. For this material, Liu et al. reported an efficiency of around 12% for the transition of mechanical into electrical energy and vice versa [88]. With a combination of PVDF, $BaTiO_3$ nanoparticles and reduced graphene oxide nanoplates, even an energy conversion efficiency of 22.5% was reached [89]. For single melt-spun PVDF, polyamide 11 and polypropylene fibers, powers in the range of 0.2 nW were generated [90]. Combining PVDF with lead zirconate titanate (PZT), a piezoelectric nanogenerator from electrospun-aligned nanofibers produced 6.35 µW [91]. Besides electrospinning and melt spinning as typical fiber production methods, it is also possible to use physical or electrochemical deposition of piezoelectric materials on textile fabrics, which is especially relevant for non-polymeric materials such as ZnO [58].

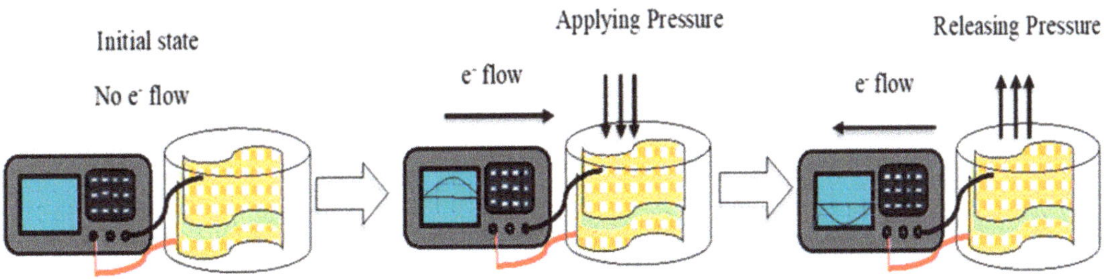

Figure 6. Working principle of a piezoelectric material when applying and releasing pressure. Reprinted from [85], originally published under a CC-BY license.

Other research groups suggested additional methods of harvesting energy by smart textiles, e.g., biological fuel cells that generate energy by biodegradation of organic matter [92,93], partly using the wearer's sweat to harvest biochemical energy [94].

Generally, as suggested by Lund et al. [78], a reasonable combination of different energy harvesting methods should be applied to optimize the overall gained energy. Depending on the environmental conditions (temperature difference, sunlight) and movements of the person wearing the respective harvesting garments, a power density of the order of magnitude 1 W/m^2 can be expected. In the next section, storing the energy harvested in a defined time in textiles will be discussed.

4. Energy-Storing Smart Textiles

Most of the aforementioned forms of energy that can be transferred into electrical energy—sunlight, body movements—are not constantly available; only the temperature difference between skin temperature and surrounding temperature, as used in thermoelectric devices, is automatically larger when more heating energy is needed. Using a combination of different energy harvesting methods, however, necessitates storing parts of the energy harvested in sunlight and during active movements for times when it is more needed.

Several researchers thus investigated methods to store energy in textile fabrics, either fiber-, yarn- or fabric-based, usually in the form of batteries or supercapacitors, as depicted in Figure 7 [95].

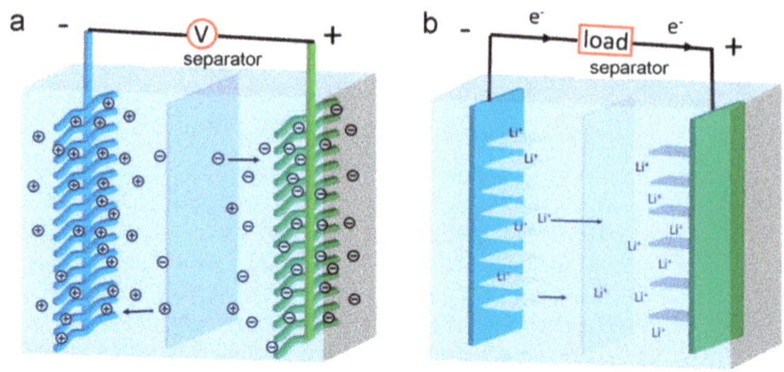

Figure 7. Schematic illustrations of energy storage mechanisms of (**a**) a supercapacitor and (**b**) a lithium-ion battery. Reprinted from [95], Copyright 2016, with permission from Elsevier.

In all these cases, several design strategies have to be taken into account, such as flexibility and washability of the materials, but also the electrical conductivity of electrodes and wires, as well as more specific physical and chemical properties of the materials used for batteries or supercapacitors [95–97]. This section discusses recent approaches to storing electrical energy in textile fabrics.

4.1. Fiber-Based Supercapacitors

Fiber-based supercapacitors can be produced, e.g., by wet spinning, dry spinning, microfluidic spinning, hydrothermal self-assembly and other methods based on graphene and other materials [98]. An example of the production of a fiber-based supercapacitor is shown in Figure 8, based on nitrogen-doped SiNs/graphene hybrid fibers [99]. More precisely, the exfoliation of $CaSi_2$ using HCl resulted in siloxene (SiNs) nanosheets after several days, which were coated with polypyrrole (Ppy) before these nanosheets were pyrolyzed, resulting in a nitrogen-doped carbon layer on the siloxene nanosheet (N-SiNs) (Figure 8a). Wet spinning from a N-SiNs/GO (graphene oxide) dispersion and chemically reducing the GO into rGO (reduced graphene oxide) resulted in fiber-based supercapacitors (Figure 8b) with an areal specific capacitance of 264 mF/cm^2 [99]. The capacitance C of a supercapacitor enables storing an energy of $E = 1/2CV2$ with voltage V.

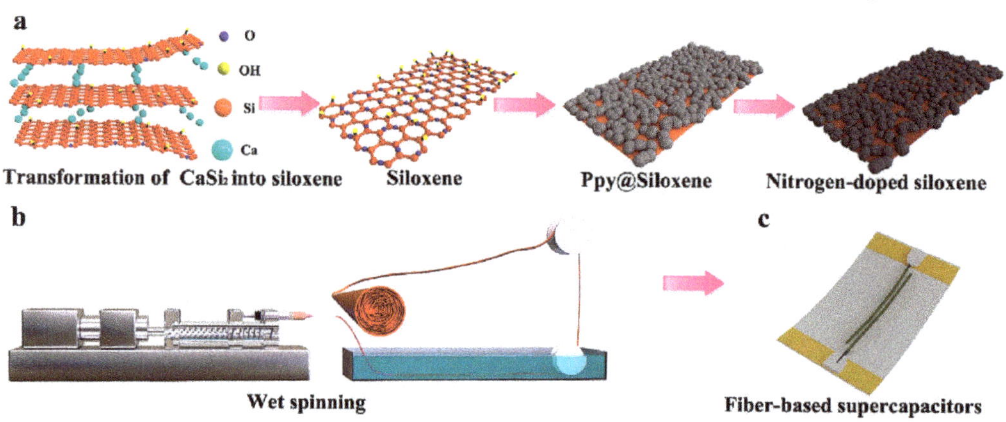

Figure 8. Schematic illustration of the fabrication of the N-SiGF sample and the fiber-based supercapacitors (**a**–**c**). Reprinted from [99], Copyright 2023, with permission from Elsevier.

When comparing different supercapacitors, it must be taken into account that sometimes only the electrodes are taken into account when calculating the capacitance for a certain mass, volume or area, while in other studies, the whole supercapacitor, including the separator and the electrolyte, is considered [100]. On the other hand, sometimes, the power density is given instead of an energy density.

A fiber-shaped supercapacitor in the form of a partially unzipped carbon nanotube/rGO hybrid fiber was produced by wet spinning and chemical reduction, leading to a high volumetric energy density of 8.63 mWh/cm^3 [101]. For an alginate/PEDOT:PSS@Ppy composite fiber, a volumetric capacity of 568 F/cm^3 and an areal capacity of 1000 mF/cm^2 were found, as well as a high energy density of 21 µWh/cm^2 [102]. A smaller areal capacitance of 115 mF/cm^2 and a similar energy density of 9 µWh/cm^2, as well as a power density of 0.11 mW/cm^2, were reported for hollow PEDOT:PSS thin-walled fibers [103]. Nanostructured MnO_2-based fiber-in-tube and particle-in-tube supercapacitors reached a capacity of 432 F/g, an energy density of 46 mWh/g and a power density of 400 mW/g [104]. Combining activated carbon fiber as substrate and a polyaniline (PAni) composite fiber coated with commercial pen ink resulted in a linear capacitance of 108 mF/cm or 68 mF/cm^2, which was coupled with MnO_2@ink/activated carbon fiber to reach an energy density of 102 µWh/cm^2 and a power density of 1 mW/cm^2 [105]. A capacitance of 17.5 F/cm^3 or 10.7 F/g, leading to an energy density of 7.88 mWh/cm^3 or 4.82 mWh/g and a power density of 2.26 W/cm^3 or 1.382 W/g, was reported for an asymmetric supercapacitor, produced by twisting a MnO_2/CNT fiber cathode and a Ppy/CNT fiber anode combined with a LiCl/poly(vinyl alcohol) (PVA) electrolyte [106].

An interesting approach based on jute fibers with PEDOT:PSS and single-wall CNTs, combined with a cellulose-based material as separator, was shown to reach an energy density of 0.71 µWh/cm^2 and a power density of 3.85 µW/cm^2, as well as a specific capacitance of 8.65 mF/cm, which is much smaller than the aforementioned values but was aiming at a more environmentally friendly approach [107].

To evaluate these different values, we can assume that the power gained with the aforementioned energy-harvesting textiles, which is in the order of magnitude 1 W/m^2, should typically be stored for approx. half a day, e.g., due to harvesting sun and movement energy during daytime and releasing it again during nighttime, so that an energy of the order of magnitude 10 Wh/m^2 should be stored if the whole textile area covering the human body is used for energy harvesting as well as for energy storing. The energy densities reported for fiber-based supercapacitors are typically in the range of 5 mWh/m^2–1 Wh/m^2, comparing the aforementioned and other values [108], which would not be sufficient to store the whole energy that can theoretically be gained during half a day. However, there are a few reports mentioning one or two orders of magnitude higher energy density, such as 147 Wh/m^2 for a core-shell MoS_2 nanosheet array/graphene hybrid fibers showing an increased specific capacitance by two orders of magnitude due to the MoS_2 nanosheet array surface deposition [109]. Apparently, it is necessary to use such specific materials to enable storing all possibly harvested energy for about a half day in fiber-based supercapacitors.

4.2. Fabric-Based Supercapacitors

Fabric-based supercapacitors can be produced by different coating, printing and lamination methods. As an example, for an inkjet-printed MnO_2–$NiCo_2O_4$/rGO (positive/negative electrode) asymmetric supercapacitor on bamboo fabric, an areal capacitance of 2.12 F/cm^2 was found, leading to an energy density of 37.8 mW/cm^3 and a power density of 2.7 W/cm^3 [110]. For a supercapacitor based on carbon fiber electrodes functionalized with vertical graphene and MnO_2 and a glass fiber separator, Sha et al. reported an areal capacitance of 31 mF/cm^2, an energy density of 12 mWh/kg and a power density of 2210 mW/kg [111]. Wen et al. showed a flexible zinc-ion supercapacitor with an energy density of 0.32 mWh/cm^2, which could also be used as a strain sensor [112]. A detailed overview of materials, production methods and energy densities of fabric-based supercapacitors can be found in [113] (Figure 9).

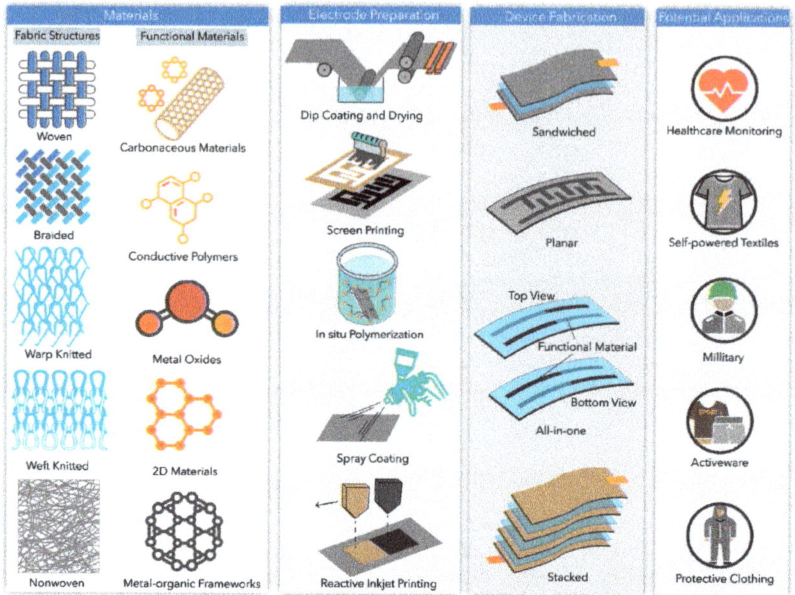

Figure 9. Materials, preparation methods and potential applications of fabric-based supercapacitors. From [113], originally published under a CC-BY license.

In his recent overview, Khadem [113] reported energy densities of up to 4.7 Wh/cm^2, i.e., 47 kWh/m^2 [114], which would be more than sufficient to store the aforementioned 10 Wh/m^2 energy for approx. half a day of energy harvesting, or even 190 kW/m^2 [115] or 1.5 MWh/m^2 [116]. However, most energy densities are several orders of magnitude smaller [113,117–120], so here again, it is important to properly choose a suitable fabric-based supercapacitor to enable sufficient energy storage capacity.

4.3. Flexible Batteries

Flexible fiber- or fabric-based batteries, such as lithium- or zinc-based batteries, often suffer from problems regarding encapsulation, high internal resistance and low durability [121]. There are, however, many more anode and cathode materials that have been investigated for their potential use in flexible batteries, such as carbon, titanium compounds, chalcogens or Na as anode materials and different oxides, sulfur or gasses as cathodes, combined with different aqueous, gel-polymer or solid-polymer electrolytes, several of which have also been investigated for fiber- or fabric-based batteries [122]. Recently, other potential electrode materials, such as vanadium nitride, have been suggested [123].

A solid-state Zn/MnO_2 fiber battery with high cycling stability was suggested by Xiao et al., who used a graphene oxide (GO)-embedded polyvinyl alcohol (PVA) hydrogel electrolyte (GPHE) (Figure 10) [124]. They found good mechanical properties, such as a stretchability of 230% without breakdown and self-healing of the gel electrolyte, as well as an energy density of 91 Wh/L, i.e., 91 mWh/cm^3 [124]. For a highly elastic fiber-based graphene/PAni-Zn@silver battery with a helical structure, a capacitance of 32.6 mAh/cm^3 and an energy density of 36 mWh/cm^3 were reported [125].

For fabric-based batteries, other problems may occur due to fractures of metalized areas [126]. Printed batteries, e.g., by screen printing or inkjet printing, are thus advantageous for the production of flexible fabric-based batteries [127]. Alternatively, batteries can be produced as (nano-)composites from fibrous materials, as depicted in Figure 11 [128]. The fabric-based battery depicted here, based on $NiCo_2S_4$@rGO nanocomposites, reached a power density of 3.2 W/g and an energy density of 0.455 Wh/g [128]. A similar energy density of 0.429 Wh/g was found for a sodium-ion with $Na_3V_2(PO_4)_2F_3$@C cathode [129]. For a potassium-ion battery with potassium nickel iron hexacyanoferrate as the cathode material, an energy density of 0.283 Wh/g was reported [130].

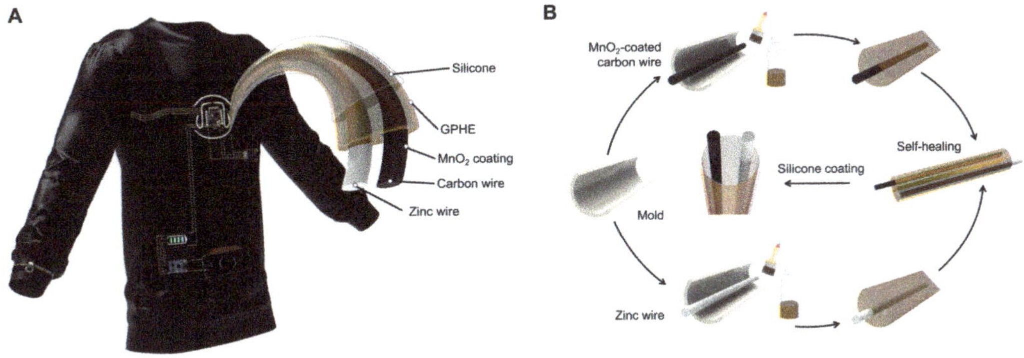

Figure 10. (**A**) Illustration and design of a flexible zinc-ion battery textile body area network. (**B**) Fabrication process of the functional fibers as the building blocks. From [124], originally published under a CC-BY-NC license.

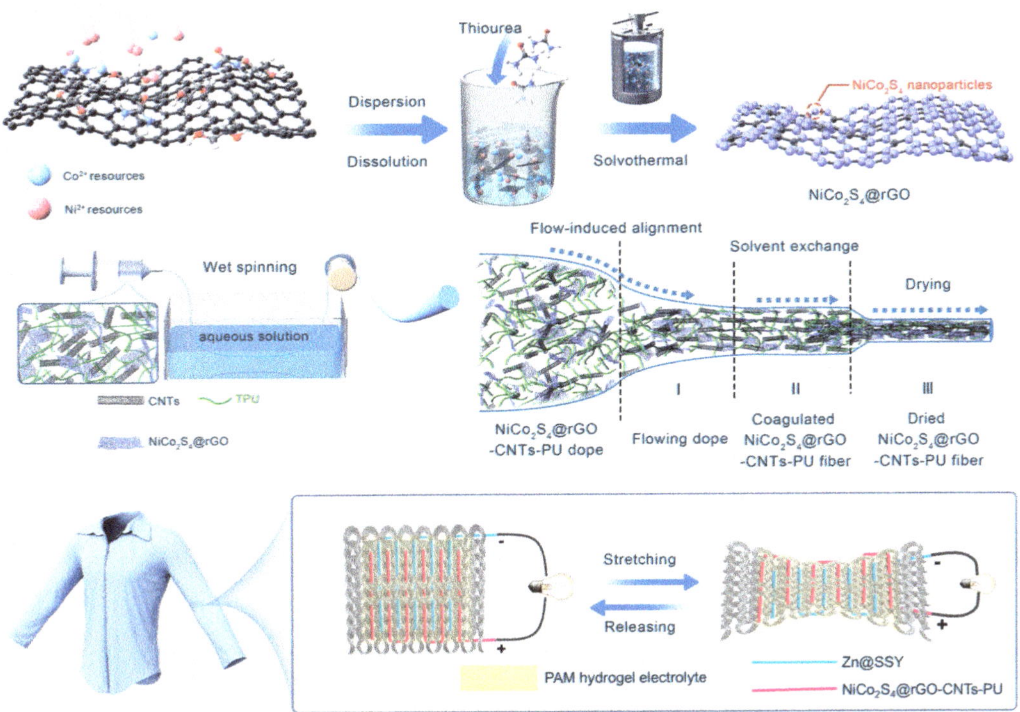

Figure 11. Schematic illustration of the fabrication of $NiCo_2S_4$@rGO nanocomposites, $NiCo_2S_4$@rGO-PU-CNTs fiber cathodes and serpentine footprint fabric Zn-based batteries. PAM: polyacrylamide, CNT: carbon nanotube, PU: polyurethane, rGO: reduced graphene oxide. Reprinted from [128], Copyright 2024, with permission from Elsevier.

For a broad range of different flexible batteries, either fiber- or fabric-based, Wang et al. reported areal energy densities around 6–60 Wh/m^2 for flexible lithium-ion batteries, 1–20 Wh/m^2 for sodium-ion batteries, and 1.5–62 Wh/m^2 for zinc-ion batteries [131]. On the one hand, these values do not differ as strongly as those for fiber- or fabric-based supercapacitors, making batteries potentially easier to build than flexible supercapacitors. On the other hand, the aforementioned value of 10 Wh/m^2, as a typical order of magnitude which energy could be harvested during half a day by textile fabrics, can be reached by all of these typical fiber- or fabric-based batteries.

As this section showed, it is thus generally possible to store the energy that could be harvested in half a day inside a textile fabric of the same area, e.g., directly below a photovoltaic textile. On the other hand, it is still necessary to find out which heating energy is necessary in different situations, either for short stays in very cold environments, such as space or cold stores, or for longer durations outside in the winter. The next section will discuss active heating by electric energy or by PCMs, where the latter will be regarded with respect to the duration of melting or crystallization at defined temperatures, while the first will be discussed in terms of necessary electric energy.

5. Energy-Self-Sufficient Heating Textiles

Some research groups have investigated energy-self-sufficient—or self-powered—heating textiles. In many cases, these studies are based on PCMs that can store a defined amount of energy and release it afterward when the environmental temperature is reduced.

5.1. PCM as Self-Sufficient Heating Textiles

To get an idea of the energy that should be stored in PCMs, it is supportive to use the heat that can be produced by a person, which is about 100 W at rest to about 600 W or even more during work or sports [132]. Typical values of heat storage capacity of phase-change materials are in the order of magnitude 200 J/g [132], meaning that energy of 1 kWh, as released by the human body at rest within ten hours, necessitates 3.6 MJ of storage capacity, i.e., phase-change material with a mass around 18 kg. This shows that PCMs from common materials, such as octadecane, hexadecane, etc., should be used more for repeatedly varying temperatures, as is the case for spacesuits in many situations, while PCMs with much higher storage capacity would be needed to support homeless people during nighttime.

It must be mentioned that there are several studies aimed at increasing the heat storage capacity of PCMs; however, for use in textile fibers or fabrics, usually an encapsulation in the form of microcapsules or core-shell fibers is needed, which reduces the mass-related heat storage capacity again. For microencapsulated PCMs with a polyurethane (PU) binder, e.g., a heat storage capacity of 0.2–7.6 J/g was reported [133].

On the other hand, the above calculation can only approximate an order of magnitude necessary to use PCMs in textiles for active heating since the necessary amount of stored energy depends strongly on the environmental temperature, isolation properties of the whole garments, activity of the wearer, etc. Additionally, thermoregulation depends on the position where the PCMs were integrated into the fabric, on the fabric construction, etc. [134]. This section thus searches for examples of measured or calculated heat retention by textile PCMs.

As an example, Zhang et al. developed a coaxially electrospun multicore-sheath nanofiber mat based on methacrylic acid (MAA), ethylene glycol dimethacrylate (EGDMA) and benzoyl peroxide (BPO) with dodecanol as a PCM, which reached a maximum latent heat value of 106 J/g [135]. Their material showed a suitable phase-change temperature of around 25–30 °C, latent heat retention of more than 98.3% after 500 thermal cycles, and nearly unchanged latent heat and tensile strength after soaking the nanofiber mat in water. The authors suggested it for potential use in space suits. Depending on the injection speed ratio of the PCM core and acrylate copolymeric shell during coaxial electrospinning, four different samples were prepared. Figure 12 shows the time-dependent temperature development of pure polyacrylate and the core-shell nanofiber mats (with an increasing amount of PCM for increasing numbers CS-1 to CS-4) after cooling the samples from 50 °C to room temperature [135]. While a clear difference from CS-1 to CS-4 is visible, it must be mentioned that the time scale used here is not sufficient to provide active warming for a longer duration. The specimens are reported to have a thickness of 85 μm, naturally leading to fast cooling when they are placed on a table at room temperature, with much higher mass and unknown thermal conductivity. Thus, these experiments can only serve as basic material comparisons but do not allow a statement about their applicability for longer active heating in garments.

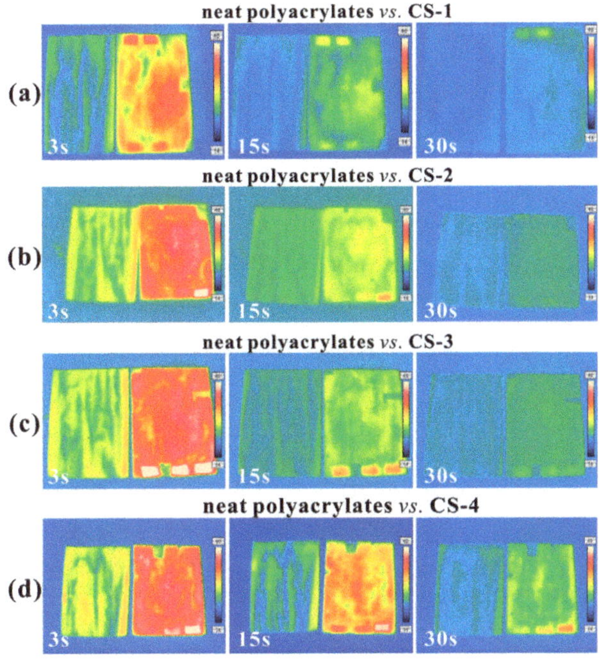

Figure 12. Thermographic images of neat polyacrylates and CS specimens after cooling from 50 °C to RT and maintaining at RT for 3 s, 15 s or 30 s. In each image, the left and the right photo, respectively, correspond to neat polyacrylates and CS specimens ((**a**) CS-1, (**b**) CS-2, (**c**) CS-3 and (**d**) CS-4). All color scales are from 14 to 40 °C. Reprinted from [135], Copyright 2022, with permission from Elsevier.

On a longer time scale of up to 60 min, Yan et al. investigated knitted fabrics, partly including hollow polypropylene (PP) fibers filled with up to 83% poly(ethylene glycol) (PEG) as the PCM [136]. The phase-change temperatures were in the range of 16–40 °C, depending on the amount of PEG in the PP fibers. Heating the samples fixed on a custom-designed sample holder up to 40 °C for 20 min by an infrared lamp and subsequently measuring the temperature of the sample with an infrared camera led to an increase in the duration that the samples needed to reach ambient temperature, from 44 min to 62 min. While this increase shows the effect of the PCM, Figure 13 [136] reveals that the difference between both investigated samples is actually not large. Besides, the inset photographs of the original and the final state show that the custom-made sample holder (of unknown material and mass) has also been heated up by the infrared lamp and is, naturally, also cooling down during the time of the experiment. Here again, the influence of the thermal contact with the unknown sample holder cannot be estimated, so the results of this study cannot be transferred to the situation of heating a person's skin in a cold environment. Generally, measuring heat retention by PCM textiles is challenging due to unclear or non-standardized factors that vary from one study to the other, such as environmental conditions (room temperature, relative humidity, underground on which the fabric is placed during measurements), fabric construction, fabric material, mass ratio of PCM to textile fabric, etc.

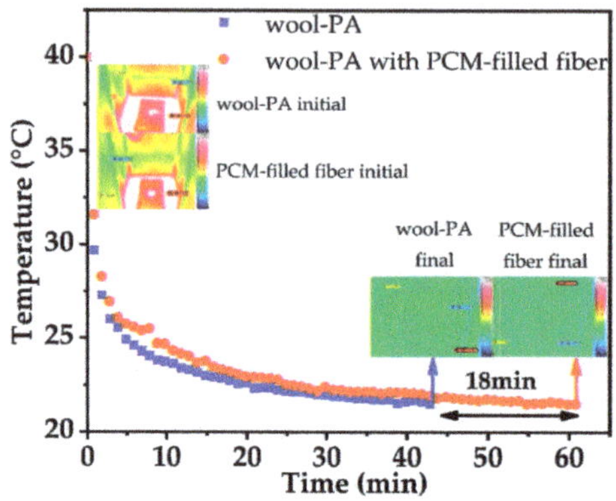

Figure 13. Temperature gradation curves of knitted wool–PA fabric with and without PEG1000-filled PP fibers. Reprinted from [136], originally published under a CC-BY license.

For the aforementioned microencapsulated PCMs with PU binder, heating durations of up to 15 min were reported in different studies [137,138]. Similarly, Ying et al. found the temperature changes between a non-PCM fabric and PCM fabrics with different PCM levels between 20 and 120 g/m^2, upon heating from 23 °C to 35 °C, vanished after less than 10 min [139]. Similar durations were found for a PCM-based cooling vest using Glauber's salt or pure gallium as the PCM in a heated chamber at 30–32 °C, starting from 22 to 25 °C [140].

Ghali et al. simulated the effect of a PCM textile upon a sudden change in the environmental temperature and found a heating effect for approx. 12.5 min, depending on the cold outdoor conditions, as well as a decrease of the clothed-body heat loss by 40–55 W/m^2 for a one-layer garment [141]. Besides oscillating temperatures, they also tested steady-state environmental conditions and stated clearly that the PCM did not influence the thermal resistance in this scenario.

Comparing the aforementioned and several other studies from the recent literature leads to the conclusion that, as discussed at the beginning of this section, PCM textiles are not suitable for long-term use at the moment but are supportive in oscillating temperatures. In addition, while only a few studies have investigated the long-term thermal performance of PCMs, the thermal characteristics of several materials were found to be unstable. Behzadi and Farid reported that Rubitherm 21, a paraffin mixture, significantly changed the peak melting point and the latent heat of fusion, while mixed esters showed nearly no such change [142]. For PCM heat storage as part of a solar combi system in a house, Johansen et al. tested sodium acetate trihydrate and found it to work properly during half a year of testing, with heating it up to 80 °C 53 times by the solar collectors [143]. For the numerical investigation of a solar greenhouse, Chen and Zhou differentiated between short- and long-term PCM storage, where an organic PCM was responsible for short-term energy storage and release during the night while no longer working in the early morning, when a mixture of salt hydrates with different melting temperatures could provide the additional long-term heating [144]. In this study, the potential change in organic and inorganic PCMs with longer use times was not investigated. Long-term stability was, however, found for TiO_2/tetradecanoic acid-based composite PCMs [145] or $LiNO_3$/KCl-expanded graphite (EG) composite PCMs [146]. Especially for textile fabrics, long-term stability was found for microcapsules loaded with *n*-docosane as the PCM [147].

While some tests of the long-term stability of PCMs have indeed been performed, as these examples show, the studies indicating energy storage for several hours [144] or even days [141] are still scarce and are usually not related to textile fabrics, where typically only relatively small masses of PCMs can be integrated, making them more suitable for short-term temperature variations than for energy storage for approx. half a day. The next sections will thus concentrate on fabrics heated by electric energy.

5.2. Self-Sufficient Joule Heating Textiles

Many research groups show the results of Joule heating textiles prepared in diverse ways. As an example, Ding et al. developed a polyester/spandex blend fabric-based Joule heater by coating it with CNT/polydopamine (PDA) [148]. Depending on the applied voltage, they reached different temperatures within approx. one minute, as depicted in Figure 14, where the measurements were not taken on a glove but on a not-defined underground with unclear thermal contact and unclear room temperature. Heating with voltages around 3–7 V should be sufficient for most environmental situations; however, without the resistance of the heating ring (which is not given in [148]), the calculation of a heating power or the energy necessary for longer heating is not possible.

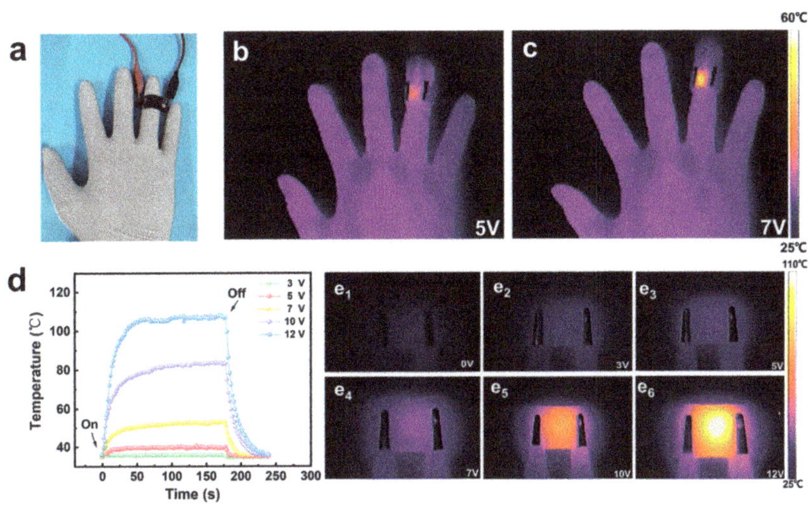

Figure 14. (**a**) Optical photograph of active heating textile wrapped around a finger; thermal infrared images at (**b**) 5 V and (**c**) 7 V; (**d**) temperature profiles of active heating textile at 3–12 V; (**e$_1$–e$_6$**) corresponding thermal infrared images at 0–12 V. Reprinted from [148], Copyright 2023, with permission from Elsevier.

For a cotton/tannic acid/Ag nanoparticle/PDMS knitted textile of dimensions 2 cm × 1 cm and a resistance of approx. 3.8 Ω/cm, Guo et al. reported saturation temperatures of 36–74 °C for voltages of 0.5–1.5 V, reached at a room temperature of around 30 °C [149]. This corresponds to a heating power of 0.07–0.6 W for the mentioned area of 2 cm^2, i.e., a necessary power density of 350–3000 W/m^2. It must be mentioned that this calculation highly overestimates the necessary energy for the given temperatures as the textile also has to heat the (unknown) underground, and no isolation from the environment is given, so this calculation cannot really be transferred to the situation of heating a human body in a cold environment in addition to common isolation clothing.

Many other groups also measured the temperature development of a Joule heated textile on an unknown underground instead of on a clothed thermal manikin or even on probands [150–152], making comparisons with other studies or even extrapolating to a real human in a cold environment quite challenging. This problem is also discussed in [8], where the authors mention the problems of transferring improvements of single thermal properties, such as thermal conductivity, towards the real warming effect for the human body and mention the necessity to perform manikin tests in simulated or real environments as well as opinion polls. Shuvo et al. also discuss the influence of characterization methods of Joule heating efficiency and boundary conditions in detail [153]. Comparing the heating efficiency of different textiles with air flow to maintain a temperature of approx. 42 °C, they found necessary power densities around 230–950 W/m^2, which are all values far above the values estimated for the power density of typical textile energy harvesting and storage methods.

One of the few studies on human probands was reported by Song et al., who measured the skin temperature on the feet inside a conventional and a heated sleeping bag on each of seven male and female volunteers [154]. The heating pads from carbon heating wires between two high-density polyester layers of area (38 cm^2) could be heated by 0–45 W, where 20 W was mostly regarded as thermo-neutral. The tests were performed in a climate chamber with air temperature −0.4 °C (for female volunteers) and −6.4 °C (for male volunteers), respectively, at an air velocity of (0.5 ± 0.1) m/s and a relative humidity of (80 ± 5)%. In all cases, a strong difference was visible for the measured temperature of the fourth toe and the left foot during 3 h of measuring, starting from approx. 24 °C (toe temperature) and nearly 30 °C (left toe) and decreasing strongly in the toe and also visibly in the foot for the conventional (unheated) sleeping bag, while the temperature slightly increased up to an approximately constant value for the heated sleeping bag, as depicted in Figure 15 [154]. Comparing the here-used heating power of 20 W with the aforementioned order of magnitude of 10 Wh/m^2 that could be harvested during half a day, however, shows that with an assumed garment surface of 2 m^2, only 1 h of using the heating pads would be possible. On the other hand, Figure 15 clearly shows that a smaller heating power would be sufficient to maintain a stable temperature of the feet.

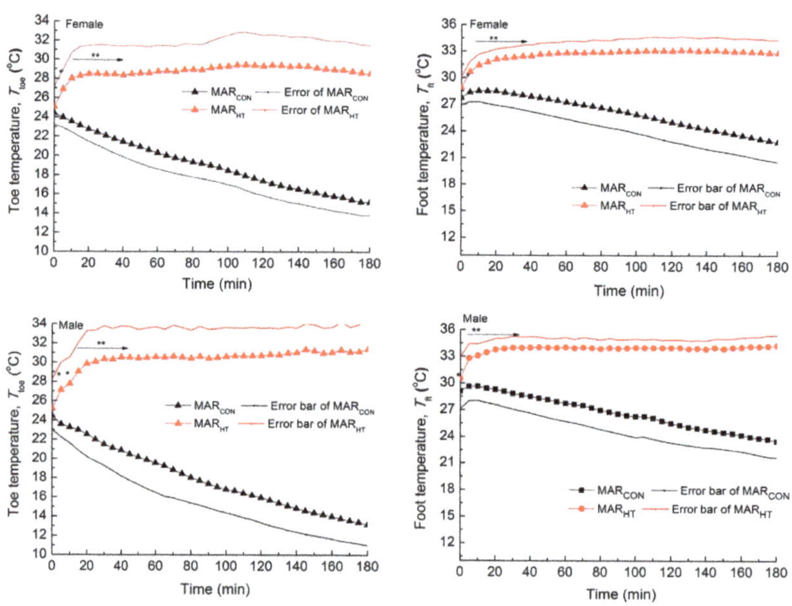

Figure 15. Evolution curves of the 4th toe and the left foot temperatures for females and males in mummy-shaped sleeping bags (conventional: MAR$_{CON}$, heated: MAR$_{HT}$). Significant deviations calculated by two-way ANOVA tests are given with significance levels $p < 0.05$ (marked * on the graphs) and < 0.01 (**), respectively. From [154], originally published under a CC-BY license.

A heated glove was used to maintain a minimum temperature of the finger surface of 15.6 °C, which had been defined as the minimum standard of spacesuit ergonomics design previously, while the environmental temperature around the clothed thermal hand manikin was −130 °C [155]. While the special extravehicular activity (EVA) glove could maintain the temperature measured at the middle finger at 20 °C for about an hour, heating was necessary after 75 min. The authors reported a heating power of 4 W to be ideal for maintaining the defined temperature of 15.6 °C [155], suggesting that either regular warming in a space vehicle or regularly taking new power supplies from the space vehicle is necessary, especially since not only the hands should be warmed during longer work in space.

Many other studies lack exact definitions and descriptions, in this way impeding extrapolations from the investigated situations to other environmental conditions. The same problem occurs for textiles heated by Peltier elements, as the next section shows.

5.3. *Self-Sufficient Peltier Heating Textiles*

Peltier modules are often integrated into smart textiles for cooling, which is easier by thermoelectric cooling than by a vest in which a coolant liquid circulates [156–158]. A jacket that can heat or cool the wearer depending on the temperature was proposed by Poikayil et al.; however, no experiments were performed with the proposed design [42]. Generating a neutral microclimate under clothing by Peltier cells was investigated by Vlad et al., who found a drop in the relative humidity upon introducing Peltier cells due to an increased temperature of the textile layers, as well as constant skin temperature during exercise and rest [159]. The general possibility of using Peltier elements for heating was mentioned in several papers [157,160–163], but no studies of this possibility were found. No reports about tests on heating a volunteer or a thermal manikin by Peltier elements were found in the literature.

6. Applications of Energy-Self-Sufficient Heating Textiles—Chances and Limits

Some of the potential applications of energy-self-sufficient heating textiles were already mentioned in the previous sections. No papers mention the potential use of active heating textiles for homeless people. Using heated textiles to survive cold environments is at least scarcely mentioned [164–167].

Some studies investigated the possibility of using heated textiles for therapeutic applications, such as relieving joint and muscle pain [168], supporting wound healing by killing bacteria [169] or thermotherapy [170]. For sports and outdoor activities, mostly cooling textiles are investigated, but cooling/heating textiles are also investigated, often based on PCMs [171–173]. Another approach is the so-called personal thermal management, which can be applied in indoor situations to heat only a person's body, while the entire indoor space can be cooler to reduce energy consumption [174–176].

More applications, however, are related to working in extreme cold, such as the aforementioned space suits [22,135,155]. Li et al. generally suggested their heated textiles in the form of a garment, a seat and an insole for outdoor workers in winter [177], while many researchers concentrated on heated gloves for protection from extreme cold [178–180]. Depending on the specific environment and work, additional requirements have to be taken into account, such as fire resistance, flame-retardant properties, defined hydrophobic/hydrophilic properties, and windproof, fast-drying, isolating and other properties [156,181–183].

Most studies, however, exclude the long-term heating applications in which self-sufficient heating would really be necessary. To support people working and living under harsh environmental conditions, especially outdoor workers or homeless people during winter, this should be changed in future research studies.

7. Conclusions and Outlook

To create energy-self-sufficient heating textiles, it is necessary to integrate energy harvesting, energy storage and active heating in clothing. Different possibilities have been investigated for energy harvesting, such as triboelectric nanogenerators (TENGs), photovoltaics, and thermoelectric or piezoelectric devices, and for energy storage, such as supercapacitors or batteries. Electric energy can be used to prepare Joule heating or Peltier heating textiles, while phase-change materials (PCMs) work independently from electric energy.

While all necessary components to prepare energy-self-sufficient active heating textiles are thus known and are being investigated by various research groups, there is still a lack of real-life tests on probands under harsh conditions or on thermal manikins under mostly realistic conditions. While most research groups aim at improving specific material parameters, only very few studies yet exist that measure the necessary energy to heat the inner part of a garment under defined environmental conditions towards a defined temperature for a given time. This makes it nearly impossible to estimate how long, e.g., the energy stored and saved in a garment during daytime could heat the clothing in the night, or how much energy must be stored in a working dress to enable active heating towards a defined temperature during a long outdoor working day under extremely cold conditions. Research on PCMs, on the other hand, clearly shows that they have recently only been suitable for short-term temperature variations, not for longer energy storage.

In the future, it is thus necessary to significantly increase the studies of active heating on a human volunteer or at least a thermal body manikin to gain more knowledge about the necessary energy for heating to a defined temperature under different environmental conditions.

To make lab experiments more comparable, heated textiles should be investigated under clearly defined, ideally standardized conditions, such as defined room temperature and relative humidity, and ideally freely held in the air instead of placed on an unknown underground to enable comparability across different studies. Since most of these experiments are performed with an infrared camera, it is necessary for all studies to mention the calibration of the emissivity, which strongly influences the measured temperatures [184]. In this way, comparability between different labs can significantly be increased, enabling more reliable measurements of material properties.

Author Contributions: Conceptualization, T.B., G.E. and E.S.-H.; methodology, T.B., I.K., E.S.-H. and A.E.; formal analysis, A.E.; investigation, M.M., I.K. and A.E.; writing—original draft preparation, T.B., M.M. and A.E.; writing—review and editing, all authors; visualization, G.E.; supervision, E.S.-H. All authors have read and agreed to the published version of the manuscript.

Funding: This research received no external funding.

Data Availability Statement: No new data were created for this review paper.

Conflicts of Interest: The authors declare no conflicts of interest.

References

1. Smith, M.H. Dress, Cloth, and the Farmer's Wife: Textiles from Ø 172 Tatsipataa, Greenland, with Comparative Data from Iceland. *J. North Atl.* **2014**, *2014*, 64–81. [CrossRef]
2. Yu, Y.F.; Zheng, G.C.; Dai, K.; Zhai, W.; Zhou, K.K.; Jia, Y.Y.; Zheng, G.J.; Zhang, Z.C.; Liu, C.T.; Shen, C.Y. Hollow-porous fibers for intrinsically thermally insulating textiles and wearable electronics with ultrahigh working sensitivity. *Mater. Horiz.* **2021**, *8*, 1037–1046. [CrossRef]
3. Jambrich, M.; Hodul, P. Textile applications of polypropylene fibers. In *Polypropylene*; Karger-Kocsis, J., Ed.; Springer: Dordrecht, The Netherlands, 1999; pp. 806–812.
4. Xie, A.-Q.; Zhu, L.L.; Liang, Y.Z.; Mao, J.; Liu, Y.J.; Chen, S. Fiber-spinning Asymmetric Assembly for Janus-structured Bifunctional Nanofiber Films towards All-Weather Smart Textile. *Angew. Chem.* **2022**, *134*, e202208592. [CrossRef]
5. Hsu, P.C.; Liu, C.; Song, A.Y.; Zhang, Z.; Peng, Y.; Xie, J.; Liu, K.; Wu, C.L.; Catrysse, P.B.; Cai, L.; et al. A dual-mode textile for human body radiative heating and cooling. *Sci. Adv.* **2017**, *3*, e1700895. [CrossRef]
6. Cai, L.; Song, A.Y.; Wu, P.; Hsu, P.C.; Peng, Y.; Chen, J.; Liu, C.; Catrysse, P.B.; Liu, Y.; Yang, A.; et al. Warming up human body by nanoporous metallized polyethylene textile. *Nat. Commun.* **2017**, *8*, 496. [CrossRef]
7. Peng, L.; Su, B.; Yu, A.; Jiang, X. Review of clothing for thermal management with advanced materials. *Cellulose* **2019**, *26*, 6415–6448. [CrossRef]
8. Peng, Y.; Cui, Y. Advanced textiles for personal thermal management and energy. *Joule* **2020**, *4*, 724–742. [CrossRef]

9. Zhang, Y.; Li, Y.; Li, K.Q.; Kwon, Y.S.; Tennakoon, T.; Wang, C.T.; Chan, K.C.; Fu, S.-C.; Huang, B.L.; Chao, C.Y.H. A large-area versatile textile for radiative warming and biomechanical energy harvesting. *Nano Energy* **2022**, *95*, 106996. [CrossRef]
10. Li, X.S.; Yang, Y.C.; Quan, Z.Z.; Wang, L.M.; Ji, D.X.; Li, F.X.; Qin, X.H.; Yu, J.Y.; Ramakrishna, S. Tailoring body surface infrared radiation behavior through colored nanofibers for efficient passive radiative heating textiles. *Chem. Eng. J.* **2022**, *430*, 133093. [CrossRef]
11. Ying, B.A.; Kwok, Y.L.; Li, Y.; Yeung, C.Y.; Zhu, Q.Y.; Li, F.Z. Computational Investigation of Thermoregulatory Effects of Multi-Layer PCM Textile Assembly. In *Computational Textile. Studies in Computational Intelligence*; Zeng, X., Li, Y., Ruan, D., Koehl, L., Eds.; Springer: Berlin/Heidelberg, Germany, 2007; Volume 55.
12. Gao, C. Phase-change materials (PCMs) for warming or cooling in protective clothing. In *Protective Clothing*; Woodhead Publishing Series in Textiles; Woodhead Publishing Ltd.: Cambridge, UK, 2014; pp. 227–249.
13. Ahrari, M.; Khajavi, R.; Dolatabadi, M.K.; Toliyat, T.; Rashidi, A. A Review on Application of Phase Change Materials in Textiles Finishing. *Int. J. Chem. Mol. Nucl. Mater. Metall. Eng.* **2017**, *11*, 400–405.
14. Choi, H.N.; Jee, S.H.; Ko, J.W.; Kim, D.J.; Kim, S.H. Properties of Surface Heating Textile for Functional Warm Clothing Based on a Composite Heating Element with a Positive Temperature Coefficient. *Nanomaterials* **2021**, *11*, 904. [CrossRef]
15. Thilagavathi, G.; Muthukumar, N.; Kannaian, T. Development and Characterization of Electric Heating Fabric Based on Silver Coated Nylon Yarn. *J. Text. Eng. Fash. Technol.* **2017**, *1*, 00036.
16. Wang, F.Q.; Liu, Y.M.; Yu, J.Y.; Li, Z.L.; Ding, B. Recent progress on general wearable electrical heating textiles enabled by functional fibers. *Nano Energy* **2024**, *124*, 109497. [CrossRef]
17. Faruk, M.O.; Ahmed, A.; Jalil, M.A.; Islam, M.T.; Shamim, A.M.; Adak, B.; Hossain, M.M.; Mukhopadhyay, S. Functional textiles and composite based wearable thermal devices for Joule heating: Progress and perspectives. *Appl. Mater. Today* **2021**, *23*, 101025. [CrossRef]
18. Hossain, M.T.; Shahid, M.A.; Ali, M.Y.; Saha, S.; Salman Ibna Jamal, M.; Habib, A. Fabrications, Classifications, and Environmental Impact of PCM-Incorporated Textiles: Current State and Future Outlook. *ACS Omega* **2023**, *8*, 45164–45176. [CrossRef]
19. Yang, K.; Venkataraman, M.; Zhang, X.L.; Wiener, J.; Zhu, G.C.; Yao, J.M.; Militky, J. Review: Incorporation of organic PCMs into textiles. *J. Mater. Sci.* **2022**, *57*, 798–847. [CrossRef]
20. Shin, Y.S.; Yoo, D.-I.; Son, K.H. Development of thermoregulating textile materials with microencapsulated phase change materials (PCM). IV. Performance properties and hand of fabrics treated with PCM microcapsules. *J. Appl. Polym. Sci.* **2005**, *97*, 910–915. [CrossRef]
21. Khadiran, T.; Hussein, M.Z.; Zainal, Z.; Rusli, R. Advanced energy storage materials for building applications and their thermal performance characterization: A review. *Renew. Sustain. Energy Rev.* **2016**, *57*, 916–928. [CrossRef]
22. Mondal, S. Phase change materials for smart textiles—An overview. *Appl. Therm. Eng.* **2008**, *28*, 1536–1550. [CrossRef]
23. Farid, M.M.; Khudhair, A.M.; Razack, S.A.K.; Al-Hallaj, S. A review on phase change energy storage: Materials and applications. *Energy Convers. Manag.* **2004**, *45*, 1597–1615. [CrossRef]
24. Bajaj, P. Thermally sensitive materials. In *Smart Fibres, Fabrics and Clothing*; Tao, X.M., Ed.; Woodhead Publishing Ltd.: Cambridge, UK, 2001; pp. 58–82.
25. Farid, M.M.; Kim, Y.; Kanzawa, A. Thermal performance of heat storage module using PCM's with different melting temperatures-experimental. *J. Sol. Energy Eng.* **1990**, *112*, 125–131. [CrossRef]
26. Feldman, D.; Shapiro, M.M. Fatty acids and their mixtures as phase-change materials for thermal energy storage. *Sol. Energy Mater.* **1989**, *18*, 201–216. [CrossRef]
27. Biswas, D.R. Thermal energy storage using sodium sulphate decahydrate and water. *Sol. Energy* **1977**, *19*, 99–100. [CrossRef]
28. Velraj, R.; Seeniraj, R.V.; Hafner, B.; Faber, C.; Schwarzer, K. Heat transfer enhancement in a latent heat storage system. *Sol. Energy* **1999**, *65*, 171–180. [CrossRef]
29. Hasnain, S.M. Review on sustainable thermal energy storage technologies, Part 1: Heat storage materials and techniques. *Energy Convers. Manag.* **1998**, *39*, 1127–1138. [CrossRef]
30. Py, X.; Olives, S.; Mauran, S. Paraffin/porous graphite matrix composite as a high and constant power thermal storage material. *Int. J. Heat Mass Transf.* **2001**, *44*, 2727–2737. [CrossRef]
31. Tan, F.L.; Hosseinizadeh, S.F.; Khodadadi, J.M.; Fan, L.W. Experimental and computational study of constrained melting of phase change materials (PCM) inside a spherical capsule. *Int. J. Heat Mass Transf.* **2009**, *52*, 3464–3472. [CrossRef]
32. Joulin, A.; Younsi, Z.; Zalewski, L.; Lassue, S.; Rousse, D.R.; Cavrot, J.-P. Experimental and numerical investigation of a phase change material: Thermal-energy storage and release. *Appl. Energy* **2011**, *88*, 2454–2456. [CrossRef]
33. Chen, C.; Guo, H.F.; Liu, Y.N.; Yue, H.L.; Wang, C.D. A new kind of phase change material (PCM) for energy-storing wallboard. *Energy and Buildings* **2008**, *40*, 882–890. [CrossRef]
34. Mazo, J.; El Badry, A.T.; Carreras, J.; Delgado, M.; Boer, D.; Zalba, B. Uncertainty propagation and sensitivity analysis of thermo-physical properties of phase change materials (PCM) in the energy demand calculations of a test cell with passive latent thermal storage. *Appl. Therm. Eng.* **2015**, *90*, 596–608. [CrossRef]
35. Tabor, J.; Chatterjee, K.; Ghosh, T.K. Smart textile-based personal thermal comfort systems: Current status and potential solutions. *Adv. Mater. Technol.* **2020**, *5*, 1901155. [CrossRef]
36. He, R.; Schierning, G.; Nielsch, K. Thermoelectric Devices: A Review of Devices, Architectures, and Contact Optimization. *Adv. Mater. Technol.* **2018**, *3*, 1700256. [CrossRef]

37. Chatterjee, K.; Ghosh, T.K. Thermoelectric Materials for Textile Applications. *Molecules* **2021**, *26*, 3154. [CrossRef] [PubMed]
38. Zhou, X.; Yan, Y.; Lu, X.; Zhu, H.; Han, X.; Chen, G.; Ren, Z. Routes for High-Performance Thermoelectric Materials. *Mater. Today* **2018**, *21*, 974–988. [CrossRef]
39. Zhao, L.-D.; Tan, G.; Hao, S.; He, J.; Pei, Y.; Chi, H.; Wang, H.; Gong, S.; Xu, H.; Dravid, V.P.; et al. Ultrahigh Power Factor and Thermoelectric Performance in Hole-Doped Single-Crystal SnSe. *Science* **2016**, *351*, 141–144. [CrossRef] [PubMed]
40. Vargiamidis, V.; Neophytou, N. Hierarchical Nanostructuring Approaches for Thermoelectric Materials with High Power Factors. *Phys. Rev. B* **2019**, *99*, 045405. [CrossRef]
41. Schmidl, G.; Gawlik, A.; Jia, G.; Andrä, G.; Richter, K.; Plentz, J. 3D spacer fabrics for thermoelectric textile cooling and energy generation based on aluminum doped zinc oxide. *Smart Mater. Struct.* **2020**, *29*, 125003. [CrossRef]
42. Poikayil, J.R.; Francis, J.; Saju, D.; Suresh, A.; Varghese, J. Peltier integrated heating & cooling jacket. In Proceedings of the 2017 International Conference of Electronics, Communication and Aerospace Technology (ICECA), Coimbatore, India, 20–22 April 2017; pp. 260–263.
43. Shi, Q.W.; Sun, J.Q.; Hou, C.Y.; Li, Y.G.; Zhan, Q.H.; Wang, H.Z. Advanced Functional Fiber and Smart Textile. *Adv. Fiber Mater.* **2019**, *1*, 3–31. [CrossRef]
44. Chen, G.R.; Li, Y.Z.; Bick, M.; Chen, J. Smart Textiles for Electricity Generation. *Chem. Rev.* **2020**, *120*, 3668–3720. [CrossRef]
45. Dong, K.; Hu, Y.F.; Yang, J.; Kim, S.-W.; Hu, W.G.; Wang, Z.L. Smart textile triboelectric nanogenerators: Current status and perspectives. *MRS Bull.* **2021**, *46*, 512–521. [CrossRef]
46. Qi, Y.; McAlpine, M.C. Nanotechnology-enabled flexible and biocompatible energy harvesting. *Energy Environ. Sci.* **2010**, *3*, 1275–1285. [CrossRef]
47. Ha, M.; Park, J.; Lee, Y.; Ko, H. Triboelectric generators and sensors for self-powered wearable electronics. *ACS Nano* **2015**, *9*, 3421–3427. [CrossRef] [PubMed]
48. Proto, A.; Penhaker, M.; Conforto, S.; Schmid, M. Nanogenerators for human body energy harvesting. *Trends Biotechnol.* **2017**, *35*, 610–624. [CrossRef]
49. Paosangthong, W.; Torah, R.; Beeby, S. Recent progress on textile-based triboelectric nanogenerators. *Nano Energy* **2019**, *55*, 401–423. [CrossRef]
50. Zhong, J.W.; Zhang, Y.; Zhong, Q.; Hu, Q.Y.; Hu, B.; Wang, Z.L.; Zhou, J. Fiber-Based Generator for Wearable Electronics and Mobile Medication. *ACS Nano* **2014**, *8*, 6273–6280. [CrossRef] [PubMed]
51. Jung, S.; Lee, J.; Hyeon, T.; Lee, M.; Kim, D.H. Fabric-based integrated energy devices for wearable activity monitors. *Adv. Mater.* **2014**, *26*, 6329–6334. [CrossRef]
52. Wang, S.; Liu, S.; Zhou, J.Y.; Li, F.X.; Li, J.; Cao, X.F.; Li, Z.Y.; Zhang, J.S.; Li, B.S.; Wang, Y.; et al. Advanced triboelectric nanogenerator with multi-mode energy harvesting and anti-impact properties for smart glove and wearable e-textile. *Nano Energy* **2020**, *78*, 105291. [CrossRef]
53. Lai, Y.-C.; Deng, J.; Zhang, S.L.; Niu, S.; Guo, H.; Wang, Z.L. Single-thread-based wearable and highly stretchable triboelectric nanogenerators and their applications in cloth-based self-powered human-interactive and biomedical sensing. *Adv. Funct. Mater.* **2017**, *27*, 1604462. [CrossRef]
54. He, T.Y.Y.; Shi, Q.F.; Wang, H.; Wen, F.; Chen, T.; Ouyang, J.Y.; Lee, C.K. Beyond energy harvesting—Multi-functional triboelectric nanosensors on a textile. *Nano Energy* **2019**, *57*, 338–352. [CrossRef]
55. Pu, X.; Song, W.; Liu, M.; Sun, C.; Du, C.; Jiang, C.; Huang, X.; Zou, D.; Hu, W.; Wang, Z.L. Wearable power-textiles by integrating fabric triboelectric nanogenerators and fiber-shaped dye-sensitized solar cells. *Adv. Energy Mater.* **2016**, *6*, 1601048. [CrossRef]
56. Tian, Z.; He, J.; Chen, X.; Zhang, Z.; Wen, T.; Zhai, C.; Han, J.; Mu, J.; Hou, X.; Chou, X.; et al. Performance-boosted triboelectric textile for harvesting human motion energy. *Nano Energy* **2017**, *39*, 562–570. [CrossRef]
57. Lee, S.; Ko, W.; Oh, Y.; Lee, J.; Baek, G.; Lee, Y.; Sohn, J.; Cha, S.; Kim, J.; Park, J.; et al. Triboelectric energy harvester based on wearable textile platforms employing various surface morphologies. *Nano Energy* **2015**, *12*, 410–418. [CrossRef]
58. Bayramol, D.V.; Soin, N.; Shah, T.; Siores, E.; Matsouka, D.; Vassiliadis, S. Energy Harvesting Smart Textiles. In *Smart Textiles. Human–Computer Interaction Series*; Schneegass, S., Amft, O., Eds.; Springer: Cham, Switzerland, 2017.
59. Kohn, S.; Wehlage, D.; Juhász Junger, I.; Ehrmann, A. Electrospinning a Dye-Sensitized Solar Cell. *Catalysts* **2019**, *9*, 975. [CrossRef]
60. Ehrmann, A.; Blachowicz, T. Recent coating materials for textile-based solar cells. *AIMS Mater.* **2019**, *6*, 234–251. [CrossRef]
61. Torah, R.; Lawrie-Ashton, J.; Li, Y.; Arumugam, S.; Sodano, H.A.; Beeby, S. Energy-harvesting materials for smart fabrics and textiles. *MRS Bull.* **2018**, *43*, 214–219. [CrossRef]
62. Plentz, J.; Andra, G.; Pliewischkies, T.; Brückner, U.; Eisenhawer, B.; Falk, F. Amorphous silicon thin-film solar cells on glass fiber textiles. *Mater. Sci. Eng. B* **2016**, *204*, 34–37. [CrossRef]
63. Jun, M.J.; Cha, S.I.; Seo, S.H.; Kim, H.S.; Lee, D.Y. Float printing deposition to control the morphology of TiO_2 photoanodes on woven textile metal substrates for TCO-free flexible dye-sensitized solar cells. *RSC Adv.* **2016**, *6*, 67331–67339.
64. Yun, M.J.; Cha, S.I.; Kim, H.S.; Seo, S.H.; Lee, D.Y. Monolithic-structured single-layered textile-based dye-sensitized solar cells. *Sci. Rep.* **2016**, *6*, 34249. [CrossRef] [PubMed]
65. Mamun, A.; Trabelsi, M.; Klöcker, M.; Sabantina, L.; Großerhode, C.; Blachowicz, T.; Grötsch, G.; Cornelißen, C.; Streitenberger, A.; Ehrmann, A. Electrospun nanofiber mats with embedded non-sintered TiO_2 for dye-sensitized solar cells (DSSCs). *Fibers* **2019**, *7*, 60. [CrossRef]

66. Sánchez-García, M.A.; Bokhimi, X.; Velázquez Martínez, S.; Jiménez-González, A.E. Dye-Sensitized Solar Cells Prepared with Mexican Pre-Hispanic Dyes. *J. Nanotechnol.* **2018**, *2018*, 1236878. [CrossRef]
67. Juhász Junger, I.; Udomrungkhajornchai, S.; Grimmelsmann, N.; Blachowicz, T.; Ehrmann, A. Effect of Caffeine Copigmentation of Anthocyanin Dyes on DSSC Efficiency. *Materials* **2019**, *12*, 2692. [CrossRef] [PubMed]
68. Desertec Foundation. An initiative of The Club of Rome. Clean Power from Deserts. The DESERTEC Concept for Energy, Water and Climate Security. WhiteBook, 4th ed. Available online: https://actionguide.info/m/pubs/158/ (accessed on 30 March 2024).
69. Chai, Z.S.; Zhang, N.N.; Sun, P.; Huang, Y.; Zhao, C.X.; Fan, H.J.; Fan, X.; Mai, W.J. Tailorable and Wearable Textile Devices for Solar Energy Harvesting and Simultaneous Storage. *ACS Nano* **2016**, *10*, 9201–9207. [CrossRef]
70. Jeong, E.G.; Jeon, Y.M.; Cho, S.H.; Choi, K.C. Textile-based washable polymer solar cells for optoelectronic modules: Toward self-powered smart clothing. *Energy Environ. Sci.* **2019**, *12*, 1878–1889. [CrossRef]
71. Satharasinghe, A.; Hughes-Riley, T.; Dias, T. An investigation of a wash-durable solar energy harvesting textile. *Prog. Photovolt.* **2020**, *28*, 578–592. [CrossRef]
72. Ilén, E.; Elsehrawy, F.; Palovuori, E.; Halme, J. Washable textile embedded solar cells for self-powered wearables. *Res. J. Text. Appar.* **2024**, *28*, 133–151. [CrossRef]
73. Biancardo, M.; West, K.; Krebs, F.C. Optimizations of large area quasi-solid-state dye-sensitized solar cells. *Sol. Energy Mater. Sol. Cells* **2006**, *90*, 2575–2588. [CrossRef]
74. Pope, J.; Lekakou, C. Thermoelectric polymer composite yarns and an energy harvesting wearable textile. *Smart Mater. Struct.* **2019**, *28*, 095006. [CrossRef]
75. Siddique, A.R.M.; Mahmud, S.; van Heyst, B. A review of the state of the science on wearable thermoelectric power generators (TEGs) and their existing challenges. *Renew. Sustain. Energy Rev.* **2017**, *73*, 730–744. [CrossRef]
76. Settaluri, K.T.; Lo, H.; Ram, R.J. Thin thermoelectric generator system for body energy harvesting. *J. Electron. Mater.* **2012**, *41*, 984–988. [CrossRef]
77. Stark, I. Converting body heat into reliable energy for powering physiological wireless sensors. In Proceedings of the 2nd Conference on Wireless Health WH '11, San Diego/La Jolla, CA, USA, 10–13 October 2011; ACM Press: New York, NY, USA, 2011.
78. Lund, A.; Tian, Y.; Darabi, S.; Müller, C. A polymer-based textile thermoelectric generator for wearable energy harvesting. *J. Power Sources* **2020**, *480*, 228836. [CrossRef]
79. Elmoughni, H.M.; Menon, A.K.; Wolfe, R.M.W.; Yee, S.K. A Textile-Integrated Polymer Thermoelectric Generator for Body Heat Harvesting. *Adv. Mater. Technol.* **2019**, *4*, 1800708. [CrossRef]
80. Finefrock, S.W.; Zhu, X.Q.; Sun, Y.M.; Wu, Y. Flexible prototype thermoelectric devices based on Ag_2Te and PEDOT:PSS coated nylon fibre. *Nanoscale* **2015**, *7*, 5598–5602. [CrossRef] [PubMed]
81. Ding, Y.F.; Qiu, Y.; Cai, K.F.; Yao, Q.; Chen, S.; Chen, L.D.; He, J.Q. High performance n-type Ag_2Se film on nylon membrane for flexible thermoelectric power generator. *Nat. Comm.* **2019**, *10*, 841. [CrossRef] [PubMed]
82. Liu, J.; Jia, Y.H.; Jiang, Q.L.; Jiang, F.X.; Li, C.C.; Wang, X.D.; Liu, P.; Liu, P.P.; Hu, F.; Du, Y.K.; et al. Highly Conductive Hydrogel Polymer Fibers toward Promising Wearable Thermoelectric Energy Harvesting. *ACS Appl. Mater. Interfaces* **2018**, *10*, 44033–44040. [CrossRef] [PubMed]
83. Lee, J.A.; Aliev, A.E.; Bykova, J.S.; Jung de Andrade, M.; Kim, D.Y.; Sim, H.J.; Lepró, X.; Zakhidov, A.A.; Lee, J.-B.; Spinks, G.M.; et al. Woven-Yarn Thermoelectric Textiles. *Adv. Mater.* **2016**, *28*, 5038–5044. [CrossRef] [PubMed]
84. Wang, L.M.; Zhang, K. Textile-Based Thermoelectric Generators and Their Applications. *Energy Environ. Mater.* **2020**, *3*, 67–79. [CrossRef]
85. Hossain, I.Z.; Khan, A.; Hossain, G. A Piezoelectric Smart Textile for Energy Harvesting and Wearable Self-Powered Sensors. *Energies* **2022**, *15*, 5541. [CrossRef]
86. Wu, J.; Mahajan, A.; Riekehr, L.; Zhang, H.; Yang, B.; Meng, N. Nano Energy for high power energy storage. *Nano Energy* **2018**, *50*, 723–732. [CrossRef]
87. Feng, Z.B.; Zhao, Z.Q.; Liu, Y.N.; Liu, Y.K.; Cao, X.Y.; Yu, D.-G.; Wang, K. Piezoelectric Effect Polyvinylidene Fluoride (PVDF): From Energy Harvester to Smart Skin and Electronic Textiles. *Adv. Mater. Technol.* **2023**, *8*, 2300021. [CrossRef]
88. Liu, Z.H.; Pan, C.T.; Lin, L.W.; Huang, J.C.; Ou, Z.Y. Direct-write PVDF nonwoven fiber fabric energy harvesters via the hollow cylindrical near-field electrospinning process. *Smart Mater. Struct.* **2013**, *23*, 025003. [CrossRef]
89. Dolez, P.I. Energy Harvesting Materials and Structures for Smart Textile Applications: Recent Progress and Path Forward. *Sensors* **2021**, *21*, 6297. [CrossRef] [PubMed]
90. Matsouka, D.; Vassiliadis, S.; Prekas, K.; Bayramol, D.V.; Soin, N.; Siores, E. On the Measurement of the Electrical Power Produced by Melt Spun Piezoelectric Textile Fibres. *J. Electron. Mater.* **2016**, *45*, 5112–5126. [CrossRef]
91. Koç, M.; Paralı, L.; Şan, O. Fabrication and vibrational energy harvesting characterization of flexible piezoelectric nanogenerator (PEN) based on PVDF/PZT. *Polym. Test.* **2020**, *90*, 106695. [CrossRef]
92. Jeerapan, I.; Sempionatto, J.R.; Wang, J. On-body bioelectronics: Wearable biofuel cells for bioenergy harvesting and self-powered biosensing. *Adv. Funct. Mater.* **2020**, *30*, 1906243. [CrossRef]
93. Gao, Y.; Cho, J.H.; Ryu, J.H.; Choi, S.H. A scalable yarn-based biobattery for biochemical energy harvesting in smart textiles. *Nano Energy* **2020**, *74*, 104897. [CrossRef]
94. Lv, J.; Jeerapan, I.; Tehrani, F.; Yin, L.; Silva-Lopez, C.A.; Jang, J.-H.; Joshuia, D.; Shah, R.; Liang, Y.Y.; Xie, L.Y.; et al. Sweat-based wearable energy harvesting-storage hybrid textile devices. *Energy Environ. Sci.* **2018**, *11*, 3431–3442. [CrossRef]

95. Zhai, S.L.; Karahan, H.E.; Wie, L.; Qian, Q.H.; Harris, A.T.; Minett, A.I.; Ramakrishna, S.; Ng, A.K.; Chen, Y. Textile energy storage: Structural design concepts, material selection and future perspectives. *Energy Storage Mater.* **2016**, *3*, 123–139. [CrossRef]
96. Duan, Y.X.; You, G.C.; Sun, K.; Zhu, Z.; Liao, X.Q.; Lv, L.F.; Tang, H.; Xu, B.; He, L. Advances in wearable textile-based micro energy storage devices: Structuring, application and perspective. *Nanoscale Adv.* **2021**, *3*, 6271–6293. [CrossRef] [PubMed]
97. Huang, Q.Y.; Wang, D.R.; Zheng, Z.J. Textile-Based Electrochemical Energy Storage Devices. *Adv. Energy Mater.* **2016**, *6*, 1600783. [CrossRef]
98. Cheng, H.Y.; Li, Q.; Zhu, L.L.; Chen, S. Graphene Fiber-Based Wearable Supercapacitors: Recent Advances in Design, Construction, and Application. *Small Methods* **2021**, *5*, 2100502. [CrossRef]
99. Bai, B.; Qiu, L.L.; Yuan, Y.F.; Song, L.X.; Xiong, J.; Du, P.F. Nitrogen doped siloxene and composite with graphene for high performance fiber-based supercapacitors. *J. Energy Storage* **2023**, *63*, 106984. [CrossRef]
100. Xu, Y.F.; Lu, W.B.; Xu, G.B.; Chou, T.-W. Structural supercapacitor composites: A review. *Compos. Sci. Technol.* **2021**, *204*, 108636. [CrossRef]
101. Ma, W.J.; Li, W.F.; Li, M.; Mao, Q.H.; Pan, Z.H.; Hu, J.; Li, X.; Zhu, M.F.; Zhang, Y.G. Unzipped Carbon Nanotube/Graphene Hybrid Fiber with Less "Dead Volume" for Ultrahigh Volumetric Energy Density Supercapacitors. *Adv. Funct. Mater.* **2021**, *31*, 2100195. [CrossRef]
102. Wang, P.Z.; Du, X.X.; Wang, X.J.; Zhang, K.W.; Sun, J.H.; Chen, Z.; Xia, Y.Z. Integrated fiber electrodes based on marine polysaccharide for ultrahigh-energy-density flexible supercapacitors. *J. Power Sourc.* **2021**, *506*, 230130. [CrossRef]
103. He, C.; Cheng, J.L.; Liu, Y.H.; Zhang, X.C.; Wang, B. Thin-walled hollow fibers for flexible high energy density fiber-shaped supercapacitors. *Energy Mater.* **2021**, *1*, 100010. [CrossRef]
104. Nie, G.D.; Luan, Y.X.; Kou, Z.K.; Jiang, J.M.; Zhang, Z.Y.; Yang, N.; Wang, J.; Long, Y.-Z. Fiber-in-tube and particle-in-tube hierarchical nanostructures enable high energy density of MnO_2-based asymmetric supercapacitors. *J. Colloid Interface Sci.* **2021**, *582*, 543–551. [CrossRef] [PubMed]
105. Dong, X.M.; Liang, J.C.; Li, H.; Wu, Z.T.; Zhang, L.; Deng, Y.G.; Yu, H.Y.; Tao, Y.; Yang, Q.-H. Matching electrode lengths enables the practical use of asymmetric fiber supercapacitors with a high energy density. *Nano Energy* **2021**, *80*, 105523. [CrossRef]
106. Ren, C.L.; Yan, Y.S.; Sun, B.Z.; Gu, B.H.; Chou, T.-W. Wet-spinning assembly and in situ electrodeposition of carbon nanotube-based composite fibers for high energy density wire-shaped asymmetric supercapacitor. *J. Collid Interface Sci.* **2020**, *569*, 298–306. [CrossRef]
107. Manjakkal, L.; Franco, F.F.; Pullanchiyodan, A.; González-Jiménez, M.; Dahiya, R. Natural Jute Fibre-Based Supercapacitors and Sensors for Eco-Friendly Energy Autonomous Systems. *Adv. Sust. Syst.* **2021**, *5*, 2000286. [CrossRef]
108. Zheng, X.H.; Hu, Q.L.; Zhou, X.S.; Nie, W.G.; Li, C.L.; Yuan, N.Y. Graphene-based fibers for the energy devices application: A comprehensive review. *Mater. Des.* **2021**, *201*, 109476. [CrossRef]
109. Tang, M.; Wu, Y.; Yang, J.; Xue, Y. Hierarchical core-shell fibers of graphene fiber/radially-aligned molybdenum disulfide nanosheet arrays for highly efficient energy storage. *J. Alloys Compd.* **2020**, *828*, 135622. [CrossRef]
110. Sundriyal, P.; Bhattacharya, S. Textile-based supercapacitors for flexible and wearable electronic applications. *Sci. Rep.* **2020**, *10*, 13259. [CrossRef] [PubMed]
111. Sha, Z.; Huang, F.; Zhou, Y.; Zhang, J.; Wu, S.Y.; Chen, J.Y.; Brown, S.A.; Peng, S.H.; Han, Z.J.; Wang, C.-H. Synergies of vertical graphene and manganese dioxide in enhancing the energy density of carbon fibre-based structural supercapacitors. *Compos. Sci. Technol.* **2021**, *201*, 108568. [CrossRef]
112. Wen, X.; Jiang, K.; Zhang, H.; Huang, H.; Yang, L.Y.; Zhou, Z.Y.; Weng, Q.H. Flexible and Wearable Zinc-Ion Hybrid Supercapacitor Based on Double-Crosslinked Hydrogel for Self-Powered Sensor Application. *Materials* **2022**, *15*, 1767. [CrossRef] [PubMed]
113. Khadem, A.H. Recent Advances in Functional Fabric-Based Wearable Supercapacitors. *Adv. Mater. Interfaces* **2024**, *11*, 2300724. [CrossRef]
114. Su, C.L.; Shao, G.W.; Yu, Q.H.; Huang, Y.L.; Jiang, J.H.; Shao, H.Q.; Chen, N.L. A flexible, lightweight and stretchable all-solid-state supercapacitor based on warp-knitted stainless-steel mesh for wearable electronics. *Text. Res. J.* **2022**, *92*, 1807–1819. [CrossRef]
115. Yu, J.H.; Wu, J.F.; Wang, H.Z.; Zhou, A.; Huang, C.Q.; Bai, H.; Li, L. Metallic Fabrics as the Current Collector for High-Performance Graphene-Based Flexible Solid-State Supercapacitor. *ACS Appl. Mater. Interfaces* **2016**, *8*, 4724–4729. [CrossRef]
116. Zhang, Q.; Liu, D.; Pei, H.Y.; Pan, W.; Liu, Y.L.; Xu, S.G.; Cao, S.K. Swelling-reconstructed chitosan-viscose nonwoven fabric for high-performance quasi-solid-state supercapacitors. *J. Colloid Interface Sci.* **2022**, *617*, 489–499. [CrossRef] [PubMed]
117. Wen, J.F.; Xu, B.G.; Zhou, J.Y. Towards 3D knitted-fabric derived supercapacitors with full structural and functional integrity of fiber and electroactive materials. *J. Power Sources* **2020**, *473*, 228559. [CrossRef]
118. Zheng, X.H.; Wang, Y.; Nie, W.Q.; Wang, Z.Q.; Hu, Q.L.; Li, C.L.; Wang, P.; Wang, W. Elastic polyaniline nanoarrays/MXene textiles for all-solid-state supercapacitors and anisotropic strain sensors. *Compos. Part A Appl. Sci. Manuf.* **2022**, *158*, 106985. [CrossRef]
119. Zuo, W.S.; Zang, L.M.; Liu, Q.F.; Qiu, J.H.; Lan, M.Y.; Yang, C. A quasi-solid-state supercapacitor based on waste surgical masks with high flexibility and designable shape. *Colloids Surf. A Physicochem. Eng. Asp.* **2022**, *634*, 128020. [CrossRef]
120. Lin, X.; Li, X.; Zhang, Z.; Li, X.; Zhang, W.; Xu, J.; Song, K. Screen-printed water-based conductive ink on stretchable fabric for wearable micro-supercapacitor. *Mater. Today Chem.* **2023**, *30*, 101529. [CrossRef]
121. Mo, F.N.; Liang, G.J.; Huang, Z.D.; Li, H.F.; Wang, D.H.; Zhi, C.Y. An Overview of Fiber-Shaped Batteries with a Focus on Multifunctionality, Scalability, and Technical Difficulties. *Adv. Mater.* **2020**, *32*, 1902151. [CrossRef]

122. Zhao, C.L.; Lu, Y.X.; Chen, L.Q.; Hu, Y.-S. Flexible Na batteries. *InfoMat* **2020**, *2*, 126–138. [CrossRef]
123. Zhou, Z.Y.; Zeng, T.T.; Zhang, H.R.; Chen, D. Mesoporous VCN Nanobelts for High-Performance Flexible Zn-Ion Batteries. *Energies* **2022**, *15*, 4932. [CrossRef]
124. Xiao, X.; Xiao, X.; Zhou, Y.; Zhao, X.; Chen, G.; Liu, Z.; Wang, Z.; Lu, C.; Hu, M.; Nashalian, A.; et al. An ultrathin rechargeable solid-state zinc ion fiber battery for electronic textiles. *Sci. Adv.* **2021**, *7*, eabl3742. [CrossRef] [PubMed]
125. Li, M.; Li, Z.Q.; Ye, X.R.; Zhang, X.J.; Qu, L.J.; Tian, M.W. Tendril-Inspired 900% Ultrastretching Fiber-Based Zn-Ion Batteries for Wearable Energy Textiles. *ACS Appl. Mater. Interfaces* **2021**, *13*, 17110–17117. [CrossRef]
126. Gao, Y.; Xie, C.; Zheng, Z.J. Textile Composite Electrodes for Flexible Batteries and Supercapacitors: Opportunities and Challenges. *Adv. Energy Mater.* **2021**, *11*, 2002838. [CrossRef]
127. Ali, A.E.; Jeoti, V.; Stojanovic, G.M. Fabric based printed-distributed battery for wearable e-textiles: A review. *Sci. Technol. Adv. Mater.* **2021**, *22*, 772–793. [CrossRef]
128. Chen, X.Y.; Gao, H.; Tian, X.J.; Wu, D.S.; Lv, P.F.; Yoon, S.S.; Yang, J.X.; Wei, Q.F. Tunable fabric zinc-based batteries utilizing core-shell like fiber electrodes with enhanced deformation durability. *Nano Energy* **2024**, *125*, 109501. [CrossRef]
129. Gu, Z.-Y.; Guo, J.-Z.; Sun, Z.-H.; Zhao, X.-X.; Li, W.-H.; Yang, X.; Liang, H.-J.; Zhao, C.-D.; Wu, X.-L. Carbon-coating-increased working voltage and energy density towards an advanced $Na_3V_2(PO_4)_2F_3$@C cathode in sodium-ion batteries. *Sci. Bull.* **2020**, *65*, 702–710. [CrossRef]
130. Chong, S.K.; Yang, J.; Sun, L.; Guo, S.W.; Liu, Y.N.; Liu, H.K. Potassium Nickel Iron Hexacyanoferrate as Ultra-Long-Life Cathode Material for Potassium-Ion Batteries with High Energy Density. *ACS Nano* **2020**, *14*, 9807–9818. [CrossRef]
131. Wang, D.H.; Han, C.P.; Mo, F.N.; Yang, Q.; Zhao, Y.W.; Li, Q.; Liang, G.J.; Dong, B.B.; Zhi, C.Y. Energy density issues of flexible energy storage devices. *Energy Storage Mater.* **2020**, *28*, 264–292. [CrossRef]
132. Mäkinen, M. Introduction to phase change materials. In *Intelligent Textiles and Clothing*; Mattila, I.H.R., Ed.; Woodhead Publishing: Cambridge, UK, 2006; pp. 21–33.
133. Tyurin, I.N.; Getmantseva, V.V.; Andreeva, E.G. Analysis of Innovative Technologies of Thermoregulating Textile Materials. *Fibre Chem.* **2018**, *50*, 1–9. [CrossRef]
134. Kizildag, N. Smart textiles with PCMs for thermoregulation. In *Multifunctional Phase Change Materials*; Woodhead Publishing: Cambridge, UK, 2023; pp. 445–505.
135. Zhang, Y.H.; Li, T.S.; Zhang, S.H.; Jiang, L.; Xia, J.; Xie, J.Y.; Chen, K.F.; Bao, L.X.; Lei, J.X.; Wang, J.L. Room-temperature, energy storage textile with multicore-sheath structure obtained via in-situ coaxial electrospinning. *Chem. Eng. J.* **2022**, *436*, 135226. [CrossRef]
136. Yan, Y.R.; Li, W.P.; Zhu, R.T.; Lin, C.; Hufenus, R. Flexible Phase Change Material Fiber: A Simple Route to Thermal Energy Control Textiles. *Materials* **2021**, *14*, 401. [CrossRef]
137. Pause, B. Nonwoven Protective Garments with Thermo-Regulating Properties. *J. Ind. Text.* **2003**, *33*, 93–99. [CrossRef]
138. Li, F.Z. Numerical simulation for effects of microcapsuled phase change material (mpcm) distribution on heat and moisture transfer in porous textiles. *Mod. Phys. Lett. B* **2009**, *23*, 501–504.
139. Ying, B.-A.; Kwok, Y.-L.; Li, Y.; Zhu, Q.-Y.; Yeung, C.-Y. Assessing the performance of textiles incorporating phase change materials. *Polym. Test.* **2004**, *23*, 541–549. [CrossRef]
140. Hamdan, H.; Ghaddar, N.; Ouahrani, D.; Ghali, K.; Itani, M. PCM cooling vest for improving thermal comfort in hot environment. *Int. J. Therm. Sci.* **2016**, *102*, 154–167. [CrossRef]
141. Ghali, K.; Ghaddar, N.; Harathani, J.; Jones, B. Experimental and Numerical Investigation of the Effect of Phase Change Materials on Clothing During Periodic Ventilation. *Text. Res. J.* **2004**, *74*, 205–214. [CrossRef]
142. Behzadi, S.; Farid, M.M. Long term thermal stability of organic PCMs. *Appl. Energy* **2014**, *122*, 11–16. [CrossRef]
143. Johansen, J.B.; Englmair, G.; Dannemand, M.; Kong, W.Q.; Fan, J.H.; Dragsted, J.; Perers, B.; Furbo, S. Laboratory testing of solar combi system with compact long term PCM heat storage. *Energy Procedia* **2016**, *91*, 330–337. [CrossRef]
144. Chen, W.; Zhou, G.B. Numerical investigation on thermal performance of a solar greenhouse with synergetic energy release of short- and long-term PCM storage. *Sol. Energy* **2024**, *269*, 112313. [CrossRef]
145. Deka, P.P.; Ansu, A.K.; Sharma, R.K.; Tyagi, V.V.; Sari, A. Development and characterization of form-stable porous TiO_2/tetradecanoic acid based composite PCM with long-term stability as solar thermal energy storage material. *Int. J. Energy Res.* **2020**, *44*, 10044–10057. [CrossRef]
146. Huang, Z.W.; Luo, Z.G.; Gao, X.N.; Fang, X.M.; Fang, Y.T.; Zhang, Z.G. Investigations on the thermal stability, long-term reliability of $LiNO_3$/KCl—Expanded graphite composite as industrial waste heat storage material and its corrosion properties with metals. *Appl. Energy* **2017**, *188*, 521–528. [CrossRef]
147. De Castro, P.F.; Minko, S.; Vinokurov, V.; Cherednichenko, K.; Shchukin, D.G. Long-Term Autonomic Thermoregulating Fabrics Based on Microencapsulated Phase Change Materials. *ACS Appl. Energy Mater.* **2021**, *4*, 12789–12797. [CrossRef]
148. Ding, Y.L.; Dong, H.H.; Cao, J.; Zhang, Z.; Chen, R.H.; Wang, Y.; Yan, J.; Liao, Y.P. A polyester/spandex blend fabrics-based e-textile for strain sensor, joule heater and energy storage applications. *Compos. Part A Appl. Sci. Manuf.* **2023**, *175*, 107779. [CrossRef]
149. Guo, Z.P.; Wang, Y.L.; Huang, J.J.; Zhang, S.Y.; Zhang, R.Q.; Ye, D.Z.; Cai, G.M.; Yang, H.J.; Gu, S.J.; Xu, W.L. Multi-functional and water-resistant conductive silver nanoparticle-decorated cotton textiles with excellent joule heating performances and human motion monitoring. *Cellulose* **2021**, *28*, 7483–7495. [CrossRef]

150. Ma, J.H.; Zhao, Q.L.; Zhou, Y.X.; He, P.X.; Pu, H.H.; Song, B.G.; Pan, S.X.; Wang, Y.W.; Wang, C. Hydrophobic wrapped carbon nanotubes coated cotton fabric for electrical heating and electromagnetic interference shielding. *Polym. Test.* **2021**, *100*, 107240. [CrossRef]
151. Zhang, D.B.; Yin, R.; Zheng, Y.J.; Li, Q.M.; Liu, H.; Liu, C.T.; Shen, C.Y. Multifunctional MXene/CNTs based flexible electronic textile with excellent strain sensing, electromagnetic interference shielding and Joule heating performances. *Chem. Eng. J.* **2022**, *438*, 135587. [CrossRef]
152. Dong, J.C.; Tang, X.W.; Peng, Y.D.; Fan, C.H.; Li, L.; Zhang, C.; Lai, F.L.; He, G.J.; Ma, P.M.; Wang, Z.C.; et al. Highly permeable and ultrastretchable E-textiles with EGaIn-superlyophilicity for on-skin health monitoring, joule heating, and electromagnetic shielding. *Nano Energy* **2023**, *108*, 108194. [CrossRef]
153. Shuvo, I.I.; Decaens, J.; Dolez, P.I. Characterization method of the Joule heating efficiency of electric textiles and influence of boundary conditions. *Flex. Print. Electron.* **2023**, *8*, 025010. [CrossRef]
154. Song, W.F.; Zhang, C.J.; Lai, D.D.; Wang, F.M.; Kuklane, K. Use of a novel smart heating sleeping bag to improve wearers' local thermal comfort in the feet. *Sci. Rep.* **2016**, *6*, 19326. [CrossRef] [PubMed]
155. Tian, Y.S.; Li, D.; Liu, H.Q. The Relationship between Active Heating Power and Temperature of the Fingers in EVA Glove. In *HCI International 2014—Posters' Extended Abstracts. HCI 2014. Communications in Computer and Information Science*; Stephanidis, C., Ed.; Springer: Cham, Switzerland, 2014; Volume 434.
156. Revaiah, R.G.; Kotresh, T.M.; Kandasubramanian, B. Technical textiles for military applications. *J. Text. Inst.* **2020**, *111*, 273–308. [CrossRef]
157. Sajjad, U.; Hamid, K.; Sultan, M.; Abbas, N.; Ali, H.M.; Imran, M.; Muneeshwaran, M.; Chang, J.Y.; Wang, C.C. Personal thermal management—A review on strategies, progress, and prospects. *Int. Commun. Heat Mass Transf.* **2022**, *130*, 105739. [CrossRef]
158. Wei, W.; Wu, B.; Guo, Y.; Hu, Y.H.; Liao, Y.H.; Wu, C.M.; Zhang, Q.H.; Li, Y.G.; Chen, J.H.; Hou, C.Y.; et al. A multimodal cooling garment for personal thermal comfort management. *Appl. Energy* **2023**, *352*, 121973. [CrossRef]
159. Vlad, L.; Mitu, S.; Buhai, C. Optimizing the Thermal Physiological Comfort by Introducing Peltier Cells in Simulated Conditions. *Appl. Mech. Mater.* **2013**, *371*, 792–796. [CrossRef]
160. Chen, X.Y.; Yang, X.N.; Han, X.; Ruan, Z.P.; Xu, J.C.; Huang, F.L.; Zhang, K. Advanced Thermoelectric Textiles for Power Generation: Principles, Design, and Manufacturing. *Glob. Chall.* **2024**, *8*, 2300023. [CrossRef] [PubMed]
161. Yang, X.N.; Zhang, K. Direct Wet-Spun Single-Walled Carbon Nanotubes-Based *p–n* Segmented Filaments toward Wearable Thermoelectric Textiles. *ACS Appl. Mater. Interfaces* **2022**, *14*, 44704–44712. [CrossRef]
162. Scott, R.A. The technology of electrically heated clothing. *Ergonomics* **1988**, *31*, 1065–1081. [CrossRef]
163. Mikolajczyk, Z.; Szalek, A. Thermal Analysis of Heating–Cooling Mat of Textile Incubator for Infants. *Autex Res. J.* **2021**, *21*, 305–322. [CrossRef]
164. Zhang, X.A.; Yu, S.; Xu, B.; Li, M.; Peng, Z.; Wang, Y.; Deng, S.; Wu, X.; Wu, Z.; Ouyang, M.; et al. Dynamic gating of infrared radiation in a textile. *Science* **2019**, *363*, 619–623. [CrossRef] [PubMed]
165. McCann, J. Identification of design requirements for smart clothes and wearable technology. In *Smart Clothes and Wearable Technology*, 2nd ed.; The Textile Institute Book Series; Elsevier: Amsterdam, The Netherlands, 2023; pp. 327–369.
166. Cui, Y.; Gong, H.X.; Wang, Y.J.; Li, D.W.; Bai, H. A Thermally Insulating Textile Inspired by Polar Bear Hair. *Adv. Mater.* **2018**, *30*, 1706807. [CrossRef] [PubMed]
167. Holmér, I. Textiles for protection against cold. In *Textiles for Protection*; Scott, R.A., Ed.; Woodhead Publishing Ltd.: Cambridge, UK, 2005; pp. 378–397.
168. Maurya, S.K.; Somkuwar, V.U.; Das, A.; Kumar, N.; Kumar, B. Influence of Knitting Engineering and Environment Conditions on the Performance of Heating Textiles for Therapeutic Applications. *ACS Appl. Eng. Mater.* **2023**, *1*, 1644–1654. [CrossRef]
169. Zhao, X.; Wang, L.-Y.; Tang, C.-Y.; Zha, X.-J.; Liu, Y.; Su, B.-H.; Ke, K.; Bao, R.-Y.; Yang, M.-B.; Yang, W. Smart $Ti_3C_2T_x$ MXene Fabric with Fast Humidity Response and Joule Heating for Healthcare and Medical Therapy Applications. *ACS Nano* **2020**, *14*, 8793–8805. [CrossRef] [PubMed]
170. Zhu, S.H.; Lou, C.-W.; Zhang, S.H.; Wang, N.; Li, J.W.; Feng, Y.J.; He, R.D.; Xu, C.G.; Lin, J.-H. Clean surface additive manufacturing of aramid paper-based electrically heated devices for medical therapy application. *Surf. Interfaces* **2022**, *29*, 101689. [CrossRef]
171. Scataglini, S.; Moorhead, A.P.; Feletti, F. A Systematic Review of Smart Clothing in Sports: Possible Applications to Extreme Sports. *Muscles Ligaments Tendons J.* **2020**, *10*, 333–342. [CrossRef]
172. McCann, J. Sports applications. In *Polyesters and Polyamides*; Woodhead Publishing Series in Textiles; Woodhead Publishing Ltd.: Cambridge, UK, 2008; pp. 505–524.
173. Mbise, E.; Dias, T.; Hurley, W.; Morris, R. The study of applying heat to enhance moisture transfer in knitted spacer structures. *J. Ind. Text.* **2017**, *47*, 1584–1608. [CrossRef]
174. Prakash, B.J.; Jahar, S.; Pralay, M. Review on passive daytime radiative cooling: Fundaments, recent researchers, challenges and opportunities. *Renew. Sustain. Energy Rev.* **2020**, *133*, 110263.
175. Zhu, F.L.; Feng, Q.Q. Recent advances in textile materials for personal radiative thermal management in indoor and outdoor environments. *Int. J. Therm. Sci.* **2021**, *165*, 106899. [CrossRef]
176. Tang, L.T.; Lyu, B.; Gao, D.G.; Jia, Z.T.; Ma, Z.J. A wearable textile with superb thermal functionalities and durability towards personal thermal management. *Chem. Eng. J.* **2023**, *465*, 142829. [CrossRef]

177. Li, S.S.; Deng, Y.; Cao, B. Study on the Performance of Personal Heating in Extremely Cold Environments Using a Thermal Manikin. *Buildings* **2023**, *13*, 362. [CrossRef]
178. Ducharme, M.B.; Brajkovic, D.; Frim, J. The effect of direct and indirect hand heating on finger blood flow and dexterity during cold exposure. *J. Therm. Biol.* **1999**, *24*, 391–396. [CrossRef]
179. Brajkovic, D.; Ducharme, M.B. Finger dexterity, skin temperature, and blood flow during auxiliary heating in the cold. *J. Appl. Physiol.* **2003**, *95*, 758–770. [CrossRef]
180. Ma, L.; Li, J. A review on active heating for high performance cold-proof clothing. *Int. J. Cloth. Sci. Technol.* **2023**, *35*, 952–970. [CrossRef]
181. Li, F.; Ren, P. Influences of Anti-G Suit Material Hygroscopicity on Thermal Stress Index of the Pilot. In Proceedings of the First International Conference on Information Sciences, Machinery, Materials and Energy, Chongqing, China, 11–13 April 2015; Atlantis Press: Beijing, China, 2015; pp. 603–608.
182. Yang, C.Q.; Yang, H. The Flame Retardant Nomex/Cotton and Nylon/Cotton Blend Fabrics for Protective Clothing. In *Advances in Modern Woven Fabrics Technology, Vassiliadis, S., Ed.*; IntechOpen: London, UK, 2011; pp. 197–210.
183. Shekar, R.I.; Kotresh, T.M.; Subbulakshmi, M.S.; Vijayalakshmi, S.N.; Prasad, A.S.K. Thermal Resistance Properties of Paratrooper Clothing. *J. Ind. Text.* **2009**, *39*, 123–148. [CrossRef]
184. Höser, D.; Wallimann, R.; Von Rohr, P.R. Uncertainty Analysis for Emissivity Measurement at Elevated Temperatures with an Infrared Camera. *Int. J. Thermophys.* **2016**, *37*, 14. [CrossRef]

Disclaimer/Publisher's Note: The statements, opinions and data contained in all publications are solely those of the individual author(s) and contributor(s) and not of MDPI and/or the editor(s). MDPI and/or the editor(s) disclaim responsibility for any injury to people or property resulting from any ideas, methods, instructions or products referred to in the content.

processes

Review

Quantitative Comparison of Personal Cooling Garments in Performance and Design: A Review

Yiying Zhou [1], Lun Lou [2] and Jintu Fan [1,3,4,*]

[1] School of Fashion and Textiles, The Hong Kong Polytechnic University, Hong Kong; yi-ying.zhou@connect.polyu.hk
[2] Energy Sector, Nano and Advanced Materials Institute, Hong Kong Science Park, Hong Kong; lawrencelou@nami.org.hk
[3] Research Center of Textiles for Future Fashion, The Hong Kong Polytechnic University, Hong Kong
[4] Research Institute of Sports Science and Technology, The Hong Kong Polytechnic University, Hong Kong
* Correspondence: jin-tu.fan@polyu.edu.hk

Abstract: Personal cooling garments (PCGs) have gained increasing attention as a promising solution to alleviate heat stress and enhance thermal comfort in hot and humid conditions. However, limited attention has been paid to the influence of clothing design on cooling performance. This review highlights the influence of design factors and provides a quantitative comparison in cooling performance for different types of PCGs, including air cooling garments, evaporative cooling garments, phase-change cooling garments, and liquid cooling garments. A detailed discussion about the relationship between design factors and the cooling performance of each cooling technique is provided based on the available literature. Furthermore, potential improvements and challenges in PCG design are explored. This review aims to offer a comprehensive insight into the attributes of various PCGs and promote interdisciplinary collaboration for improving PCGs in both cooling efficiency and garment comfort, which is valuable for further research and innovation.

Keywords: personal cooling garments; cooling technique; cooling performance; clothing design; quantitative comparison

Citation: Zhou, Y.; Lou, L.; Fan, J. Quantitative Comparison of Personal Cooling Garments in Performance and Design: A Review. *Processes* **2023**, *11*, 2976. https://doi.org/10.3390/pr11102976

Academic Editors: Iqbal M. Mujtaba and Blaž Likozar

Received: 11 September 2023
Revised: 10 October 2023
Accepted: 12 October 2023
Published: 14 October 2023

Copyright: © 2023 by the authors. Licensee MDPI, Basel, Switzerland. This article is an open access article distributed under the terms and conditions of the Creative Commons Attribution (CC BY) license (https://creativecommons.org/licenses/by/4.0/).

1. Introduction

Heat stress has become a growing concern with the impact of climate change. The high air temperatures, high relative humidity, high radiant temperature, and high activity levels or a combination of these factors present a significant challenge to both public health and occupational productivity [1–4]. Since humans can only regulate their core body temperature within a narrow range, excessive body heat beyond the capacity of physiological thermoregulation can result in heat accumulation. It may lead to a higher risk of heat-related illness and conditions such as dehydration, heat stroke, chronic kidney injuries, or even mortality [5–9].

Thermal comfort, as defined by the International Organization for Standardization [10], refers to the subjective state of an individual's satisfaction with the thermal environment. Individual differences in thermal comfort may exist in different climates as well as different physiological, behavioral, and cultural elements. Consequently, even when exposed to the same environmental conditions, individuals may perceive thermal comfort differently [11,12]. Finding acceptable temperatures for individuals depends on their opportunities to adjust conditions, such as changing their clothing or workplace to feel more comfortable. Although heating, ventilation, and air conditioning (HVAC) systems are commonly employed in indoor environments to lower temperatures and alleviate heat stress, they are primarily designed to cater to groups of occupants within buildings. Additionally, they are limited by energy requirements and operational costs, making them unsuitable for outdoor settings [13,14]. In contrast, personal cooling garments (PCGs) aim to provide direct thermal management to individuals, having been proven as an effective measure to manage the thermal state, satisfy individual thermal comfort needs, and improve working performance [15,16]. By focusing on individuals and offering personalized cooling solutions, PCGs offer a promising approach to address thermal comfort challenges in various situations.

Various types of PCGs have been investigated and reviewed in the past decade [17–24]. However, prior reviews primarily focused on the classification of PCG types, the introduction of materials, and potential applications; there is limited discussion about the quantitative comparison of different PCGs. Moreover, the factors of garment design that are closely related to the cooling performance are mentioned less in previous reviews. This paper aims to provide a comprehensive overview of the current state of the art, examine design factors associated with cooling efficiency, and conduct a quantitative comparison of different PCGs in critical aspects.

2. Methodology

2.1. Literature Search Strategy

Database searches were conducted in Google Scholar and Web of Science, covering the period from 1966 to 2023. All the terms that related to 'personal cooling garment' and 'cooling power' were searched through the Google Scholar system. The search strategy incorporated keywords and synonyms, which were searched in the title, abstract, or keyword fields of the databases. Moreover, to find more relevant studies, we surveyed journals using keywords that included 'air cooling garment', 'evaporative cooling garments', 'phase change cooling garments', 'liquid cooling garments', 'thermoelectric cooling garments', and 'radiative cooling garments'. We also assessed all full-text papers for eligibility.

2.2. Selection Criteria

A systematic literature search was conducted to identify relevant articles, research papers, and studies related to personal cooling garments. The following criteria were used for source selection.

Inclusion criteria:

1. Peer-reviewed journal articles, conference papers, and academic studies.
2. Publications addressing personal cooling garments, technologies, and related topics.
3. Publications addressing cooling performance (e.g., cooling power).
4. Publications addressing the clothing design which affected the cooling performance.

Exclusion criteria:

1. Non-peer-reviewed sources, such as blog posts.
2. Publications that do not directly pertain to personal cooling.
3. Publications that do not assess the cooling performance (e.g., cooling power) of personal cooling garments.
4. Publications that do not related to garments.

3. Classification and Design Parameters

Personal cooling garments can be categorized into six different types based on the cooling technique used, namely, air cooling garments, evaporative cooling garments, phase-change cooling garments, liquid cooling garments, thermoelectric cooling garments, and radiative cooling garments. Detailed discussions such as the definition, principle, and design factors affecting the effectiveness of each category will be discussed in this section.

- Air cooling garments

Air cooling garments (ACGs) which improve the convective and evaporative heat dissipation of the human body can be classified into two types: active and passive ACGs. The active ACGs require batteries or external energy supplies to promote ventilative heat exchange within the clothing microenvironment. A forced convection can be introduced by electronic devices such as fans [25] (Figure 1a), vortex tubes [26] (Figure 1b), blowers [27], etc. In terms of garment design, to create sufficient air space between the garment and the human body, the outer layer of most active ACGs is usually made of wind-proof fabric to prevent air from escaping to the environment. In contrast, passive ACGs rely on air ventilation triggered by body movements and natural convection as well as the internal characteristics of textiles. The cooling effect can be regulated by the opening and mesh or spacer and mesh structures of a garment [28–31].

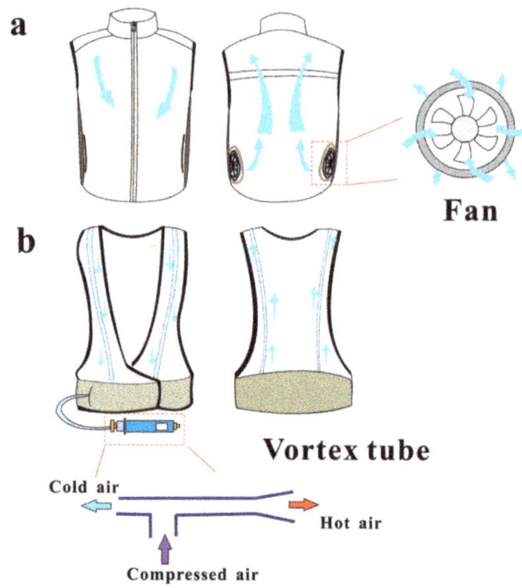

Figure 1. The schematic diagram of active air cooling garments (ACGs) (e.g., cooling achieved by fan (**a**) and vortex tube (**b**)).

The cooling effectiveness of ACGs can be affected by many factors, such as clothing design, environmental conditions, and the users. From the aspect of clothing design, these factors are related to the clothing size, the type of devices, the placement site of devices, the opening site, and the clothing eyelet. Zhao et al. [32] studied the design of locations for fans and openings at different torso sites. They found that the lower front placement site with both front and back openings achieved the best cooling performance. For the localized cooling, the ventilation location had more influence than the design of the opening. The ventilation units can be placed anywhere requiring more evaporative cooling. The adjustable openings (closed or opened) are helpful for the wearers' comfort but have no significant difference in the cooling performance under the same flow rate. Zhao et al. [33] investigated the effect of clothing eyelet designs (sizes and positions) for the air ventilation system and found that the eyelet could reduce the clothing bulkiness ($p < 0.05$) but had no significant impact on the cooling effectiveness ($p > 0.05$). Ho et al. [28] designed 10 different opening and mesh styles of T-shirts and found that the openings applied at two vertical side panels along the side seams of t-shirts were the most effective to release heat and moisture from the human body. Yang et al. [34] found that the upper body heat loss of ACGs was related to the combined influences of ventilation rate and clothing size. The ventilation rate can increase upper body heat loss, while the clothing size has almost no impact on the effectiveness in high ventilation. Yi et al. [25] evaluated different ventilation units for ACGs and found that ACGs with higher flow rates perform better, which can achieve a higher cooling power. Lou et al. [35] investigated the effects of garment design on cooling performance and recommended that the inner space between ACGs and the human body that may influence the efficiency of heat exchange is important for garment design.

In brief, ACGs provide an innovative approach to enhance thermal comfort and mitigate physiological strain. These cooling garments, which create an air gap between the body and clothing, coupled with active cooling mechanisms such as fans or blowers, offer effective heat dissipation. For the design of active ACGs, the eyelet design and adjustable openings are helpful for the bulkiness problem caused by the impermeable outer fabric; at the same time, the inner space between ACGs and the human body also needs to be considered in the design process. Placing devices with a higher flow rate at the sweating region is preferred for improving cooling efficiency, while the direct cooling of fans at their location may decrease the localized comfort sensation.

- Evaporative cooling garments

Evaporative cooling garments (ECGs) are based on the phase change of liquid (mostly water or sweat generated by the human body) from liquid to vapor state that can absorb heat for personal thermal management. As a passive cooling garment, it is energy-saving and environment-friendly. There are two mainly used ECGs (Figure 2): One is a liquid-soaked (water mainly) garment that provides cooling by direct contact with the skin surface; it should be dipped in liquid to reserve the cooling liquid in the garment before use. Another is dry evaporative cooling garments that need to be filled with water before use.

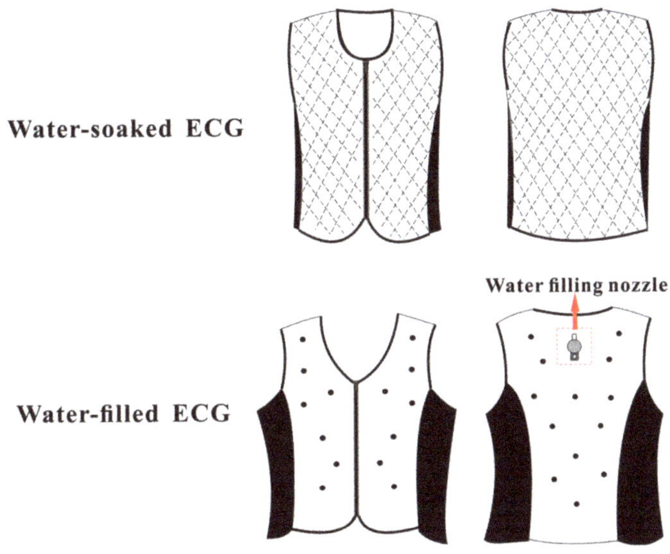

Figure 2. The schematic diagram of evaporative cooling garments (ECGs).

The cooling performance of the ECG is related to the properties of the clothing fabric, fitness, and ambient environment, such as the moisture evaporation and water vapor permeability of fabric and the air gap between the human body and clothing. For this type of garment, the cooling only happened on the outer clothing surface, not the inner surface. Hes et al. [36] studied the cooling effect of wet fabric with the fabric worn with or without an air gap between the skin and the fabric. They found that the cooling efficiency increased with the increase of the fabric moisture content without air layers, and if 2 mm and 4 mm thick air layers were involved, the total cooling efficiency would not be influenced by the water content. The better cooling performance happened when the fabric had direct contact with the skin. Guan et al. [37] studied the cooling power of ECGs that provide cooling by sweating from the human body and showed that the evaporative area and locus (in-plane and trans-plane moisture transfer) were determinants of the cooling efficiency. The cooling efficiency was negatively correlated with the evaporative resistance and the thickness of the fabric. Gillis et al. [38] assessed an ECG saturated with combined menthol and ethanol for evaporative cooling performance versus that with water and found that menthol and ethanol caused cooler sensations and a heat storage response.

ECGs utilize the natural phase change of water from liquid to vapor to provide effective personal cooling. They offer an energy-efficient and environmentally friendly approach to maintaining thermal comfort. To improve the cooling performance of ECGs, the first step is to increase the liquid content and ensure that the fabric has direct contact with skin. The second is to decrease the evaporative resistance and thickness of the fabric for ECGs.

- Phase-change cooling garments

Phase-change cooling garments (PCCGs) provide cooling through phase-change materials (PCMs) which can store latent heat for thermal energy transfer. PCMs can absorb heat from the body surface during the phase-change process that can increase the heat loss of the human body. PCMs can absorb or release heat at a constant temperature. Over 150 phase-change materials are used in scientific research [39]. Some commonly used PCMs are ice, frozen gel, and paraffin waxes. They have different phase-change temperatures (from 0–40 °C) and can be applied for various heat storage capacities [40]. There are two primary technologies used for the development of PCCGs. One is the phase-change material packs which can be placed inside the garment's pockets (Figure 3). The other is to incorporate PCMs into daily clothing, such as ordinary fabrics or fibers. (1) Phase-change microcapsules are PCMs packaged within a suitable wall material which can be impregnated or coated on the surface of the garment [41,42]. (2) Phase-change fiber is the PCM packaged inside a fiber that can be used in garment textiles [43,44]. However, the preparation process of phase-change microcapsules and fiber is complicated and costly, and most studies focus on the development of materials; there are limited studies about the cooling garments that are made by phase-change microcapsules and fiber.

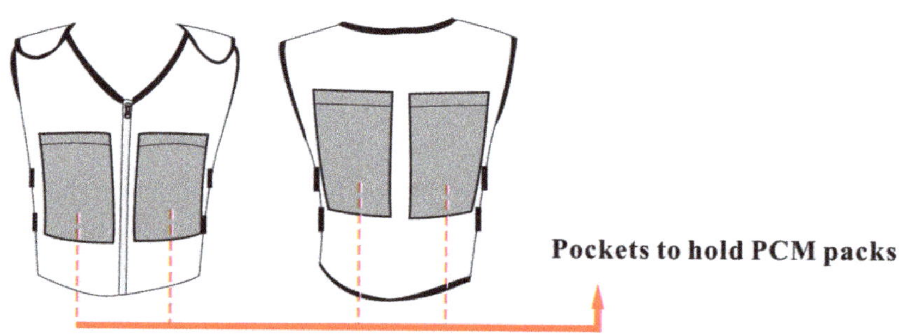

Figure 3. The schematic diagram of phase-change cooling garments (PCCGs).

The cooling effect of PCCGs mainly depends on the clothing design as well as the form and selection of PCMs. Yoo et al. [45] investigated the effects of the number and position of PCM-treated fabrics and found that the treated fabric with a greater amount of PCMs performed better. Moreover, if there is more than one layer, the outer layer of the garment is more appropriate for the effects of the PCMs. According to Mokhtari et al. [23], the PCM microcapsules can only provide a small capacity of heat absorption (about 15 W) with the limitation of the weight of materials that can be coated on the garment. Hence, they are not applicable in hot environments, where ice packs are more effective. House et al. [46] assessed the performance of four PCCGs containing different PCMs, which melted at 0 °C, 10 °C, 20 °C, and 30 °C separately, and the PCCGs were worn under firefighters' protective clothing. They found that 10 °C has the best cooling efficiency when combining work and rest periods, and they also stated that a cooling vest containing ice packs (melting at 0 °C) could be used only if the thermal resistance between the ice packs and the skin was higher. Gao et al. [47] evaluated PCCGs with different melting temperatures, masses, and covering areas and found that the PCCG with the lower melting temperature (24 °C) performed better than those with higher melting temperatures. They also found that the cooling performance was mainly determined by the covering area, while the cooling duration depended on the PCM mass. In hot climates, the temperature gradient for PCCGs was suggested to be equal to or greater than 6 °C.

PCCGs represent an innovative and promising solution for achieving effective cooling and enhancing thermal comfort in various environments. By harnessing the properties of PCMs that can change from solid to liquid and vice versa to create a cooling effect, these garments offer a unique approach to regulating body temperature. The design and composition of PCCGs play an important role in their cooling efficiency. A higher mass of PCMs or PCM microcapsules is positive for cooling efficiency, and the lower melting temperature can provide a higher cooling rate. The higher phase-transition temperature can provide a longer cooling duration. A temperature gradient over 6 °C is recommended. Ice packs can only be used when there is sufficient thermal resistance between the skin and ice. At the same time, the covering area needs to be considered in the design of PCCGs.

- Liquid cooling garments

Liquid cooling garments (LCGs) provide cooling by the pump-driven circulation of liquid coolant inside the tubes embedded in the garment (Figure 4). The tubes may be sewn into the fabric layer, and the pump is powered by electricity. The circulating liquid may be cold or icy water, liquid metal, or a mixture of water and propylene or ethylene glycol [48,49]. The liquid microclimate cooling systems produce a temperature gradient that makes the conductive and convective heat transfer take place among the coolant, the human body, and the environment.

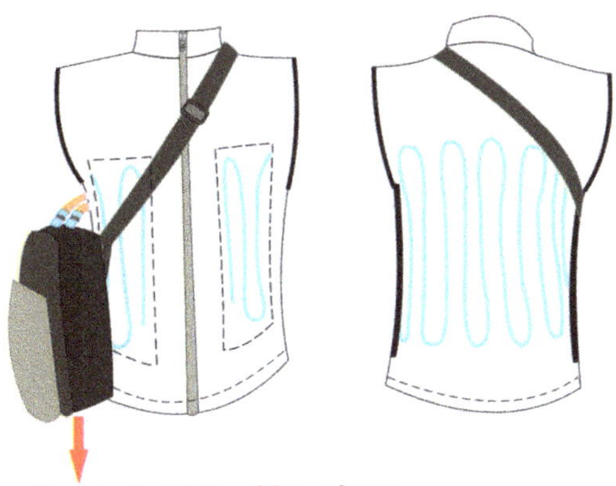

Figure 4. The schematic diagram of liquid cooling garments (LCGs).

The cooling performance of LCGs depends on the design of the clothing, the ambient condition, and the controller, which determines the liquid flow rate, coolant temperature, and intermittent or regional cooling control [50–54]. In terms of clothing design, it is related to the fitness, tubing characteristics, different body areas, and fabric [55–58]. Kayacan et al. [51] compared two different designed LCGs and found that the fabrics with a natural curved structure performed better than the plain fabrics; they also found that when the liquid inlet temperature was decreased, the effective cooling increased slightly. Burton et al. [52] found that the heat removal rate of LCGs was not linearly related to the flow rate when the flow was relatively slow. When the flow rate exceeds 1 L/min, little further improvement in the cooling performance can be made with the increase in flow rate. These findings were also shown by Frim et al. [50]. Dionne et al. [59] evaluated PCGs with different densities of tubing through thermal manikin testing. They found that under the condition of the same flow rate and liquid inlet temperature, the heat removal rate increased with the density of the tubing. Branson et al. [55] used a 3D body scanner to compare the fit analyses of LCGs and indicated that from the aspect of different users, it is better to develop an adjustable cooling garment. Cao et al. [58] investigated the effect of inner fabric layers of LCGs on heat exchange and cooling efficiency. They found that an inner layer with good thermal conductivity, good moisture management, and good tactile properties is desirable.

LCGs offer a promising approach to regulating body temperature and enhancing comfort. These garments utilize the circulation of cold liquid through tubes in clothing to facilitate heat exchange between the body, the liquid, and the environment. For the design of LCGs, a lower liquid temperature with a higher flow rate (no more than 1 L/min) is beneficial to improve cooling performance. As conductive heat transfer is the primary cooling mechanism for LCGs, a good fitness of LCGs can also increase cooling efficiency. An adjustable (fitness and flows) LCG could be designed to reduce the discomfort caused by the coolant having direct contact with skin, and a three-layer system may be better for wearer comfort. The comparative analysis of different LCGs demonstrates their varied cooling powers, durations, and temperature ranges, allowing for tailored choices based on specific cooling needs. Further exploration and real-world studies will contribute to a deeper understanding of LCGs' performance and their utility in diverse settings.

- Thermoelectric cooling garments

Thermoelectric cooling garments (TCGs) are based on the Peltier effect that can convert electrical energy into thermal power. The single-stage thermoelectric (TE) module is composed of type-N and type-P semiconductors which connected electrically in series. The Peltier effect can transfer heat from one side to the other; therefore, one face is heated and the opposite is cooled, which can be changed by the direction of the electric current [60,61]. A single TE device can be embedded in a clothing system to provide cooling by direct contact with the skin [62] (Figure 5). However, there are some apparent weaknesses in the design of such TCGs, for example, the poor coefficient of performance and low efficiency, so there is nearly no cooling garment designed solely using the Peltier effect. Some enhanced cooling garments, such as the combination of the air and TE cooling techniques which enhance air cooling through the TE unit [63,64] and the combination of the liquid and TE cooling techniques which enhance liquid cooling through the TE unit [64] were developed for better cooling performance.

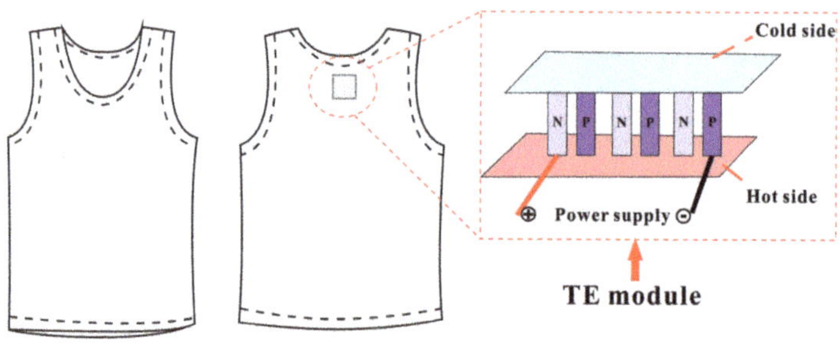

Figure 5. The schematic diagram of thermoelectric cooling garments (TCGs).

- Radiative cooling garments

Radiative cooling garments (RCGs) provide thermal comfort through enhanced radiative heat loss from the human skin or the clothing outer surface. Human skin can be regarded as a near-black radiating source with an emissivity value over 0.95 [65,66], which makes it a near-perfect emitter of thermal infrared radiation. Based on this concept, some fabrics, textiles, and garments were designed for personal thermal management. The cooling garment with mid-IR transparency [67], mid-IR emissivity [68], or solar-reflecting mid-IR emissivity [69] are all potential solutions to achieve better cooling efficiency. RCGs as passive cooling systems contribute to energy saving. The innovation of RCGs mainly focuses on the material design [70–73]; there is limited knowledge about the cooling performance of this kind of garment which may need more attention.

4. Comparative Study

Several methods have been employed to evaluate PCGs. Human trials, thermal manikin tests, and model simulation were the three mainly used methods for the assessment of PCGs [74]. Human testing provides practical data for the physiological impact of PCGs on the human body with the work intensity and specific test conditions set [75–77]. A sweating, heated manikin (ASTM F2371-16) can provide objective and reproducible results to assess the effectiveness of PCGs [78,79], while thermal manikins cannot simulate realistic thermo-physiological responses, such as changes in sweat and heart rate. Thermal manikins can be used as an alternative method when human trials are not applicable, while it cannot replace the human subject tests. Generally, human trials can be conducted independently or combined with thermal manikins to evaluate PCGs. Thermoregulation models could be a good alternative to human trial studies to avoid the ethical problems and operating limitations. Thermoregulation models that can simulate physiological responses and the complex heat transfer of the human body were developed to predict human thermal responses in different environmental and activity conditions [80–82]. However, thermoregulation models need to be combined with human trials to demonstrate their reliability and applicability. The evaluation of different types of PCGs plays an important role in the development and improvement of cooling technology. In this section, the separate and overall quantitative comparisons in cooling performance for different types of PCGs will be discussed.

Table 1 provides a comparative analysis of different ACGs based on key parameters such as cooling power, weight, duration, and flow rate. The cooling powers of the listed ACGs in Table 1 were measured through thermal manikin tests (ASTM-F2371-16). ACG 1, designed by Zhao et al. [32], demonstrates the highest estimated cooling power range of 71.5–106 W, while ACG 2 [83] offers a cooling power of 79.5–97.6 W, and the cooling temperature is 30 °C. ACG 3 [25] has a cooling power of approximately 67.72 W. ACG 4 [63] was proposed as a thermoelectric ACG, providing a personal cooling of 15.5 W. ACG 5 [84] was designed for healthcare workers and provides a constant cooling temperature of 22.85 °C and exhibits a cooling power of 51.7 W. Notably, these ACGs were tested under different environmental settings. ACG 4 [63] and ACG 5 [84] as the thermoelectric ACGs that can provide a lower cooling temperature through a thermoelectric unit are heavier than normal ACGs. The cooling duration of active ACGs mainly relies on the capacity of the cooling device's battery. Research by Yi et al. [25] highlights the positive impact of higher airflow rates on ACGs' performance. Additionally, the cooling coverage area is influenced by the positioning and outlet of the cooling mechanism.

Table 1. Quantitative analysis for air cooling garments (ACGs).

No.	Cooling Power	Weight	Duration	Testing Scenario	Flow Rate	Cooling Area
ACG 1	71.5–106 W	N/A	N/A	34 °C, 60% RH, 0.4 m/s	12 L/s	N/A
ACG 2	79.5–97.6 W	N/A	N/A	30–34 °C, 90% RH	10 m^3/h or 14 m^3/h.	0.53 m^2
ACG 3	67.72 ± 0.74 W	98 g (Fan only)	7.05 h	34 °C, 60% RH, 0.4 ± 0.1 m/s	8–22 L/s	N/A
ACG 4	15.5 W	994 g	N/A	26.1 °C, 50% RH	1×10^{-3}~1.17×10^{-3} m^3/s	N/A
ACG 5	51.7 W	1.2 kg	N/A	23 °C, 50% RH	70 L/min	N/A

Table 2 offers a comprehensive insight into the characteristics and performances of different PCCGs as assessed in various studies. The cooling power of PCCGs 1—3 were measured through thermal manikin tests (ASTM-F2371-16), the value of PCCG 4 was calculated from the average evaporation heat flux using an equation, and the value of PCCG

5 was simulated through Stolwijk's mathematical modeling [85]. A notable observation is the variability in cooling power across the examined cooling garments, ranging from 6.0 to 98.6 W. Regarding cooling duration, the assessed garments demonstrate a range from 65 to 360 min, suggesting varying capacities for sustained cooling during use. In addition, the garments were evaluated under different testing scenarios, encompassing diverse conditions such as hot climates and specific temperature gradients, which are significant factors influencing their efficacy. PCCG 1 [47] exhibits varying cooling powers, ranging from 6 to 28.4 W based on different phase-transition temperatures. The lower phase-transition temperature can provide a longer cooling duration, and the duration of the cooling is from 2.5 to 6 h. PCCG 2 [76] achieves cooling powers of 8.4 W and 10.5 W, with a weight of 1296 g. It operates for durations of 65 min and 125 min which depends on the different phase-transition temperatures. The cooling mechanism involves a dual-phase transition at 15 °C + 15 °C and 15 °C + 23 °C, contributing to the flexibility of the garment in addressing different thermal requirements. PCCG 3 [86] emphasizes its cooling power from 10–20 W and its weight of 2224 g. It sustains cooling for an extended duration of 210 min under specific testing conditions. PCCG 4, designed by Yang et al. [77], is based on the concept of vacuum desiccant cooling and emphasizes an average cooling power of 89.6 ± 9 W with a weight of 3400 g. It can achieve a minimum cooling temperature of 16 °C. Hou et al. [87] proposed a PCM-liquid cooling garment (PCCG 5) to enhance PCCGs; it demonstrates a cooling power ranging from 13.4 to 19.4 W, optimized for a prolonged usage of over 2 h under a testing scenario of 30 °C and 45% relative humidity.

Table 2. Quantitative analysis for phase-change cooling garments (PCCGs).

NO.	Cooling Power	Weight	Duration	Testing Scenario	Phase-Transition Temperature	Cooling Area	Cooling Temperature
PCCG 1	19.2–28.4 W 12.9–21.2 W 6.0–14.2 W	2224 g 2226 g 1973 g	150 min 288 min 360 min	Hot climates (the required temperature gradient is suggested to be greater than 6 °C)	24 °C 28 °C 32 °C	0.2054 m^2	≥24 °C ≥28 °C ≥32 °C
PCCG 2	8.4 W 10.5 W	1296 g	65 min 125 min	30 ± 0.5 °C, 80 ± 5% RH, <0.1 m/s	15°C + 15 °C 15°C + 23 °C	0.1404 m^2	≥15 °C ≥15 °C
PCCG 3	10–20 W	2224 g	210 min	34 ± 0.5 °C, 60% RH, 0.4 m/s	21 °C	0.57 m^2	≥21 °C
PCCG 4	89.6 ± 9 W	3400 g	N/A	37 °C, 50% RH	N/A	0.4 m^2	≥16 °C
PCCG 5	13.4–19.4 W	1800 g	≥2 h	30 °C, 45% RH	24–26 °C	N/A	≥24 °C

Table 3 provides a comparison of four LCGs based on these parameters including cooling power, weight, duration, testing scenario, flow rate, cooling area, and cooling temperature. The cooling power of the LCG 1 and LCG 4 is calculated through equations, the value of LCG 2 is estimated by analysis, and the value of LCG 3 is estimated through the previous study [88]. LCG 1, designed by Guo et al. [89], demonstrates the estimated cooling power range of 67.2–138.1 W with a weight of 1.5 kg, and it offers a cooling duration of 0.79 to 3.36 h, tested within a temperature range of 40 °C to 50 °C. Meanwhile, LCG 2 [90] offers a relatively higher cooling power of 300 W, and its weight is less than 10 kg; it provided cooling for a minimum of 1 h under varying conditions of 30–45 °C in temperature and 20–80% relative humidity. LCG 3 [91] has a cooling power of 90.8–96.5 W with a weight of 3 kg, offering a cooling duration of 1.5 h within the specific conditions of 35.89 °C ± 1.25 °C in temperature and 35% relative humidity. Notably, LCG 2 [90] achieves a cooling temperature range of 22 to 27 °C, whereas LCG 3 [91] achieves a lower cooling temperature range of 10 to 15 °C. LCG 4 which was designed by Xu et al. [64] and based on thermoelectric refrigeration exhibits a high cooling power of 340.4 W, thereby providing effective cooling in a 30 °C environment.

Table 3. Quantitative analysis for liquid cooling garments (LCGs).

NO.	Cooling Power	Weight	Duration	Testing Scenario	Flow Rate	Cooling Area	Cooling Temperature
LCG 1	67.2–138.1 W	1.5 kg	0.79–3.36 h	40 °C–50 °C	224.5–544.2 mL/min	0.568 m^2	N/A
LCG 2	300 W	<10 kg	≥1 h	30–45 °C, 20–80% RH	3.8 L/min	N/A	22–27 °C
LCG 3	90.8–96.5 W	3 kg	90 min	35.89 ± 1.25 °C, 35% RH	N/A		10–15 °C
LCG 4	340.4 W	N/A	N/A	30 °C	N/A	N/A	15.7 °C

Table 4 offers a comparative analysis of four different types of PCGs based on these parameters including cooling power, weight, duration, testing scenario, cooling area, and cooling temperature. The results show that the cooling powers of the different PCGs range from 6 to 340.4 W. The enhanced LCG which is based on thermoelectric refrigeration demonstrates the highest cooling power, while the PCCG exhibits the lowest. The estimated cooling power for ACGs, PCCGs, LCGs, and ECGs ranges from 15.5 to 106 W, 6–98.6 W, 67.2–340.4 W, and 48–57 W, respectively. Weight is an important consideration for portable cooling systems, and lighter systems are usually preferred [92]. The weight of the personal cooling garments listed in this table varies from less than 100 g to over 3000 g, and ACGs present a relatively lower weight among these cooling garments. The testing scenarios, including temperature, humidity, and air velocity, listed in the table are important for the evaluation and application of different cooling systems. Most cooling garments are designed for use in hot climates which are higher than 30 °C, and the recommended temperature gradient for PCCGs is suggested to be greater than 6 °C.

Table 4. Comparative analysis for different types of personal cooling garments (PCGs).

Ref.	Cooling Technique	Cooling Power	Weight	Duration	Testing Scenario	Cooling Area	Cooling Temperature
[32]	ACG	71.5–106 W	N/A	N/A	34 °C, 60% RH, 0.4 m/s	N/A	N/A
[83]		79.5–97.6 W	N/A	N/A	30–34 °C, 90% RH	0.53 m^2	30 °C
[25]		67.72 ± 0.74 W	98g (Fan only)	7.05 h	34 °C, 60% RH, 0.4 ± 0.1 m/s	N/A	N/A
[63]		15.5 W	994 g	N/A	26.1 °C, 50% RH	N/A	19.8–26.5 °C
[84]		51.7 W	1.2 kg	N/A	23 °C, 50% RH	N/A	N/A
[47]	PCCG	19.2–28.4 W 12.9–21.2 W 6.0–14.2 W	2224 g 2226 g 1973 g	2.5 h 4.8 h 6 h	Hot climates (the required temperature gradient is suggested to be greater than 6 °C)	0.2054 m^2	≥24 °C ≥28 °C ≥32 °C
[76]		8.4 W 10.5 W	1296 g	65 min 125 min	30 ± 0.5 °C, 80 ± 5% RH, <0.1m/s	0.1404 m^2	≥15 °C ≥15 °C
[86]		10–20 W	2224 g	210 min	34 ± 0.5 °C, 60% RH, 0.4 m/s	0.57 m^2	≥21 °C
[77]		89.6 ± 9 W	3400 g	N/A	37 °C, 50% RH	0.4 m^2	N/A
[87]		13.4–19.4 W	1800 g	≥2 h	30 °C, 45% RH	N/A	N/A
[89]	LCG	67.2–138.1 W	1500 g	0.79–3.36 h	40 °C–50 °C	0.568 m^2	N/A
[90]		300 W	<10 kg	≥1 h	30–45 °C, 20–80% RH	N/A	22–27 °C
[91]		90.8–96.5 W	3000 g	90 min	35.89 ± 1.25 °C, 35% RH		10–15 °C
[64]		340.4 W	N/A	N/A	30 °C		≥15.7 °C
[93]	ECG	48–57 W	N/A	N/A	40 °C, 10% RH	0.6 m^2	N/A

The duration of cooling is an important consideration when choosing a cooling system for a particular application. The cooling duration needs to be considered when selecting a cooling system for a particular application. The duration of cooling provided by each cooling technique ranges from less than an hour to several hours. For active ACGs, the cooling duration is dependent on the battery provided for the cooling device. For PCCGs, it is related to the weight, covering area, and cooling temperature of the PCM. For LCGs, it is influenced by the flow rate, weight, and temperature of the cooling liquid.

Additionally, the cooling area is more important for designers to achieve better cooling performance. The cooling temperature for active ACGs is related to the environmental temperature and cannot be controlled by users; PCCGs and LCGs' cooling temperature is related to the phase-transition temperature of the PCMs and the temperature of the cooling liquid, respectively, and both cannot be controlled manually; and for ECGs, water evaporation is highly wind-speed dependent.

5. Conclusions

This paper presents a literature review on the design and comparison of personal cooling garments. Based on the cooling technique, PCGs are divided into six different types, including ACGs, PCCGs, LCGs, ECGs, TCGs, and RCGs. A brief introduction of each kind of cooling system has been presented, and the cooling efficiency, design-affecting parameters were also discussed.

The results of this review reveal that the cooling power exhibited by different types of cooling garments varies. It seems that these cooling garments are effective at reducing the physiological strain in a hot environment. LCGs demonstrate a relatively higher cooling power, ranging from 67.2–340.4 W.

As a relatively lightweight cooling garment, the cooling efficiency of ACGs is sensitive to environmental factors such as humidity levels and air temperature. In addition, ACGs rely on the exchange of the air from the inside to the outside of clothing; it cannot work when worn underneath personal protective equipment (PPE). The active ACGs require a power source to operate, and this can be problematic in situations where a continuous power source is not readily available. The passive ACGs show limited cooling efficiency, but they are lightweight and no extra devices needed, which is suitable for sportswear. For the design of active ACGs, the fabric of the outer layer should be considered for the air circulation inside the clothing, and the incorporating features such as the eyelet design and adjustable openings can prove beneficial in mitigating the bulkiness problem. Furthermore, optimizing the placement of devices with higher flow rates in the sweating region is recommended to enhance cooling efficiency.

The suitability of LCGs and PCCGs as normal clothing is limited by their excessive weight, which in turn affects the duration of cooling provided. The weight can add bulk, making them less comfortable and practical for everyday use, especially in athletic settings. The LCGs and PCCGs typically have a limited cooling duration, and their cooling sources need to be recharged or replaced to maintain their function. This limitation can be a concern for individuals requiring long-lasting cooling, such as athletes during extended competitions. In terms of LCGs, optimizing the cooling performance involves utilizing a lower liquid temperature coupled with a higher flow rate (not exceeding 1 L/min). Additionally, the design of adjustable (fitness and flows) LCGs can minimize the discomfort resulting from direct contact between the coolant and the skin. For PCCGs, enhanced cooling efficiency can be achieved by increasing the mass of the PCMs or phase-change microcapsules. Lower melting temperatures of the PCM provide higher cooling rates. Finally, achieving a proper fit of LCGs and PCCGs is essential to enhance cooling efficiency.

The performance of ECGs is influenced by ambient temperature and humidity. In extremely hot or humid conditions, they may struggle to provide adequate cooling, especially for individuals engaged in strenuous activities. To enhance the cooling performance of direct ECGs, firstly, increasing the liquid content and ensuring direct contact between the fabric and the skin can enhance cooling efficiency. Secondly, reducing the evaporative resistance and the thickness of the fabric used in ECGs can promote improved cooling effects.

However, the detailed exploration of other cooling garments, such as TCGs and RCGs, remains relatively limited in current research. Thus, further investigations in these categories are warranted to expand our understanding and guide future advancements in this field.

Further research is needed to balance the function, ergonomics, and aesthetic design for different applications. The efficiency of these PCGs not only depends on the effect of the cooling performance but also the effects of ambient temperature and experimental protocol. Therefore, different designs based on the consideration of end-use scenarios is necessary for the improvement of PCGs. Moreover, the development of these types of PCGs requires the impact of human factors, including metabolic rates and physical activities, on individuals' responses during the use of cooling garments under normal working conditions and should be studied in the future.

Author Contributions: Conceptualization, J.F., L.L., and Y.Z.; methodology, J.F. and Y.Z.; resources, Y.Z.; data curation, Y.Z.; writing—original draft preparation, Y.Z.; writing—review and editing, Y.Z. and L.L.; supervision, J.F. All authors have read and agreed to the published version of the manuscript.

Funding: The funding from Research Centre of Textiles for Future Fashion is acknowledged.

Data Availability Statement: No new data were created or analyzed in this study. Data sharing is not applicable to this article.

Conflicts of Interest: The authors declare no conflict of interest.

References

1. Sherwood, S.C. Adapting to the challenges of warming. *Science* **2020**, *370*, 782–783. [CrossRef]
2. Pennisi, E. *Living with Heat*; American Association for the Advancement of Science: Washington, DC, USA, 2020.
3. Habibi, P.; Moradi, G.; Dehghan, H.; Moradi, A.; Heydari, A. The impacts of climate change on occupational heat strain in outdoor workers: A systematic review. *Urban Clim.* **2021**, *36*, 100770. [CrossRef]
4. Patz, J.A.; Campbell-Lendrum, D.; Holloway, T.; Foley, J.A. Impact of regional climate change on human health. *Nature* **2005**, *438*, 310–317. [CrossRef] [PubMed]
5. Gao, C.; Kuklane, K.; Östergren, P.-O.; Kjellstrom, T. Occupational heat stress assessment and protective strategies in the context of climate change. *Int. J. Biometeorol.* **2018**, *62*, 359–371. [CrossRef]
6. Flouris, A.D.; Dinas, P.C.; Ioannou, L.G.; Nybo, L.; Havenith, G.; Kenny, G.P.; Kjellstrom, T. Workers' health and productivity under occupational heat strain: A systematic review and meta-analysis. *Lancet Planet. Health* **2018**, *2*, e521–e531. [CrossRef]
7. Ioannou, L.G.; Mantzios, K.; Tsoutsoubi, L.; Panagiotaki, Z.; Kapnia, A.K.; Ciuha, U.; Nybo, L.; Flouris, A.D.; Mekjavic, I.B. Effect of a simulated heat wave on physiological strain and labour productivity. *Int. J. Environ. Res. Public Health* **2021**, *18*, 3011. [CrossRef] [PubMed]
8. Ioannou, L.G.; Mantzios, K.; Tsoutsoubi, L.; Nintou, E.; Vliora, M.; Gkiata, P.; Dallas, C.N.; Gkikas, G.; Agaliotis, G.; Sfakianakis, K. Occupational heat stress: Multi-country observations and interventions. *Int. J. Environ. Res. Public Health* **2021**, *18*, 6303. [CrossRef] [PubMed]
9. Chan, A.P.; Yi, W. *Heat Stress and Its Impacts on Occupational Health and Performance*; SAGE Publications: London, UK, 2016; Volume 25, pp. 3–5.
10. Zare, S.; Hasheminezhad, N.; Sarebanzadeh, K.; Zolala, F.; Hemmatjo, R.; Hassanvand, D. Assessing thermal comfort in tourist attractions through objective and subjective procedures based on ISO 7730 standard: A field study. *Urban Clim.* **2018**, *26*, 1–9. [CrossRef]
11. Wang, Z.; de Dear, R.; Luo, M.; Lin, B.; He, Y.; Ghahramani, A.; Zhu, Y. Individual difference in thermal comfort: A literature review. *Build. Environ.* **2018**, *138*, 181–193. [CrossRef]
12. Wang, Z.; Zhang, H.; He, Y.; Luo, M.; Li, Z.; Hong, T.; Lin, B. Revisiting individual and group differences in thermal comfort based on ASHRAE database. *Energy Build.* **2020**, *219*, 110017. [CrossRef]
13. Guo, W.; Zhou, M. Technologies toward thermal comfort-based and energy-efficient HVAC systems: A review. In Proceedings of the 2009 IEEE International Conference on Systems, Man and Cybernetics, San Antonio, TX, USA, 11–14 October 2009; pp. 3883–3888.
14. Karjalainen, S.; Koistinen, O. User problems with individual temperature control in offices. *Build. Environ.* **2007**, *42*, 2880–2887. [CrossRef]
15. Golbabaei, F.; Heydari, A.; Moradi, G.; Dehghan, H.; Moradi, A.; Habibi, P. The effect of cooling vests on physiological and perceptual responses: A systematic review. *Int. J. Occup. Saf. Ergon.* **2022**, *28*, 223–255. [CrossRef]
16. Morris, N.B.; Jay, O.; Flouris, A.D.; Casanueva, A.; Gao, C.; Foster, J.; Havenith, G.; Nybo, L. Sustainable solutions to mitigate occupational heat strain–an umbrella review of physiological effects and global health perspectives. *Environ. Health* **2020**, *19*, 95.
17. Peng, Y.; Cui, Y. Advanced textiles for personal thermal management and energy. *Joule* **2020**, *4*, 724–742. [CrossRef]
18. Lou, L.; Chen, K.; Fan, J. Advanced materials for personal thermal and moisture management of health care workers wearing PPE. *Mater. Sci. Eng. R Rep.* **2021**, *146*, 100639. [CrossRef]
19. Sajjad, U.; Hamid, K.; Sultan, M.; Abbas, N.; Ali, H.M.; Imran, M.; Muneeshwaran, M.; Chang, J.-Y.; Wang, C.-C. Personal thermal management-A review on strategies, progress, and prospects. *Int. Commun. Heat Mass Transf.* **2022**, *130*, 105739. [CrossRef]

20. Song, W.; Zhang, Z.; Chen, Z.; Wang, F.; Yang, B. Thermal comfort and energy performance of personal comfort systems (PCS): A systematic review and meta-analysis. *Energy Build.* **2022**, *256*, 111747. [CrossRef]
21. Ren, S.; Han, M.; Fang, J. Personal Cooling Garments: A Review. *Polymers* **2022**, *14*, 5522. [CrossRef] [PubMed]
22. Lou, L.; Wu, Y.S.; Shou, D.; Fan, J. Thermoregulatory clothing for personal thermal management. *Annu. Rev. Heat Transf.* **2018**, *21*, 205–244. [CrossRef]
23. Mokhtari Yazdi, M.; Sheikhzadeh, M. Personal cooling garments: A review. *J. Text. Inst.* **2014**, *105*, 1231–1250. [CrossRef]
24. Sarkar, S.; Kothari, V. Cooling garments—A review. *Indian J. Fibre Text. Res.* **2014**, *39*, 450–458.
25. Yi, W.; Zhao, Y.; Chan, A.P. Evaluation of the ventilation unit for personal cooling system (PCS). *Int. J. Ind. Ergon.* **2017**, *58*, 62–68. [CrossRef]
26. Zhai, X. Research on the application of vortex tube type of cooling jacket in coal mine. *AIP Conf. Proc.* **2017**, *1864*, 020220.
27. Al Sayed, C.; Vinches, L.; Dupuy, O.; Douzi, W.; Dugue, B.; Hallé, S. Air/CO_2 cooling garment: Description and benefits of use for subjects exposed to a hot and humid climate during physical activities. *Int. J. Min. Sci. Technol.* **2019**, *29*, 899–903. [CrossRef]
28. Ho, C.; Fan, J.; Newton, E.; Au, R. Effects of athletic T-shirt designs on thermal comfort. *Fibers Polym.* **2008**, *9*, 503–508. [CrossRef]
29. Sun, C.; Au, J.S.-C.; Fan, J.; Zheng, R. Novel ventilation design of combining spacer and mesh structure in sports T-shirt significantly improves thermal comfort. *Appl. Ergon.* **2015**, *48*, 138–147. [CrossRef]
30. Wang, W.; Yao, L.; Cheng, C.-Y.; Zhang, T.; Atsumi, H.; Wang, L.; Wang, G.; Anilionyte, O.; Steiner, H.; Ou, J. Harnessing the hygroscopic and biofluorescent behaviors of genetically tractable microbial cells to design biohybrid wearables. *Sci. Adv.* **2017**, *3*, e1601984. [CrossRef]
31. Chai, J.; Kang, Z.; Yan, Y.; Lou, L.; Zhou, Y.; Fan, J. Thermoregulatory clothing with temperature-adaptive multimodal body heat regulation. *Cell Rep. Phys. Sci.* **2022**, *3*, 100958. [CrossRef]
32. Zhao, M.; Gao, C.; Wang, F.; Kuklane, K.; Holmér, I.; Li, J. A study on local cooling of garments with ventilation fans and openings placed at different torso sites. *Int. J. Ind. Ergon.* **2013**, *43*, 232–237. [CrossRef]
33. Zhao, M.; Yang, J.; Wang, F.; Udayraj; Chan, W.C. The cooling performance of forced air ventilation garments in a warm environment: The effect of clothing eyelet designs. *J. Text. Inst.* **2022**, *114*, 378–387. [CrossRef]
34. Yang, J.; Wang, F.; Song, G.; Li, R.; Raj, U. Effects of clothing size and air ventilation rate on cooling performance of air ventilation clothing in a warm condition. *Int. J. Occup. Saf. Ergon.* **2022**, *28*, 354–363. [CrossRef]
35. Lou, L.; Wu, Y.S.; Zhou, Y.; Fan, J. Effects of body positions and garment design on the performance of a personal air cooling/heating system. *Indoor Air* **2022**, *32*, e12921. [CrossRef] [PubMed]
36. Hes, L.; de Araujo, M. Simulation of the effect of air gaps between the skin and a wet fabric on resulting cooling flow. *Text. Res. J.* **2010**, *80*, 1488–1497. [CrossRef]
37. Guan, M.; Annaheim, S.; Li, J.; Camenzind, M.; Psikuta, A.; Rossi, R.M. Apparent evaporative cooling efficiency in clothing with continuous perspiration: A sweating manikin study. *Int. J. Therm. Sci.* **2019**, *137*, 446–455. [CrossRef]
38. Gillis, D.J.; Barwood, M.; Newton, P.; House, J.; Tipton, M. The influence of a menthol and ethanol soaked garment on human temperature regulation and perception during exercise and rest in warm, humid conditions. *J. Therm. Biol.* **2016**, *58*, 99–105. [CrossRef] [PubMed]
39. Sharma, A.; Tyagi, V.V.; Chen, C.R.; Buddhi, D. Review on thermal energy storage with phase change materials and applications. *Renew. Sustain. Energy Rev.* **2009**, *13*, 318–345. [CrossRef]
40. Pause, B. Driving more comfortably with phase change materials. *Tech. Text. Int.* **2002**, *11*, 24–27.
41. Zhang, W.; Hao, S.; Zhao, D.; Bai, G.; Zuo, X.; Yao, J. Preparation of PMMA/SiO2 PCM microcapsules and its thermal regulation performance on denim fabric. *Pigment Resin Technol.* **2020**, *49*, 491–499. [CrossRef]
42. Geng, X.; Li, W.; Wang, Y.; Lu, J.; Wang, J.; Wang, N.; Li, J.; Zhang, X. Reversible thermochromic microencapsulated phase change materials for thermal energy storage application in thermal protective clothing. *Appl. Energy* **2018**, *217*, 281–294. [CrossRef]
43. Chen, C.; Wang, L.; Huang, Y. Electrospun phase change fibers based on polyethylene glycol/cellulose acetate blends. *Appl. Energy* **2011**, *88*, 3133–3139. [CrossRef]
44. Golestaneh, S.; Mosallanejad, A.; Karimi, G.; Khorram, M.; Khashi, M. Fabrication and characterization of phase change material composite fibers with wide phase-transition temperature range by co-electrospinning method. *Appl. Energy* **2016**, *182*, 409–417. [CrossRef]
45. Yoo, H.; Lim, J.; Kim, E. Effects of the number and position of phase-change material-treated fabrics on the thermo-regulating properties of phase-change material garments. *Text. Res. J.* **2013**, *83*, 671–682. [CrossRef]
46. House, J.R.; Lunt, H.C.; Taylor, R.; Milligan, G.; Lyons, J.A.; House, C.M. The impact of a phase-change cooling vest on heat strain and the effect of different cooling pack melting temperatures. *Eur. J. Appl. Physiol.* **2013**, *113*, 1223–1231. [CrossRef]
47. Gao, C.; Kuklane, K.; Holmér, I. Cooling vests with phase change material packs: The effects of temperature gradient, mass and covering area. *Ergonomics* **2010**, *53*, 716–723. [CrossRef]
48. Richardson, G.; Cohen, J.; McPhate, D.; Hayes, P. A personal conditioning system based on a liquid-conditioned vest and a thermoelectric supply system. *Ergonomics* **1988**, *31*, 1041–1047. [CrossRef] [PubMed]
49. Vallerand, A.; Michas, R.; Frim, J.; Ackles, K. Heat balance of subjects wearing protective clothing with a liquid-or air-cooled vest. *Aviat. Space Environ. Med.* **1991**, *62*, 383–391. [PubMed]
50. Frim, J.; Michas, R.D.; Cain, B. Personal cooling garment performance: A parametric study. In *Environmental Ergonomics. Recent Progress and New Frontiers*; Freund Publishing House, Ltd.: London, UK; Tel Aviv, Israel, 1996.

51. Kayacan, Ö.; Kurbak, A. Effect of garment design on liquid cooling garments. *Text. Res. J.* **2010**, *80*, 1442–1455. [CrossRef]
52. Burton, D. Performance of water conditioned suits. *Aerosp. Med May* **1966**, *1966*, 500–504.
53. Cheuvront, S.N.; Kolka, M.A.; Cadarette, B.S.; Montain, S.J.; Sawka, M.N. Efficacy of intermittent, regional microclimate cooling. *J. Appl. Physiol.* **2003**, *94*, 1841–1848. [CrossRef]
54. Xu, X. Multi-loop control of liquid cooling garment systems. *Ergonomics* **1999**, *42*, 282–298. [CrossRef]
55. Branson, D.H.; Cao, H.; Jin, B.; Peksoz, S.; Farr, C.; Ashdown, S. Fit analysis of liquid cooled vest prototypes using 3D body scanning technology. *J. Text. Appar. Technol. Manag.* **2005**, *4*, 1–15.
56. Burton, D. Engineering aspects of personal conditioning. In Proceedings of the Symposium on Individual Cooling, Manhattan, KS, USA, 17–18 March 1969; pp. 33–49.
57. Young, A.J.; Sawka, M.N.; EpStein, Y.; Decristofano, B.; Pandolf, K.B. Cooling different body surfaces during upper and lower body exercise. *J. Appl. Physiol.* **1987**, *63*, 1218–1223. [CrossRef] [PubMed]
58. Cao, H.; Branson, D.H.; Peksoz, S.; Nam, J.; Farr, C.A. Fabric selection for a liquid cooling garment. *Text. Res. J.* **2006**, *76*, 587–595. [CrossRef]
59. Dionne, J.; Semeniuk, K.; Makris, A.; Teal, W.; Laprise, B. Thermal manikin evaluation of liquid cooling garments intended for use in hazardous waste management. In Proceedings of the Waste Management 2003 Symposium, Tucson, AZ, USA, 3–27 February 2003.
60. Enescu, D.; Virjoghe, E.O. A review on thermoelectric cooling parameters and performance. *Renew. Sustain. Energy Rev.* **2014**, *38*, 903–916. [CrossRef]
61. Riffat, S.B.; Ma, X. Thermoelectrics: A review of present and potential applications. *Appl. Therm. Eng.* **2003**, *23*, 913–935. [CrossRef]
62. Hong, S.; Gu, Y.; Seo, J.K.; Wang, J.; Liu, P.; Meng, Y.S.; Xu, S.; Chen, R. Wearable thermoelectrics for personalized thermoregulation. *Sci. Adv.* **2019**, *5*, eaaw0536. [CrossRef] [PubMed]
63. Lou, L.; Shou, D.; Park, H.; Zhao, D.; Wu, Y.S.; Hui, X.; Yang, R.; Kan, E.C.; Fan, J. Thermoelectric air conditioning undergarment for personal thermal management and HVAC energy saving. *Energy Build.* **2020**, *226*, 110374. [CrossRef]
64. Xu, Y.; Li, Z.; Wang, J.; Zhang, M.; Jia, M.; Wang, Q. Man-portable cooling garment with cold liquid circulation based on thermoelectric refrigeration. *Appl. Therm. Eng.* **2022**, *200*, 117730. [CrossRef]
65. Sanchez-Marin, F.J.; Calixto-Carrera, S.; Villaseñor-Mora, C. Novel approach to assess the emissivity of the human skin. *J. Biomed. Opt.* **2009**, *14*, 024006. [CrossRef]
66. Tong, J.K.; Huang, X.; Boriskina, S.V.; Loomis, J.; Xu, Y.; Chen, G. Infrared-transparent visible-opaque fabrics for wearable personal thermal management. *Acs Photonics* **2015**, *2*, 769–778. [CrossRef]
67. Hsu, P.-C.; Song, A.Y.; Catrysse, P.B.; Liu, C.; Peng, Y.; Xie, J.; Fan, S.; Cui, Y. Radiative human body cooling by nanoporous polyethylene textile. *Science* **2016**, *353*, 1019–1023. [CrossRef] [PubMed]
68. Zeng, S.; Pian, S.; Su, M.; Wang, Z.; Wu, M.; Liu, X.; Chen, M.; Xiang, Y.; Wu, J.; Zhang, M. Hierarchical-morphology metafabric for scalable passive daytime radiative cooling. *Science* **2021**, *373*, 692–696. [CrossRef] [PubMed]
69. Sun, Y.; Ji, Y.; Javed, M.; Li, X.; Fan, Z.; Wang, Y.; Cai, Z.; Xu, B. Preparation of passive daytime cooling fabric with the synergistic effect of radiative cooling and evaporative cooling. *Adv. Mater. Technol.* **2022**, *7*, 2100803. [CrossRef]
70. Zhu, F.; Feng, Q. Recent advances in textile materials for personal radiative thermal management in indoor and outdoor environments. *Int. J. Therm. Sci.* **2021**, *165*, 106899. [CrossRef]
71. Zhu, B.; Li, W.; Zhang, Q.; Li, D.; Liu, X.; Wang, Y.; Xu, N.; Wu, Z.; Li, J.; Li, X. Subambient daytime radiative cooling textile based on nanoprocessed silk. *Nat. Nanotechnol.* **2021**, *16*, 1342–1348. [CrossRef]
72. Iqbal, M.I.; Lin, K.; Sun, F.; Chen, S.; Pan, A.; Lee, H.H.; Kan, C.-W.; Lin, C.S.K.; Tso, C.Y. Radiative cooling nanofabric for personal thermal management. *ACS Appl. Mater. Interfaces* **2022**, *14*, 23577–23587. [CrossRef]
73. Wu, X.; Li, J.; Jiang, Q.; Zhang, W.; Wang, B.; Li, R.; Zhao, S.; Wang, F.; Huang, Y.; Lyu, P. An all-weather radiative human body cooling textile. *Nat. Sustain.* **2023**, *2023*, 1–9.
74. Bogerd, N.; Psikuta, A.; Daanen, H.; Rossi, R. How to measure thermal effects of personal cooling systems: Human, thermal manikin and human simulator study. *Physiol. Meas.* **2010**, *31*, 1161. [CrossRef]
75. Ouahrani, D.; Itani, M.; Ghaddar, N.; Ghali, K.; Khater, B. Experimental study on using PCMs of different melting temperatures in one cooling vest to reduce its weight and improve comfort. *Energy Build.* **2017**, *155*, 533–545. [CrossRef]
76. Zheng, Q.; Ke, Y.; Wang, H. Design and evaluation of cooling workwear for miners in hot underground mines using PCMs with different temperatures. *Int. J. Occup. Saf. Ergon.* **2022**, *28*, 118–128. [CrossRef]
77. Yang, Y.; Stapleton, J.; Diagne, B.T.; Kenny, G.P.; Lan, C.Q. Man-portable personal cooling garment based on vacuum desiccant cooling. *Appl. Therm. Eng.* **2012**, *47*, 18–24. [CrossRef]
78. Jetté, F.-X.; Dionne, J.-P.; Rose, J.; Makris, A. Effect of thermal manikin surface temperature on the performance of personal cooling systems. *Eur. J. Appl. Physiol.* **2004**, *92*, 669–672. [CrossRef]
79. Miura, K.; Takagi, K.; Ikematsu, K. Evaluation of two cooling devices for construction workers by a thermal manikin. *Fash. Text.* **2017**, *4*, 23. [CrossRef]
80. Schellen, L.; Loomans, M.; Kingma, B.; De Wit, M.; Frijns, A.; van Marken Lichtenbelt, W. The use of a thermophysiological model in the built environment to predict thermal sensation: Coupling with the indoor environment and thermal sensation. *Build. Environ.* **2013**, *59*, 10–22. [CrossRef]

81. Zolfaghari, A.; Maerefat, M. A new simplified thermoregulatory bioheat model for evaluating thermal response of the human body to transient environments. *Build. Environ.* **2010**, *45*, 2068–2076. [CrossRef]
82. Kang, Z.; Wang, F. An advanced three-dimensional thermoregulation model of the human body: Development and validation. *Int. Commun. Heat Mass Transf.* **2019**, *107*, 34–43. [CrossRef]
83. Wu, G.; Liu, H.; Wu, S.; Liu, Z.; Mi, L.; Gao, L. A study on the capacity of a ventilation cooling vest with pressurized air in hot and humid environments. *Int. J. Ind. Ergon.* **2021**, *83*, 103106. [CrossRef]
84. Lou, L.; Zhou, Y.; Yan, Y.; Hong, Y.; Fan, J. Wearable cooling and dehumidifying system for personal protective equipment (PPE). *Energy Build.* **2022**, *276*, 112510. [CrossRef]
85. Stolwijk, J.A. *A Mathematical Model of Physiological Temperature Regulation in Man*; NASA: Washington, DC, USA, 1971.
86. Gao, C.; Kuklane, K.; Wang, F.; Holmér, I. Personal cooling with phase change materials to improve thermal comfort from a heat wave perspective. *Indoor Air* **2012**, *22*, 523–530. [CrossRef]
87. Hou, J.; Yang, Z.; Xu, P.; Huang, G. Design and performance evaluation of novel personal cooling garment. *Appl. Therm. Eng.* **2019**, *154*, 131–139.
88. Nag, P.; Pradhan, C.; Nag, A.; Ashtekar, S.; Desai, H. Efficacy of a water-cooled garment for auxiliary body cooling in heat. *Ergonomics* **1998**, *41*, 179–187. [CrossRef] [PubMed]
89. Guo, T.; Shang, B.; Duan, B.; Luo, X. Design and testing of a liquid cooled garment for hot environments. *J. Therm. Biol.* **2015**, *49*, 47–54. [CrossRef] [PubMed]
90. Bartkowiak, G.; Dabrowska, A.; Marszalek, A. Assessment of an active liquid cooling garment intended for use in a hot environment. *Appl. Ergon.* **2017**, *58*, 182–189. [CrossRef]
91. Shirish, A.; Kapadia, V.; Kumar, S.; Kumar, S.; Mishra, S.; Singh, G. Effectiveness of a cooling jacket with reference to physiological responses in iron foundry workers. *Int. J. Occup. Saf. Ergon.* **2016**, *22*, 487–493. [CrossRef]
92. Butts, C.L.; Torretta, M.L.; Smith, C.R.; Petway, A.J.; McDermott, B.P. Effects of a phase change cooling garment during exercise in the heat. *Eur. J. Sport Sci.* **2017**, *17*, 1065–1073. [CrossRef]
93. Rykaczewski, K. Rational design of sun and wind shaded evaporative cooling vests for enhanced personal cooling in hot and dry climates. *Appl. Therm. Eng.* **2020**, *171*, 115122. [CrossRef]

Disclaimer/Publisher's Note: The statements, opinions and data contained in all publications are solely those of the individual author(s) and contributor(s) and not of MDPI and/or the editor(s). MDPI and/or the editor(s) disclaim responsibility for any injury to people or property resulting from any ideas, methods, instructions or products referred to in the content.

processes

Review

Current Research Status and Development Trends of Cooling Suits in High-Temperature Mine Environments: A Review

Yu Ma [1], Qing Wan [1], Zidan Gong [2], Yiwei Wu [1] and Jie Zhou [1,*]

[1] School of Apparel and Art Design, Xi'an Polytechnic University, Xi'an 710048, China; 20120710@xpu.edu.cn (Y.M.); 15150957135@163.com (Q.W.); jackzxc603@gmail.com (Y.W.)
[2] Sino German College of Intelligent Manufacturing, Shenzhen Technology University, Shenzhen 518118, China; gongzidan@sztu.edu.cn
* Correspondence: zhoujie@xpu.edu.cn

Abstract: To gain a deeper understanding of the current research status of cooling suits in high-temperature mines, this paper provides separate introductions to vest-type cooling suits and full-body cooling suits. It summarizes the categories of cooling suits based on different cooling media and systematically elucidates the advantages and disadvantages of each type. The paper also analyzes the current application status of cooling suits in mine environments. It suggests that the future research directions for cooling suits in mines include the miniaturization of components, intelligent temperature control, optimization of new phase-change materials, development of cooling fabrics, and research in smart fibers.

Keywords: cooling suit; thermal comfort; phase-change materials; high-temperature mine environment

Citation: Ma, Y.; Wan, Q.; Gong, Z.; Wu, Y.; Zhou, J. Current Research Status and Development Trends of Cooling Suits in High-Temperature Mine Environments: A Review. *Processes* **2023**, *11*, 3256. https://doi.org/10.3390/pr11113256

Academic Editors: Alessandro Tonacci and Sergio Bobbo

Received: 16 October 2023
Revised: 16 November 2023
Accepted: 18 November 2023
Published: 20 November 2023

Copyright: © 2023 by the authors. Licensee MDPI, Basel, Switzerland. This article is an open access article distributed under the terms and conditions of the Creative Commons Attribution (CC BY) license (https://creativecommons.org/licenses/by/4.0/).

1. Introduction

Currently, China possesses billions of tons of coal resources in need of extraction. Coal resources have made a significant contribution to China's economic and social development. However, as the depth of mining increases year by year, reaching from several hundred meters to a thousand meters, the temperature in mines is also rising. It can reach up to 40 °C, which greatly impacts the miners [1]. Due to limited space underground, the heat generated by surrounding rocks, compressed air, and mechanical and electrical equipment cannot be dissipated, resulting in a gradual deterioration of the thermal environment in mines. Thermal hazards in mines are widespread. Therefore, it is crucial to lower the temperature in the mining environment.

Methods for preventing and controlling heat hazards in mines can be broadly categorized into two main types: traditional cooling techniques and individual cooling suits. Traditional cooling techniques primarily rely on ventilation to achieve lower temperatures. This is achieved by controlling parameters such as the airflow speed, volume, and temperature underground to achieve the desired cooling effect [2]. However, this method has certain limitations, and the cooling effect may not be very pronounced. To meet the demand for thermal comfort of personnel working in high-temperature mine environments, improve their work efficiency, and ensure work quality, the development of individual cooling suits has come into the view of researchers [3]. Individual cooling suits have the capability to regulate the microclimate temperature between the human body and the clothing. They exhibit high cooling efficiency, leading to effective temperature reduction and ultimately enhancing the thermal comfort of mine workers. Therefore, herein existing individual cooling suits have been classified and summarized. Initially, from a structural perspective, they are divided into vest-type cooling suits and full-body cooling suits. Upon analysis and comparison, it is evident that vest-type cooling suits are more suitable for wearing in mining operations. They are easy to put on and take off, facilitate movement, and provide good cooling effects. Based on their different cooling principles, the cooling suits are further categorized into gas-based cooling suits [4], liquid-based cooling suits [5], phase change cooling suits [6], and hybrid cooling suits [7]. This paper concludes and summarizes the working principles, advantages, disadvantages, and feasibility of using various types of cooling suits in underground mines. The strengths and weaknesses of different cooling suits are analyzed. Finally, recommendations are made for the future development of cooling suits for mining applications.

2. Human Thermal Comfort

After a person puts on clothing, a small climate is formed between the surface of the body and the outermost layer of clothing, known as the clothing microclimate. The comfort of this microclimate is closely related to human comfort. Temperature and humidity are important indicators for evaluating thermal comfort. The relationship between the clothing microclimate and human thermal comfort is shown in Figure 1. As shown in the figure, when the temperature of the clothing microclimate is between 31 °C and 33 °C, and the relative humidity is between 40% and 60%, the person feels comfortable. According to the ISO 7730-2005 "Ergonomics of the Thermal Environment" standard, human thermal comfort is defined as the subjective thermal environment evaluation made by a person in the surrounding environment. Thermal comfort occurs when the heat production and dissipation of the human body are in balance. However, in mining environments, the airflow between the underground environment and the outside world is weak, which severely affects evaporative and convective heat dissipation. Therefore, heat-related issues in mines significantly impact the thermal stability of mine workers, posing a threat to their health. To address the issue of cooling the human body in a hot environment, a method is proposed for mine workers to wear cooling suits in high-temperature mining environments. By regulating the heat in the clothing microclimate, the goal is to achieve thermal balance and thermal comfort for the human body. Factors affecting human thermal comfort include aspects related to the human body, clothing, and the environment [8]. The relationships between the human body, the environment, and the cooling suit during wearing are illustrated in Figure 2. For mine workers, it is challenging to control the intensity of activities and the environmental temperature underground. However, utilizing cooling suits to reduce the generation of body heat and facilitate heat dissipation is an effective approach.

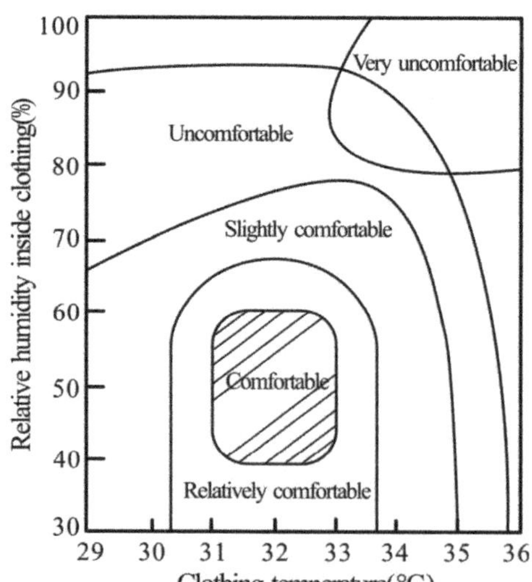

Figure 1. Relationship between clothing microclimate regions and human thermal comfort.

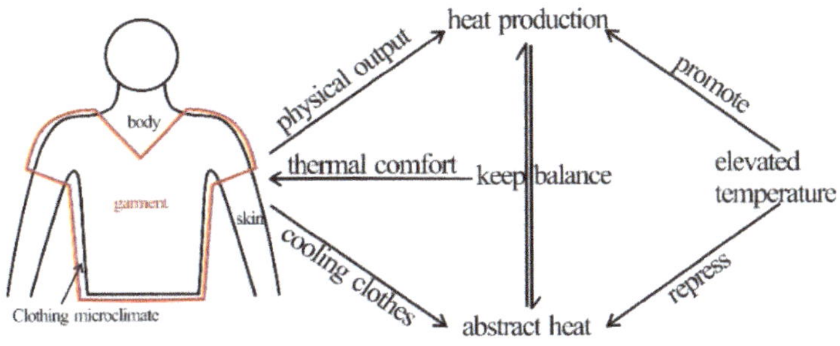

Figure 2. The relationship between the human body, environment, and cooling clothing.

3. Working Principle of Cooling Clothes

The heat generated during the human body's metabolic processes is dissipated to the surrounding environment through the skin. When the surrounding temperature is too high or heat cannot be effectively dissipated, heat accumulates in the microclimate zone on the surface of the skin, causing discomfort. Therefore, the primary purpose of cooling garments is to regulate the temperature in the microclimate zone on the surface of the skin. The human body dissipates heat through various methods, including convective heat dissipation, radiative heat dissipation, conductive heat dissipation, and evaporative heat dissipation. The cooling effect of cooling garments on the human body primarily involves the transfer of heat between the skin surface, clothing, and the environment, as illustrated in Figure 3.

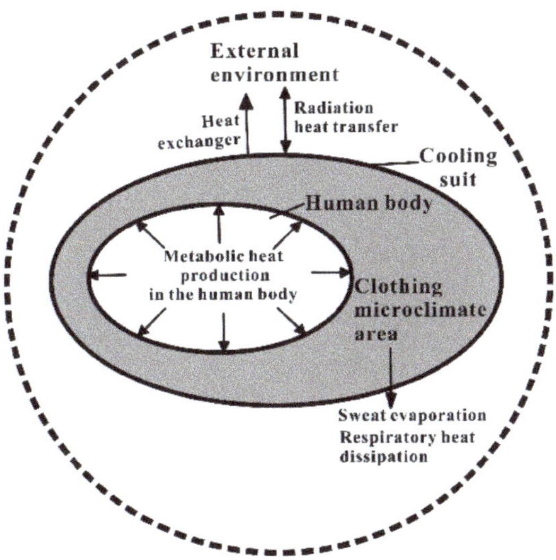

Figure 3. Heat exchange diagram between human body and environment.

As seen in Figure 3, when the heat exchange between the human body and the environment reaches dynamic equilibrium, the production and dissipation of heat by the body are balanced. Wearing cooling garments involves two main paths of heat transfer between the human body and the external environment: the first path involves the transfer of heat generated by metabolic processes through the skin surface to the cooling garment in a process of thermal equilibrium; the second path involves the transfer of heat between the external environment and the cooling garment. In harsh mining environments, where the heat generated by the body's metabolism is constant, the cooling effect of cooling garments directly influences the heat transfer between the body and the environment. When the body is in a high-temperature environment, cooling garments absorb excess heat produced by the body to maintain thermal comfort.

4. Research Status of Cooling Clothing

Utilizing the Web of Science database for retrieval, we gathered 118 articles on cooling garments from the past 20 years, as depicted in Figure 4. The linear trend chart indicates an increasing focus by researchers on the innovation and application of cooling garments. The prominent areas of interest include liquid-cooled garments, evaporative cooling, cooling performance, cooling textiles, protective devices, phase-change materials, and others.

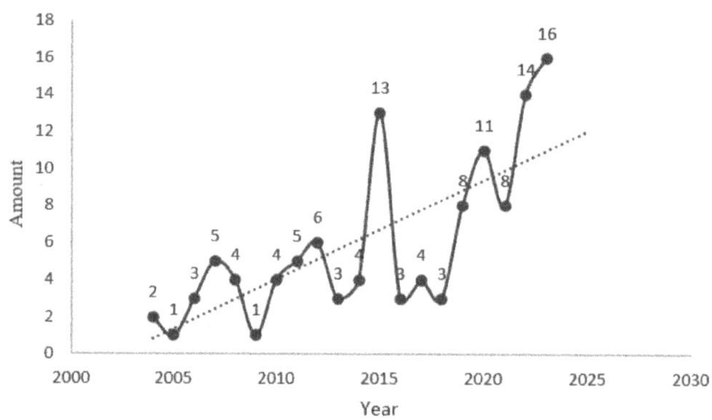

Figure 4. Number of publications on temperature-reducing clothing.

The effectiveness of cooling suits is influenced by factors such as environmental temperature and humidity, cool fabric, cooling clothing structure, cooling clothing design, and level of physical activity, as shown in Figure 5. Among these, factors that can be improved to enhance the cooling effect of cooling suits include the cool fabric, cooling clothing structure, and cooling clothing design.

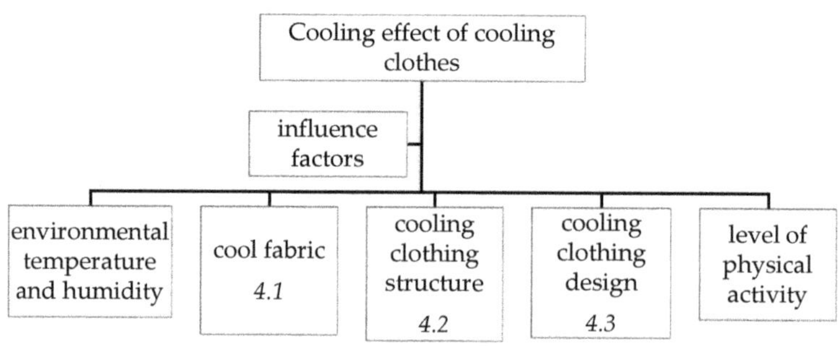

Figure 5. Factors affecting the cooling effect of cooling clothing.

4.1. Cool Fabric

In the high-temperature environment of mines, the use of cooling fabrics can enhance heat conduction and facilitate more effective convective evaporation, thus increasing comfort when the fabric is in close contact with the body. By increasing the thermal conductivity of fibers, regulating fiber cross-sectional structures, introducing phase-change materials, and employing post-fabric processing techniques, the fabric can rapidly transfer heat and accelerate perspiration, resulting in a cooling effect. This, in turn, enhances the thermal comfort of the human body. In recent years, experts and scholars have paid considerable attention to research on fiber thermal conductivity and the introduction of phase-change materials. Elevating thermal conductivity can enhance the fabric's ability to dissipate heat to the environment. Among these, phase-change fibers utilize the properties of phase-change materials to absorb and store body heat, thereby achieving a cooling effect [9].

Thermal conductivity is a metric used to assess the ability of clothing fibers to conduct heat. The thermal conductivity of fibers primarily depends on factors such as the internal molecular arrangement of the material, fiber morphology, and the microstructure within the fibers. Research conducted by You B et al. [10] through experimental comparative analysis indicates that modal fibers are better suited for mining clothing due to their high thermal conductivity, leading to enhanced cooling effects. To increase fiber thermal conductivity, Gao et al. [11] prepared boron nitride/polyvinyl alcohol composite fiber fabrics. The high thermal conductivity of the boron nitride nanosheets resulted in a thermal conductivity coefficient of the composite fiber fabric that was 2.22 times that of cotton fabric. Yuan et al. [12] utilized copper nanowires coated with polydopamine to prepare electrically insulating epoxy resin nanocomposite fibers with a thermal conductivity of up to 2.87 W/(m K). Wu et al. [13] developed a regenerated cellulose/BNNS (boron nitride nanosheets) composite fiber with a high BNNS content of up to 60% (by weight percentage). The thin and transparent two-dimensional structure of BNNS was particularly conducive to heat conduction. Chien et al. [14] fabricated a semi-crystalline polyamide (nylon) nanofiber. Annealing treatment was employed to enhance the configuration orderliness of the polymer, resulting in a high thermal conductivity of 59.1 W/(m K). This fiber can be employed to enhance the human thermal comfort performance.

Phase-change cooling fabrics integrate fibers with phase-change materials, leveraging the capacity of these materials to absorb, store, and release heat during phase transitions. This enables the regulation of skin temperature. In order to achieve an optimal thermal regulation response with phase-change fibers, it is important to enhance the thermal conductivity of both the phase-change material and the polymer substrate. Liu et al. [15] utilized wet spinning and freeze-drying processes to prepare Kevlar aerogel fibers with high porosity and surface area. After filling them with phase-change material, they obtained phase-change fibers with a phase-change enthalpy of 162 J/g. Wu et al. [16] used silk fibroin as a raw material and employed a freeze-spinning method to create microstructured fibers with high porosity. They then filled these fibers with polyethylene glycol (PEG) and coated them with polydimethylsiloxane (PDMS) to prevent leakage. The phase-change enthalpy of these fibers was approximately 118.1 J/g.

With the development of smart fibers, material intelligence has gradually come into the view of researchers. Wang Wen et al. [17] used deformable bacteria on fabrics to make them sensitive to humidity. They placed the bacteria on both sides of latex and kept them stable at room temperature. When one side was exposed to high temperatures, the expansion of the bacteria caused the latex to bend outward. Sweat evaporates through the opened ventilation port, achieving the goal of cooling. Lao S et al. [18] developed a solar cooling suit. They installed a semi-flexible solar panel on the back of the suit and supplied power to two battery packs through a controller. This provided a continuous power source for a fan, enabling sustained cooling. This cooling suit is environmentally friendly. Ke et al. [19] designed a ladies' blouse made from nano-porous polyethylene material. The study found that when using this cooling suit, increasing the set air conditioning temperature from 25.5 °C to 27.0 °C could save 9% to 15% of cooling energy. Wei et al. [20] applied a layer of Al_2O_3-dispersed cellulose acetate on the textile. This increased the textile's solar reflectance from 62.6% to 80.1%. As a result, the wearer absorbs less radiant heat, achieving the goal of cooling.

4.2. Cooling Clothing Structure

Experimental measurements have revealed that different parts of the human body exhibit varying microclimates within clothing, and their relationship and response to the external environment are distinct. Areas such as the armpits, sides, and waist have relatively little correlation with environmental climate variations (changing by approximately 5 °C over the course of a year). Conversely, regions like the chest, back, shoulders, upper and lower limbs, including the forearms and thighs, have a closer relationship with external climate changes (varying by 5 °C to 10 °C for the forearms and thighs, and by 10 °C to 15 °C for the calves). Additionally, during physical activity, the trunk of the body has the highest basal metabolic rate and heat capacity [4]. Based on the relationship between different parts of the human body and environmental variations, various types of cooling suits with different structural designs have been developed. Figure 6a depicts a vest-type cooling suit. Vest-type cooling suits are widely used in various fields of production and daily life due to their simple structure, ease of wear, effective cooling, and high practicality. Figure 6b shows a full-body cooling suit. The advantage of a full-body cooling suit lies in its ability to comprehensively cool the trunk, limbs, and other areas, ensuring that the entire body remains in a relatively comfortable state. Gerrett et al. [21] conducted a study analyzing the thermal sensitivity of different areas on the human body. They found that the abdominal area and the sides of the lower back are more sensitive to cold sensations. Therefore, cooling the lower half of the body is more conducive to enhancing thermal comfort. In comparison to the upper body, the legs generate more heat during walking. Cooling suits need to absorb a significant amount of heat. As a result, the legs may not achieve the desired cooling effect in the later stages of exercise and during periods of rest [22]. However, for workers in high-temperature mines with extensive work areas and high intensity, it may affect their work efficiency and overall comfort of wearing.

(**a**)

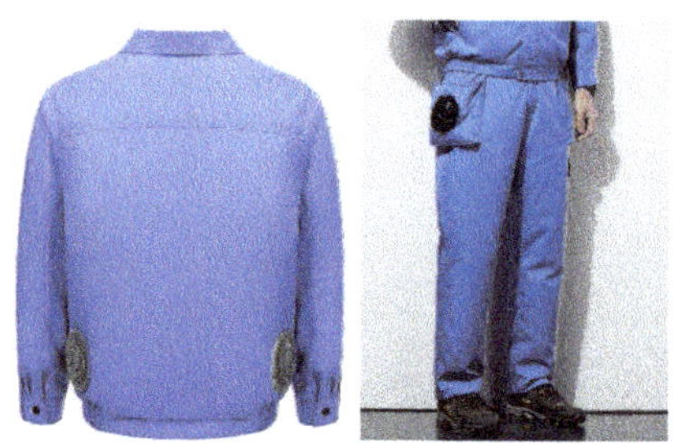

(**b**)

Figure 6. Structural types of cooling suits. (**a**) Tank top cooling suit; (**b**) full-body cooling suit.

4.3. Cooling Clothing Design

To date, researchers have conducted extensive studies on different types of cooling suits and their performance. Based on different cooling media, cooling suits can be categorized as gas-based cooling suits, liquid-based cooling suits, phase-change cooling suits, and hybrid cooling suits. According to the source of the cooling power, they can be classified as active cooling suits or passive cooling suits. Additionally, based on different cooling methods, they can further be divided into cold storage cooling suits, vapor compression cooling suits, vortex tube cooling suits, and thermoelectric refrigeration cooling suits [23–25].

4.3.1. Gas Cooling Suit

Gas-based cooling suits blow air into the microclimate of the clothing, dissipating heat through evaporation and convection. This falls under the category of active cooling [26]. The principle of heat dissipation involves accelerating the evaporation of sweat and enhancing air convection, thereby achieving the goal of cooling. Depending on the different methods of heat dissipation, gas-based cooling suits can be further divided into convective gas-based cooling suits and evaporative gas-based cooling suits. Additionally, based on different structural designs, gas-based cooling suits can be categorized as fan-type cooling suits and duct-type cooling suits [27].

Fan-based cooling suits primarily operate by using an external power source to drive a fan or blower device, creating a flow of air between the clothing and the skin surface. This accelerates the evaporation of sweat, thereby cooling the microclimate, as illustrated in Figure 7. Currently, some construction workers and traffic control personnel have adopted these suits. Mengmeng Zhao [4] conducted experiments under different fan placement configurations and found that placing fans along the spine and lower back yielded the best results. Yi et al. [28] added multiple settings to the fan in the cooling suit to ensure a stable airflow. Each setting provided a relatively consistent volume of air. Gas-based cooling suits offer excellent cooling effects. Underground environments are rich in gas resources, allowing for extended cooling durations. These suits are compact, lightweight, and can alleviate the burden on workers. However, fan-based cooling suits can be cumbersome to wear, potentially hindering the wearer's mobility. Moreover, there may be risks of explosion in underground mines due to the presence of hazardous gases like methane. And its cooling effect is not significant in high humidity or extremely hot environments [29].

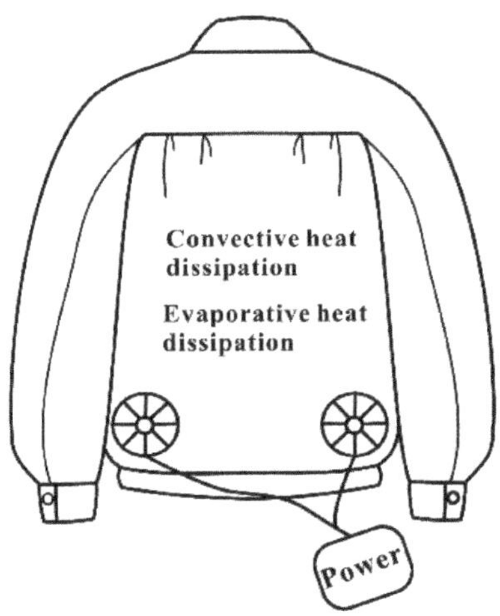

Figure 7. Fan-style cooling suit.

In 2005, Zheng Xiang Pu [30] introduced the duct-type gas-based cooling suit. In this design, an air compressor compresses air and pre-cools it through a vortex tube. The compressed air is then routed through a system of tubes distributed throughout the suit. This promotes heat dissipation on the body surface through the evaporation of sweat and air convection. The gas, having completed its heat exchange, is expelled through the exhaust port. This technology applies the cooling effect of the vortex tube to provide cooling protection for mine workers. During use, temperature output within the range of human thermal comfort can be ensured simply by controlling the adjustment valve. The suit features a straightforward structure, convenient operation, and occupies minimal space, making it suitable for the complex environments found in mines. Compared to traditional fan-based cooling suits, vortex-based cooling suits do not require power sources or refrigerants and can operate continuously for extended periods. However, their cooling effect is relatively weaker and is influenced by the intake pressure and flow rate. Guo et al. [31] developed a tubular ventilation suit with an open front and unsealed ventilation. The ventilation device introduces natural air or cool air into the duct, which is then blown onto the inner clothing through holes. This design offers characteristics such as ease of wear, being lightweight, and low resistance.

4.3.2. Liquid Cooling Suit

Liquid-based cooling suits circulate a cooling liquid within the microclimate, dissipating the body's metabolic heat to maintain thermal balance. This falls under active cooling. The cooling mediums for liquid cooling suits mainly include water, a mixture of water and ice, phase-change emulsions, and microcapsule emulsions. The design styles are depicted in Figure 8. Since liquid cooling suits rely on conductive heat dissipation, they are typically designed in a form-fitting style to ensure effective heat conduction by keeping the clothing close to the skin [32]. However, to reduce the initial discomfort when the skin surface comes into direct contact with the low-temperature pipes, McLellan [33] suggests wearing a T-shirt or long-sleeved shirt underneath the liquid cooling suit. Guo et al. [5] developed a detailed theoretical model for heat transfer from the liquid cooling suit to the environment and determined its maximum working duration to be 3.36 h, which is relatively short. Liquid cooling suits generally require additional equipment such as power sources, water pumps, and refrigeration devices, making them heavy and less portable. To address these issues, Rahman et al. [34] designed an active cooling system that uses non-toxic gallium-based liquid metal (Galinstan) as the coolant. Compared to other active cooling systems, the liquid-metal cooling system extended the cooling time by four times, reduced the weight to one-third of the original weight, and lowered the cost to one-eighth of the original cost, significantly improving upon the drawbacks of traditional liquid cooling suits. However, further research is needed to address issues such as preventing leaks in liquid cooling suits and improving freedom of movement while wearing them.

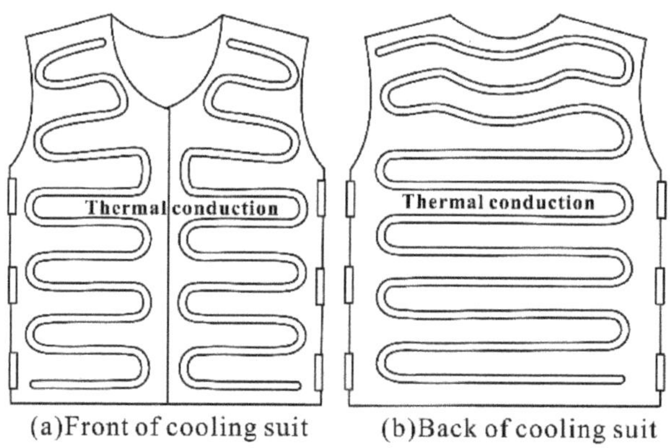

Figure 8. Style diagram of liquid cooling suit. (**a**) Front of cooling suit; (**b**) Back of cooling suit.

4.3.3. Phase-Change Cooling Suit

According to the cooling principle, phase-change cooling suits fall under passive cooling technology. Due to their simple system, user-friendly operation, lack of power requirements, absence of explosion-proof demands, wide range of phase-change materials, effective cooling performance, and reusability, phase-change cooling suits possess extensive application prospects. They have become a focal point of research for international scholars.

Phase-change cooling suits operate on a passive cooling principle [35]. Typically, multiple pockets are incorporated into the clothing to hold phase-change materials. Through the phase transition of these materials, they absorb heat from the body, achieving the cooling effect. Common phase-change materials include ice, dry ice, gels, crystalline salts, and phase-change material capsules [36]. Ali et al. [37] explored the melting points of different phase-change materials, ultimately choosing hexadecane, with a melting point of 18-20 °C, as the phase-change material to develop a phase-change cooling suit. Itani et al. [38] tested the metabolic rate of individuals wearing three different cooling vests at different environmental temperatures to evaluate their respective cooling performance. The study found that using a mixed-phase-change cooling suit with a desiccant vest had a good effect on maintaining a stable metabolic rate. Ji Changfa et al. [39] conducted experimental research on the characteristics of phase-change storage materials in cooling suits. They found that inorganic salts have good storage effects when used as coolants, demonstrating that an inorganic salt solution with a certain concentration ratio can be used as a phase-change refrigerant to fill the cooling suit.

However, phase-change cooling suits do have some drawbacks, such as a non-adjustable temperature, heavy weight, limited duration, and potential for leakage. To address these issues, Yuan Pei et al. selected a phase-change material with a transition temperature close to body temperature and used a vortex tube as a cold source to cool the phase-change material, resolving the problem of repeated cooling. Sun Wenjuan et al. [40] proposed a composite phase-change latent heat storage material based on paraffin, which was blended with a polymer material in a proportionate melt. When the paraffin-based phase-change material underwent solid–liquid phase transition, the surface of the composite phase-change material remained unchanged, effectively solving the issues of leakage and the encapsulation of phase-change latent heat storage materials in individual cooling in mining. Liang Y et al. [41] selected the TH-SL-23 hybrid solid–liquid phase-change material, which has a high latent heat of fusion for an organic phase-change material, low undercooling ability in the liquid phase, and stable thermal and chemical properties without reaction or corrosion. The total mass of the working material was only 618 g. Today, various critical issues in phase-change cooling suits have been resolved. However, the future focus of research lies in how to integrate and apply these key technologies. This includes the development of a new type of comprehensive cooling suit based on phase-change materials that not only provides effective cooling but also offers temperature control. Additionally, there is a need for improved encapsulation materials and methods, enhanced thermal conductivity of phase-change materials, the development of composite phase-change materials with excellent heat dissipation and long endurance, as well as the creation of new materials or devices for rapid activation of phase-change materials, all of which will drive the advancement of phase-change cooling suits.

4.3.4. Hybrid Cooling Suit

A hybrid cooling suit refers to a cooling suit that combines two or more cooling methods, taking advantage of their respective strengths and compensating for their weaknesses. Song et al. [42] developed a hybrid cooling suit equipped with both a ventilation fan and phase-change materials. Their research results indicate that compared to a single cooling suit, wearing this hybrid cooling suit during indoor work for 90 min demonstrates better cooling effects and a longer cooling duration. Kang et al. [43], on the basis of a hybrid cooling suit combining phase-change materials and a fan, added an insulating layer to the outer surface of the phase-change cooling suit to reduce the absorption of heat from the external environment, significantly enhancing the cooling effectiveness and the duration of effective cooling for the hybrid cooling suit.

5. The Application of Cooling Clothes in Mine Environments

In hot underground mining environments, the initial investment for cooling suits is only 1/25 of that for traditional cooling methods, and the operational maintenance costs are 1/16 of the original. Mining cooling suits are flexible and convenient to use, not limited by the workplace, and can be moved according to changes in the mining face. It can be foreseen that in the hot environment of mines, cooling suits have a broad application prospect. The characteristics of various types of cooling suits are compared in Table 1.

Table 1. A comparison of characteristics of various cooling suits.

Types of Cooling Suits	Cooling Methods/Principles	Advantages	Disadvantages	Is It Suitable for the Mine Environment?
Vest-style cooling suits	Cooling vest designed for the torso	Simple structure, easy to wear, effective cooling, practical.	The application is limited, as it may not be suitable for certain specialized positions that require full-body protection.	Suitable
Full-body cooling suits	Comprehensive cooling for the trunk and limbs	Provides comprehensive cooling protection for all parts of the body, especially suitable for special operations or special environments.	The range of motion and comfort during activity are limited, which can affect operational and work efficiency. It is not easy to put on, and it is heavy.	Not suitable
Gas cooling suit	Convective cooling and evaporative cooling	Abundant gas source, lightweight, effective cooling, and sustainable cooling.	The volume of the clothing after ventilation is relatively large. The cooling effect is poorer in hot and humid environments. It is not suitable for environments above 45 °C.	It is not advisable for use in environments with poor air quality and hazardous gases. Additionally, it can be bulky, restricting the wearer's movement and potentially impacting work efficiency.
Liquid cooling suit	Conductive cooling	Temperature controllable, high cooling efficiency, and sustainable cooling.	There is a relatively high risk of liquid leakage. It requires a power source and a refrigeration unit, making it heavy and less portable.	Liquid leakage is a concern, and the cooling device is complex, heavy, and difficult to carry. The comfort during movement while wearing it is also poor.
Phase-change cooling suit	Conductive cooling	No electronic devices, easy to operate, minimal risk of overcooling or overheating, effective cooling, lightweight and portable.	There are potential issues with phase-change material leakage, limited cooling duration, uncontrollable temperatures, and the need for repeated material cooling.	It provides short-term continuous cooling. However, it has poor breathability in the covered areas, requires repeated cooling, has uncontrollable temperature, and may face leakage issues.
Hybrid cooling suit	Multiple cooling methods	Promotes strengths and avoids weaknesses, complements each other's strengths.		Indeterminacy

Based on the above comparative analysis, it is evident that existing cooling suits still face challenges when applied in high-temperature mines, including limited cooling duration, material leakage, and inconvenience in wearability. When selecting cooling suits for use in high-temperature mines, various factors must be taken into consideration, such as the hot and humid working environment, underground safety conditions, permitted battery types and CO_2 volume fractions, comfort and flexibility in wear, as well as overall cost-effectiveness. Therefore, the issues that need to be addressed for different types of cooling suits in mining applications include:

(1) The use of fan-cooled suits carries a risk of electrical leakage and explosion. Phase-change cooling suits are susceptible to deformation and potential leakage of phase-change materials when miners engage in high-intensity physical labor, which can lead to skin corrosion, resulting in lower safety levels.

(2) While gas cooling and liquid cooling suits provide effective cooling, they rely on refrigeration units to supply the cooling medium. This results in suits with a large size and weight, which may impede the movement of underground workers.

(3) Gas cooling suits are suitable for mines with low pollution levels, low cooling requirements, low humidity, and no electrical explosion-proof requirements. Liquid cooling suits are suitable for large spaces in high-temperature mines. Phase-change cooling suits are suitable for short-duration work in high-temperature mines.

6. Conclusions and Outlook

The fabric of the suit and the design of the cooling suit structure both impact the cooling effectiveness of the cooling suit. Cooling suits have already been applied in high-temperature mining environments to improve human thermal comfort.

With the advancement of technology, cooling fibers have emerged. Cooling suits are no longer limited to conventional materials. Cooling-fabric-based cooling suits are comfortable to wear and environmentally friendly. The development of new cooling fabrics has improved the thermal conductivity and convective evaporation of cooling suits, enhancing comfort when the suit is in close contact with the body. In the future, interdisciplinary collaboration in fields such as textiles, materials, optics, thermodynamics, and mechanics could lead to the development of intelligent cooling suits with new materials that meet the basis of human thermal comfort.

Gas cooling suits, liquid cooling suits, phase-change cooling suits, and hybrid cooling suits each have their own strengths and weaknesses. To be effectively used in mining environments, further improvements are needed. In addition to ensuring safety and comfort, factors such as cooling duration, weight, leakage issues, ease of wearing, power requirements, and aesthetics need to be comprehensively considered. Continued optimization in design is necessary to meet the needs of mining personnel. Among them, hybrid cooling suits exhibit excellent comprehensive performance and high practicality, making them a key focus for future development.

Taking into consideration the unique environment and conditions in high-temperature mines, the future development trends of cooling suits for mining will focus on the miniaturization of components, intelligent temperature control, optimization of new phase-change materials, development of cooling fabrics, and research in smart fibers. Among these, phase-change cooling suits and composite cooling suits show significant potential and are likely to be the best-suited cooling solutions for miners.

Author Contributions: Y.M. and Q.W. designed the study. Z.G. and Y.W. collected the related materials. J.Z. re-examined the related materials. Y.M. and Q.W. wrote the manuscript. Z.G. and Y.W. drew pictures. Q.W. and J.Z. reviewed and revised the manuscript. All authors have read and agreed to the published version of the manuscript.

Funding: This research was funded by Shaanxi Provincial Key Research and Development Program 2023, grant number 2023-YBSF-203.

Institutional Review Board Statement: Not applicable.

Informed Consent Statement: Not applicable.

Data Availability Statement: Data are contained within the article.

Conflicts of Interest: The authors declare no conflict of interest.

References

1. Xie, Y.; Lu, Y.; Shen, H. Effect of human age on thermo physiology and behavior ability in a high temperature environment. *Basic Sci. J. Text. Univ.* **2021**, *34*, 32–39. [CrossRef]
2. Liu, Y.; Li, M.; Cao, Y.; Jia, M.T. Development of local cooling equipment for deep mines. *Mod. Min.* **2019**, *35*, 177–181. [CrossRef]
3. Tian, S.; Zhou, R.; Yang, J. Effects of cooling suits on firefighters' thermal responses under high temperature. *China Saf. Sci. J.* **2020**, *30*, 166–171. [CrossRef]
4. Zhao, M.M.; Gao, C.S.; Wang, F.M.; Kuklane, K.; Holmér, I.; Li, J. A study on local cooling of suits with ventilation fans and openings placed at different torso sites. *Int. J. Ind. Ergon.* **2013**, *43*, 232–237. [CrossRef]
5. Guo, T.; Shang, B.; Duan, B.; Luo, X.B. Design and testing of a liquid cooled suit for hot environments. *J. Therm. Biol.* **2015**, *49–50*, 47–54. [CrossRef]

6. Hunter, I.; Hopkins, J.T.; Case, D.J. Warming up with an ice vest: Core body temperature before and after cross country racing. *J. Athl. Train.* **2005**, *41*, 371–374. [CrossRef]
7. Lu, Y.; Wei, F.; Lai, D.; Wen, S.; Wang, F.; Gao, C.S. A novel personal cooling system (PCS) incorporated with phase change materials (PCMs) and ventilation fans. *J. Thermal. Biol.* **2015**, *52*, 137–146. [CrossRef]
8. Wu, T.; Lu, Y. Analytical investigation on human thermoregulation model based on CiteSpace. *Basic Sci. J. Text. Univ.* **2023**, *36*, 1–7. [CrossRef]
9. Zeng, S.N.; Hu, J.Y.; Zhang, M.N.; Xiang, Y.Z.; Wu, J.W.; Su, M.Y.; Zhang, Y.Q.; Shen, M.; Hong, P.; Huang, Z.L.; et al. Cooling textiles for personal thermal management. *Chin. Sci. Bull.* **2022**, *67*, 1167–1179. [CrossRef]
10. You, B.; Liu, J.F.; Zhang, Y.X.; Shi, S.L.; Liu, H.Q.; Lu, Y. Experimental of influence of clothing materials on cooling effect of mine ventilation clothing. *China Saf. Sci. J.* **2022**, *32*, 174–180. [CrossRef]
11. Gao, T.T.; Yang, Z.; Chen, C.J.; Li, Y.J.; Fu, K.; Dai, J.Q.; Hitz, E.M.; Xie, H.; Liu, B.Y.; Song, J.W.; et al. Three-Dimensional Printed Thermal Regulation Textiles. *ACS Nano* **2017**, *11*, 11513–11520. [CrossRef] [PubMed]
12. Yuan, H.; Wang, Y.; Li, T.; Ma, P.; Zhang, S.; Du, M.; Chen, M.; Dong, W.; Ming, W. Highly thermal conductive and electrically insulating polymer composites based on polydopamine-coated copper nanowire. *Compos. Sci. Technol.* **2018**, *164*, 153–159. [CrossRef]
13. Wu, K.; Yu, L.P.; Lei, C.X.; Huang, J.X.; Fu, Q. Green Production of Regenerated Cellulose/Boron Nitride Nanosheet Textiles for Static and Dynamic Personal Cooling. *ACS Appl. Mater. Interfaces* **2019**, *11*, 40685–40693. [CrossRef] [PubMed]
14. Chien, H.-C.; Peng, W.-T.; Chiu, T.-H.; Wu, P.-H.; Liu, Y.-J.; Tu, C.-W.; Wang, C.-L.; Lu, M.-C. Heat Transfer of Semicrystalline Nylon Nanofibers. *ACS Nano* **2020**, *14*, 2939–2946. [CrossRef] [PubMed]
15. Liu, Z.; Lyu, J.; Fang, D.; Zhang, X. Nanofibrous Kevlar Aerogel Threads for Thermal Insulation in Harsh Environments. *ACS Nano* **2019**, *13*, 5703–5711. [CrossRef] [PubMed]
16. Wu, J.; Hu, R.; Zeng, S.N.; Xi, W.; Huang, S.Y.; Deng, J.H.; Tao, G.M. Flexible and Robust Biomaterial Microstructured Colored Textiles for Personal Thermoregulation. *ACS Appl. Mater. Interfaces* **2020**, *12*, 19015–19022. [CrossRef] [PubMed]
17. Anonymous. Biointelligent clothing that can automatically ventilate during sweating. *Text. Test. Stand.* **2017**, *3*, 4.
18. Lao, S.J.Y.; Xu, D. Design and research of solar powered cooling suits. *Explor. Sci.* **2016**, 191.
19. Ke, Y.; Wang, F.; Xu, P.J.; Yang, B. On the use of a novel nanoporous polyethylene (nanoPE) passive cooling material for personal thermal comfort management under uniform indoor environments. *Build. Environ.* **2018**, *145*, 85–95. [CrossRef]
20. Wei, W.; Zhu, Y.; Li, Q.; Cheng, Z.; Yao, Y.; Zhao, Q.; Zhang, P.; Liu, X.; Chen, Z.; Xu, F. An Al_2O_3-cellulose acetate-coated textile for human body cooling. *Sol. Energy Mater. Sol. Cells* **2020**, *211*, 110525. [CrossRef]
21. Gerrett, N.; Ouzzahra, Y.; Havenith, G. Distribution of skin thermal sensitivity. In *Agache's Measuring the Skin: Non-invasive Investigations, Physiology, Normal Constants*, 2nd ed.; Springer: Cham, Switzerland, 2017; pp. 1285–1301. [CrossRef]
22. Zheng, Q.; Wang, H.F.; Ke, Y.; Li, S. Design and evaluation of cooling clothing by phase change materials for miners. *J. Text. Res.* **2020**, *41*, 124–129. [CrossRef]
23. Li, L.N.; Qian, X.M.; Xu, J. The current international research situation and application of cooling suit. *China Pers. Prot. Equip.* **2008**, *2*, 24–28. [CrossRef]
24. Han, Z.; Tang, S.; Lai, J. The current international research situation and key techniques of cooling suit. *China Pers. Prot. Equip.* **2009**, 11–14. [CrossRef]
25. Sheng, W.; Zheng, H. Literature review on application of body cooled suits in mine thermal environment. *J. Saf. Sci. Technol.* **2013**, 95–101. [CrossRef]
26. Chinevere, T.D.; Cadarette, B.S.; Goodman, D.A.; Ely, B.R.; Cheuvront, S.N.; Sawka, M.N. Efficacy of body ventilation system for reducing strain in warm and hot climates. *Eur. J. Appl. Physiol.* **2008**, *103*, 307–314. [CrossRef]
27. Bartkowiak, G.; Dąbrowska, A.; Włodarczyk, B. Construction of a suit for an integrated liquid cooling system. *Text. Res. J.* **2015**, *85*, 1809–1816. [CrossRef]
28. Yi, W.; Zhao, Y.; Chan, A.P.C. Evaluation of the ventilation unit for personal cooling system (PCS). *Int. J. Ind. Ergon.* **2017**, *58*, 62–68. [CrossRef]
29. Jing, R.; Liang, J. Research progress of active cooling suits for outdoor workers in summer. *Basic Sci. J. Text. Univ.* **2023**, *36*, 29–35. [CrossRef]
30. Zheng, X.P. *A Dynamic Model of the Human/Cooling System/Clothing/Environment System*; University of Central Florida: Orlando, FL, USA, 2005.
31. Guo, X.M.; Zhang, X.; Yuan, X.G. Design and development of novel ventilated clothing. In Proceedings of the 13th International Conference on Man-Machine-Environment System Engineering; Long, S., Dhillon, B.S., Eds.; Springer: Berlin/Heidelberg, Germany, 2014; Volume 259, pp. 451–459.
32. Branson, D.H.; Cao, H.; Jin, B.; Peksoz, S.; Farr, C.; Ashdown, S. Fit analysis of liquid cooled vest prototypes using 3D body scanning technology. *J. Text. Appar. Technol. Manag.* **2005**, *4*, 1–13.
33. Mcleilan, T.M. *Cooling Options for Shipboard Personnel Operating in Hot Environments*; Defense Technical Information Center: Fort Belvoir, VA, USA, 2002.
34. Rahman, M.A.; Elassy, K.S.; Roose, L.R.; Shiroma, W.A. A Lightweight, Low-Cost Liquid-Metal Personal Cooling System for Prolonged Cooling. In Proceedings of the 2020 IEEE International Conference on Consumer Electronics (ICCE), Las Vegas, NV, USA, 4–6 January 2020.

35. Zhu, F.L.; Fan, J.B.; Feng, Q.Q.; Zhou, Y. Application and feasibility analysis of phase change materials for fire-fighting suit. *J. Text. Res.* **2014**, *35*, 124–132. [CrossRef]
36. Shen, T.; Lu, S.; Xin, C. Preparation of thermal energy storage microcapsule by phase of change and the application in intelligent textiles. *J. Xi'an Polytech. Univ.* **2017**, *31*, 306–314. [CrossRef]
37. Ali, K.; Gul, R.M.; Arshad, S.N.; Kamran, M.A. Personalized Cooling System Using Phase Change Materials. *Key Eng. Mater.* **2021**, *875*, 184–192. [CrossRef]
38. Itani, M.; Bachnak, R.; Ghaddar, N.; Ghali, K. Evaluating Performance of Hybrid PCM-Fan and Hybrid PCM-Desiccant Vests in Moderate and Hot Climates. *J. Build. Eng.* **2019**, *22*, 383–396. [CrossRef]
39. Ji, C.F.; Ji, C.Y.; Wang, Z.R.; Li, S.J. Experimental Study on Properties of Phase Change Cold Storage Materials Filled in Cooling Suits. *Coal Mine Saf.* **2019**, *50*, 21–25+30. [CrossRef]
40. Sun, W.; Liang, G. Research and development of individual cooling composite shaped paraffin based phase change cold storage materials for mining. *Chem. Eng. Equip.* **2021**, 43–44+47. [CrossRef]
41. Liang, Y.H.; Li, W.L.; Zhu, J.Y.; Liu, C.H.; Ning, B.S. Optimal design and thermal comfort evaluation of ultralight phase change cooling suit. *Wool Text. J.* **2022**, *50*, 97–103. [CrossRef]
42. Song, W.; Wang, F.; Wei, F. Hybrid cooling clothing to improve thermal comfort of office workers in a hot indoor environment. *Build. Environ.* **2016**, *100*, 92–101. [CrossRef]
43. Kang, Z.; Udayraj; Wan, X.; Wan, F. A new hybrid personal cooling system (HPCS) incorporating insulation pads for thermal comfort management: Experimental validation and parametric study. *Build. Environ.* **2018**, *145*, 276–289. [CrossRef]

Disclaimer/Publisher's Note: The statements, opinions and data contained in all publications are solely those of the individual author(s) and contributor(s) and not of MDPI and/or the editor(s). MDPI and/or the editor(s) disclaim responsibility for any injury to people or property resulting from any ideas, methods, instructions or products referred to in the content.

processes

MDPI

Review

Systematic Evaluation of Research Progress in the Textile Field over the Past 10 Years: Bibliometric Study on Smart Textiles and Clothing

Ting Wang [1], Changqing Liu [1], Jun Zhang [2] and Aosi Wang [2,*]

[1] School of Apparel and Art Design, Xi'an Polytechnic University, Xi'an 710048, China; wangting331@126.com (T.W.); changqingliu1999@gmail.com (C.L.)
[2] School of Fashion and Textiles, The Hong Kong Polytechnic University, Hong Kong, China; alicezhang34@hotmail.com
* Correspondence: 22124308r@connect.polyu.hk

Abstract: Intelligent textile clothing is one of the most popular topics in the field. In recent decades, rapid advances have been made in the area of intelligent textile clothing research, and the intellectual structure pertaining to this domain has significantly evolved. We used CiteSpace 6.2.R4, VOSviewer 1.6.19, to evaluate and visualize the results, analyzing articles, countries, regions, institutions, authors, journals, citations, and keywords. Both a macroscopic sketch and a microscopic characterization of the entire knowledge domain were realized. The aim of this paper is to utilize bibliometric and knowledge mapping theories to identify relevant research papers on the subject of smart textiles and clothing that have been published by the China Knowledge Network Web of Science (WOS) within the last decade. It is concluded that the main topics of smart textile and garment research can be divided into nine categories: wearable electronics, smart textiles, flexible antennas, energy storage, textile actuators, mechanical properties, asymmetric supercapacitors, carbon nanotubes, and fiber extrusion. In addition to the latter analysis, emerging trends and future research foci were predicted. This review will help scientists discern the dynamic evolution of intelligent textile clothing research as well as highlight areas for future research.

Keywords: smart textiles and clothing; smart fibers; CiteSpace; intellectual structure; knowledge mapping; bibliometrics

Citation: Wang, T.; Liu, C.; Zhang, J.; Wang, A. Systematic Evaluation of Research Progress in the Textile Field over the Past 10 Years: Bibliometric Study on Smart Textiles and Clothing. *Processes* **2023**, *11*, 2797. https://doi.org/10.3390/pr11092797

Academic Editors: Zhanxiao Kang and Qing Chen

Received: 3 August 2023
Revised: 6 September 2023
Accepted: 9 September 2023
Published: 20 September 2023

Copyright: © 2023 by the authors. Licensee MDPI, Basel, Switzerland. This article is an open access article distributed under the terms and conditions of the Creative Commons Attribution (CC BY) license (https://creativecommons.org/licenses/by/4.0/).

1. Introduction

With the increasing popularity and advancement of smart textile clothing, a wide range of smart clothing products are gaining traction in the market. As a result, people have developed multifaceted demands for smart textile clothing with respect to its functionalities [1]. Due to the interesting electrical, thermal, and optical properties of smart textile garments, among others, a high degree of innovation and smart garment potential can be realized, e.g., they can be used simultaneously as sensors, heaters, energy generators, and storage devices. Smart textiles and clothing products utilize feedback mechanisms to sense changes in both the external environment and the human body, allowing for a triangular interaction between the wearer, their surroundings, and their garments. This creates an interdependent organism. The development of intelligent textile garments requires multidisciplinary technical support and the integration of advanced technologies from various fields, including biotechnology, sensor technology, computer science, microelectronics, polymer chemistry, and materials science and technology [2–4].

There have been numerous studies and reports on wearable technology [5–7], garment design [8–10], and smart clothing [11–13] in the context of smart textile garments. The aim of this paper is to utilize bibliometric and knowledge mapping theories to identify relevant research papers on the subject of smart textiles and clothing that have been published by the China Knowledge Network Web of Science (WOS) within the last decade. We used CiteSpace 6.2.R4 and VOSviewer 1.6.19 software to analyze the data, aiming to understand the research hotspots and trends in the field of smart textiles and clothing in China in recent years. Additionally, we provided references for future research on this topic.

2. Research Methodology and Data Analysis

2.1. Research Methodology

The CiteSpace 6.2.R4 and VOSviewer 1.6.19 software were used to analyze the knowledge structure, developmental patterns, and distribution of intelligent textiles and garments. A knowledge graph is presented to illustrate the relationship between the annual article volume, organizational affiliations, and author collaborations in this research field. This graph intuitively reflects the internal relationships and trajectories within this field of study while also sorting out its research trajectory and future trends [14].

2.2. Data Analysis

The data sources utilized in this study comprise the Web of Science (WOS) core collection database, which was searched from 1 January 2012 to 31 December 2022.

A precise search was conducted in the WOS core collection database for articles with subject terms (TS) = "Smart clothing*" or "Smart textiles*". The retrieval strategy is shown in Figure 1. The literature category included papers, reviews, and online publications in the English language. The retrieved articles were exported in plain text format, including both full records and cited references, resulting in 2456 articles. After deduplication processing, 2398 valid pieces of literature were obtained.

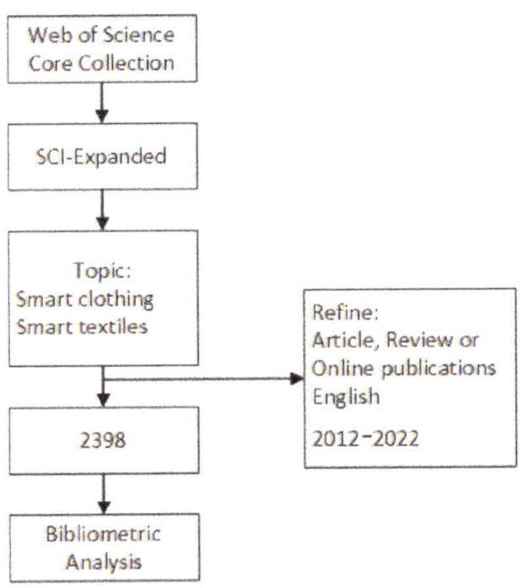

Figure 1. Flowchart steps of the search strategy.

3. Results and Analysis

3.1. Volume of Literature

The annual article count serves as a visual indicator of the current research status and facilitates the identification of developmental trends, which is crucial for predicting future directions. Based on the data from the WOS, Figure 2 depicts the trend graph of the article counts.

Figure 2 showed an upward trend in the overall number of WOS database papers from 2012 to 2022. Specifically, the average annual number of WOS articles is approximately 218, which has been growing rapidly since 2015, exceeding 200 articles for the first time in 2018 and reaching 505 articles in 2021. The growth trend in the WOS's postings since 2017 is apparent, with an average annual growth of 27.7%.

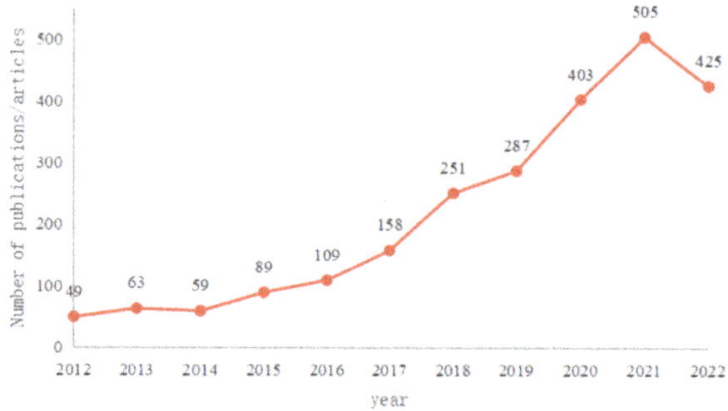

Figure 2. Statistical chart of documents issued.

3.2. *Network Analysis of the Country Based on Published Articles*

National cooperation network mapping can be used to visualize the degree of connection and social relationships between different countries in this research field, providing a novel perspective for evaluating the academic influence and research capacity of nations. WOS documents were used to construct the cooperation network mapping between countries and regions, as presented in Table 1 and Figure 3.

Table 1. Top five countries in terms of publication volume and centrality.

Ranking	Number of Essay	Centrality	Ranking	Number of Essay
	Country	Number	Country	Value
1	China	909	United States	0.24
2	United States	338	China	0.21
3	Korea	246	Italy	0.19
4	England	187	Germany	0.15
5	Australia	105	England	0.12

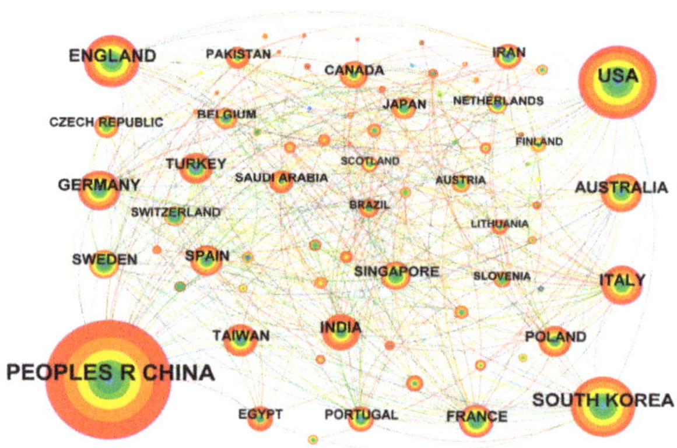

Figure 3. Map of national and regional cooperation networks.

In the field of smart textiles and apparel, the WOS database contains 2398 papers published in 83 countries and regions. China leads with 909 papers, followed by the United States with 338 papers. Korea ranks third with 246 papers, and England and Australia follow closely behind with 187 and 105 papers, respectively. In Figure 2, China and the United States are the two dominant nodes, indicating their significant research leadership in this field compared to other countries. The United States has the highest cooperation centrality at 0.24, demonstrating its close collaboration with a total of 42 countries and regions. The quantity of literature published in China and indexed in WOS has significantly grown in recent years, with a cooperation centrality of 0.21, ranking second. This indicates the high level of international collaboration among Chinese scholars.

3.3. Institutions Analysis

Analyzing research institutions can facilitate our comprehension of the key players in the field and offer guidance for scholars in selecting collaborative partners. The top 10 institutions, ranked by the number of their published articles in the WOS database [15], are tabulated as shown in Table 2.

Table 2. Top 10 institutions in terms of publication volume and centrality.

	WOS Database	
	Scientific Research Institution	**Frequency**
1	Donghua University	143
2	Chinese Academy of Sciences	124
3	Hong Kong Polytechnic University	86
4	Jiangnan University	74
5	Qingdao University	67
6	University of Boras	53
7	Soochow University	46
8	University of Chinese Academy of Sciences	45
9	University of Manchester	40
10	Georgia Institute of Technology	31

The VOSviewer 1.6.19 software was used to visualize collaborative networks among research institutions in the WOS literature. Among these, 224 foreign research institutions were identified, including Donghua University (141 articles), the Chinese Academy of Sciences (124 articles), Hong Kong Polytechnic University (86 articles), Jiangnan University (74 articles), Qingdao University (67 articles), the University of Boras (53 articles), Soochow University (46 articles), the University of Chinese Academy of Sciences (45 articles), the University of Manchester (40 articles) and Georgia Institute of Technology (31 articles). The number of articles issued by these institutions is substantial. This analysis indicates that Donghua University and the Chinese Academy of Sciences have the highest number of publications in English. Please refer to Table 2 and Figure 4 for details.

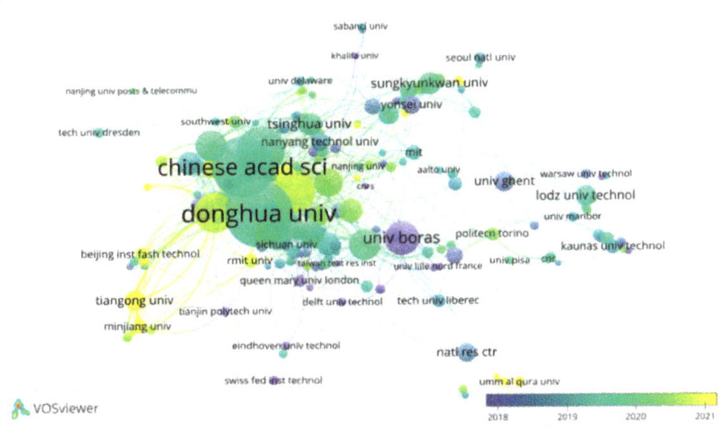

Figure 4. WOS Collaborative network of research institutions based on the literature.

3.4. Author Group Analysis

To gain a deeper understanding of the fundamental structure of this field, it is crucial to identify its core authors. Table 3 presents the top six author groups ranked by their publication output. The statistical analysis was conducted in accordance with the Price formula $M = 0.749N_{max}^{1/2}$, where M represents the number of publications by core authors and N_{max} represents the number of publications by the author with the highest output in that year. The selection criteria must be greater than or equal to the number of publications by core authors [15,16]. In the WOS database, there were 71 core authors who had published more than five articles. Only one author, Li Yi's group from the University of Manchester, had published more than 20 articles. Li Yi is considered a prolific author, with a total of 35 publications.

Table 3. Top six authors in terms of publication volume.

WOS Database		
Volume of Literature/Articles	**Author Group**	**Number of People**
35	Li, Yi, Liu Zekun, Zheng Zijian et al.	43
23	Wang, Zhong Lin, Dong Kai, Zhang Yang, Liu Mengmeng, Hu Weiguo et al.	48
20	Qu, Lijun, Tian, Mingwei, Zhang Xueji, Zhao Hongtao, Liu Xuqing et al.	25
18	Chen, Jun, Chen Guorui, Zhou Yihao, Yang Jin, Liu Jun, Zhao Xun et al.	33
15	Zheng, Zijian, Gao Yuan, Huang Qiyao, Hu Hong, Xie Chuan et al.	33
14	Zhu Meifang, Chen Yanhua, Yang Shenyuan, Ramakrishna Seeram et al.	29

In the WOS database, there are 54 distinct clusters of core authors based on publication volume (see Figure 5), and the number of papers published by a maximum of one author per cluster was selected as the standard. Li Yi's group (35 articles), Wang Zhong Lin's group (23 articles), Qu Lijun's group (20 articles), Chen Jun's group (18 articles), and Zheng Zijian's group (15 articles) each form a cluster. Each cluster contains more than two authors. There are nine closely cooperative teams, several of which have long-term collaborations with high publication volumes. This forms a stable cross-team communication and cooperation network. The most prominent team is Li Yi's group at the University of Manchester, whose collaborative network is the largest in this field. Their research focuses on characterizing wearable tensile strain sensors [17].

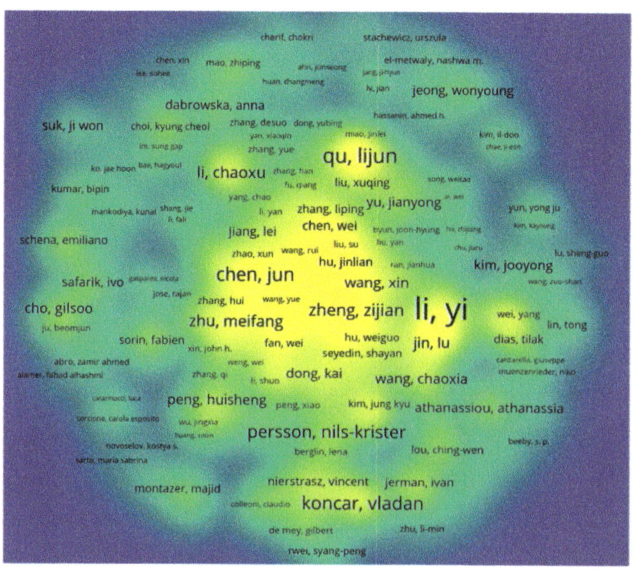

Figure 5. Density view of literature authors.

3.5. Journals Analysis

Papers related to intelligent textile clothing were published in 296 journals from 2012 to 2022. We listed the top ten journals in this field based on the number of publications, as demonstrated in Table 4. ADV MATER was in first place with 447 published articles, ACS APPL MATER INTER was in second place with 169 articles, and ADV FUNCT MATER was in third place with 124 articles.

Table 4. Top 10 most productive journals.

Journal	Count	IF (2021)	Quartile in Category (2021)	H-Index
ADV MATER	1332	4.18	Q1	447
ACS APPL MATER INTER	1195	1.48	Q2	169
ADV FUNCT MATER	1058	2.66	Q1	269
ACS NANO	969	2.51	Q1	310
NAT COMMUN	761	3.13	Q1	248
SCIENCE	743	10.15	Q1	1058
SENSORS-BASEL	712	0.9	Q3	132
NANO LETT	672	1.95	Q1	430
SCI REP-UK	670	1.05	Q3	149
TEXT RES J	661	0.64	Q3	74

The JIF of a journal is another important parameter to evaluate the value of the journal itself and the publications included in it. Among the top 10 academic journals, Science had the highest JIF at 10.15, followed by ADV MATER at 4.18, and both were classified as Q1. Additionally, the top three H-index journals were SCIENCE (1058), ADV MATER (447), and NANO LETT (430). Based on the above, we believe that Science and Ad Mater are the more authoritative journals in this field.

3.6. Citations Analysis

Citation analysis is an important indicator in bibliometric studies [18,19]. A total of 254 co-cited references were visualized using CiteSpace 6.2.R4, with the time slice set as one year and the time span spanning from 2012 to 2022. Table 5 showed the top ten most cited papers, including four review studies (Stoppa M et al., 2014 [20]; Zeng W et al., 2014 [21]; Heo JS et al., 2018 [22]; and Amjadi M, 2016 [23]) and six experimental studies (Dong K et al., 2020 [24]; Weng W et al., 2016 [25]; Shi JD et al., 2020 [26]; Pu X et al., 2016 [27]; Lee J et al., 2015 [28]; Chen J et al., 2016 [29]), which focused on the electronic components and multi-functions of smart textiles. The most-cited papers were written by Stoppa M's group, who work at the University of Istituto Italiano di Tecnologia (IIT), reflecting their great influence in the field. Citation bursts refer to references that caught the attention of scholars in a specific field at a specific time interval and whose analysis can be used to observe the evolution of a field of knowledge and to predict frontier trends. In Table 6, the timeline is shown in a circle, and the burst time interval is shown in a solid black circle, indicating the start year, end year, and duration of the burst. Of these burst citations, the shortest burst duration for intelligent textiles and garments was one year, and the longest was four years. Notably, 40% of the citation bursts ended close to 2019, focusing on advances in flexible sensitive strain sensors and supercapacitors (Amjadi M et al., 2016 [23]; Wen Z et al., 2016 [30]; Pu X et al., 2016 [27]; Ren JS et al., 2017 [31]; Kou L et al., 2014 [32]). Additionally, 15% of the citation bursts ending in 2022 or later focused on smart textiles that integrate microelectronic systems and functional textiles (Hsu PC et al., 2016 [33]; Zhao ZZ et al., 2016 [34]; Shi JD et al., 2020 [26]), suggesting that these research topics have been receiving attention in recent years and are expected to be a focus of research in the future.

Table 5. Top 10 most cited references.

Rank	Title	References	Author	Year	Citation
1	Wearable Electronics and Smart Textiles: A Critical Review	[20]	Stoppa M	2014	143
2	Fiber-Based Wearable Electronics: A Review of Materials, Fabrication, Devices, and Applications	[21]	Zeng W	2014	77
3	Fiber/Fabric-Based Piezoelectric and Triboelectric Nanogenerators for Flexible/Stretchable and Wearable Electronics and Artificial Intelligence	[24]	Dong K	2020	73
4	Recent Progress of Textile-Based Wearable Electronics: A Comprehensive Review of Materials, Devices, and Applications	[22]	Heo JS	2018	70
5	Smart Electronic Textiles	[25]	Weng W	2016	67
6	Smart Textile-Integrated Microelectronic Systems for Wearable Applications	[26]	Shi JD	2020	67
7	Wearable Self-Charging Power Textile Based on Flexible Yarn Supercapacitors and Fabric Nanogenerators	[27]	Pu X	2016	56
8	Conductive Fiber-Based Ultrasensitive Textile Pressure Sensor for Wearable Electronics	[28]	Lee J	2015	56
9	Micro-cable structured textile for simultaneously harvesting solar and mechanical energy	[29]	Chen J	2016	52
10	Stretchable, Skin-Mountable, and Wearable Strain Sensors and Their Potential Applications: A Review	[23]	Amjadi M	2016	52

Table 6. Top 25 references with the strongest citation bursts.

References	Cited References	Year	Strength	Begin	End	2012–2022
Hu LB, 2010, NANO LETT, V10, P708,	[35]	2010	13.37	2012	2015	●●●●○○○○○○○
Cherenack K, 2012, J APPL PHYS, V112, P0	[36]	2012	18.28	2014	2017	○○●●●●○○○○○
Meng YN, 2013, ADV MATER, V25, P2326	[37]	2013	12.1	2014	2018	○○●●●●●○○○○
Lee YH, 2013, NANO LETT, V13, P5753	[38]	2013	9.05	2014	2018	○○○●●●●●○○○
Zeng W, 2014, ADV MATER, V26, P5310	[21]	2014	25.01	2015	2019	○○○●●●●●○○○
Castano LM, 2014, SMART MATER STRUCT, V23, P0	[39]	2014	19.37	2015	2019	○○○●●●●●○○○
Kou L, 2014, NAT COMMUN, V5, P0	[32]	2014	12.82	2015	2019	○○○●●●●●○○○
Lee JA, 2013, NAT COMMUN, V4, P0	[40]	2013	10.68	2015	2018	○○○●●●●○○○○
Wang K, 2013, ADV MATER, V25, P1494	[41]	2013	8.44	2015	2018	○○○●●●●○○○○
Fu YP, 2012, ADV MATER, V24, P5713	[42]	2012	7.83	2015	2016	○○○●●○○○○○○
Zhong JW, 2014, ACS NANO, V8, P6273	[43]	2014	7.66	2015	2017	○○○●●●○○○○○
Stoppa M, 2014, SENSORS-BASEL, V14, P11957	[20]	2014	48.36	2016	2019	○○○○●●●●○○○
Lee J, 2015, ADV MATER, V27, P2433	[28]	2015	13.92	2017	2020	○○○○○●●●●○○
Amjadi M, 2014, ACS NANO, V8, P5154	[44]	2014	9.49	2017	2019	○○○○○●●●○○○
Wen Z, 2016, SCI ADV, V2, P0	[30]	2016	8.57	2017	2019	○○○○○●●●○○○
Pu X, 2016, ADV MATER, V28, P98	[27]	2016	8.53	2017	2019	○○○○○●●●○○○
Weng W, 2016, ANGEW CHEM INT EDIT, V55, P6140	[25]	2016	8.2	2017	2019	○○○○○●●●○○○
Cheng Y, 2015, ADV MATER, V27, P7365	[45]	2015	9.37	2018	2020	○○○○○○●●●○○
Seyedin S, 2015, ACSAPPL MATER INTER, V7, P21150	[46]	2015	8.15	2018	2020	○○○○○○●●●○○
Ren JS, 2017, CARBON, V111, P622	[31]	2017	7.48	2018	2019	○○○○○○●●○○○
Liu MM, 2017, ADV MATER, V29, P0	[47]	2017	8.83	2019	2020	○○○○○○○●●○○
Ryu S, 2015, ACS NANO, V9, P5929	[48]	2015	8.1	2019	2020	○○○○○○○●●○○
Hsu PC, 2016, SCIENCE, V353, P1019	[33]	2016	11.2	2020	2022	○○○○○○○○●●●
Zhao ZZ, 2016, ADV MATER, V28, P10267	[34]	2016	7.75	2020	2022	○○○○○○○○●●●
Shi JD, 2020, ADV MATER, V32, P0	[26]	2020	7.69	2020	2022	○○○○○○○○●●●

3.7. Keywords Analysis

This paper used VOSviewer 1.6.19 to visualize keywords in the literature and employed a keyword co-occurrence analysis to explore research hotspots in smart textiles and apparel. To gain a deeper understanding of this field, we constructed visual maps for 2398 English literature, as shown in Figure 6.

The keywords were analyzed based on literature indexed in the Web of Science database. The most frequently occurring terms included smart textiles, fibers, performance, sensors, composites, textile design, fabrication techniques, and nanocomposites. After excluding self-referential terms, the frequently occurring keywords included fibers, performance, sensors, composites, design, fabrication, and nanocomposites. Based on these identified keywords, it is evident that research on smart textiles and clothing can be classified into two distinct directions. The first direction pertains to material development, which includes smart fiber materials [49,50] such as composite materials, nanocomposites, phase change fibers, and shape memory fibers. The chemical fibers can be purified and deodorized by adding nano-level ZnO, SiO_2, or other chemicals. Adding nanometer-sized ZnO to polyester fiber can increase the material's anti-ultraviolet and anti-bacterial abilities. Adding nano-sized metal particles to the chemical fibers can enhance the antistatic ability of the material. Adding nanoscale silver ions to the chemical fibers can enhance the material's own bactericidal ability. In addition, adding carbon black nanoparticles to rubber materials can also greatly improve the strength and anti-wear properties of rubber materials, thereby improving their service life. The second aspect concerns garment design, which involves dividing smart garment design into areas such as garment structure and fabric elasticity to create functional garments that address specific issues [51].

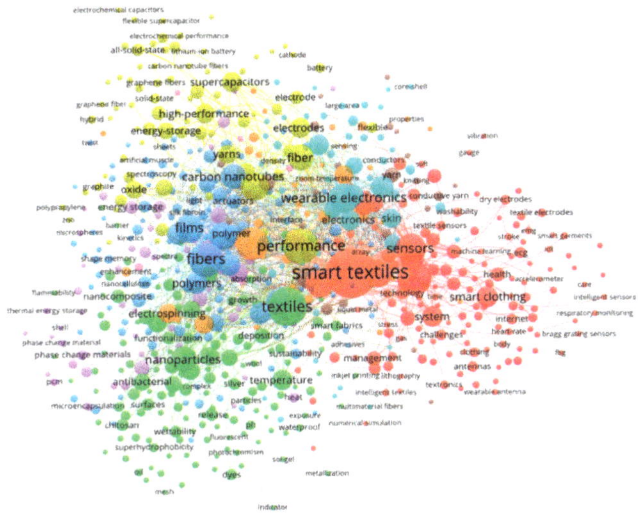

Figure 6. WOS keyword co-occurrence map.

A keyword clustering analysis facilitates the identification of relevant studies in a given field. A visual clustering analysis of these keywords was conducted using the LLR test algorithm within the CiteSpace 6.2.R4 software. We identified the key research areas in intelligent textiles and clothing from 2012 to 2021, and their clustering is illustrated in Figure 7.

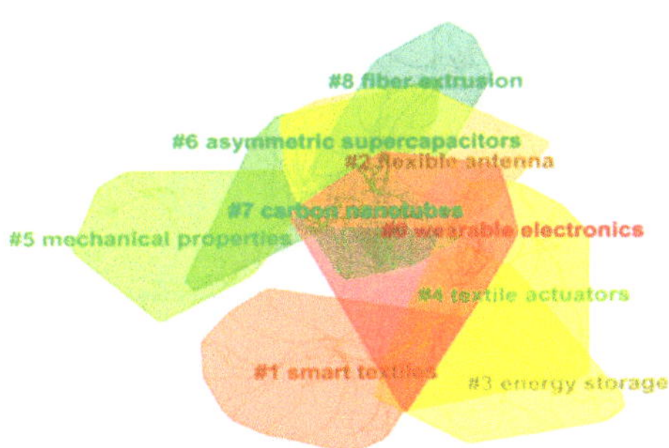

Figure 7. Keyword clustering map.

The clustering module (Q) for the WOS literature keywords is 0.768, with an average contour value (S) of 0.906, as depicted in Figure 7. The top nine keywords with the highest frequency are "wearable electronics", "smart textiles", "flexible antennas", "energy storage", "textile actuators", "mechanical properties", "asymmetric supercapacitors", "carbon nanotubes," and "fiber extrusion". It can be seen that the English literature also focuses on smart wearable products and smart textiles. English literature places more emphasis on mechanical properties, capacitors, and other electronic devices in smart textiles and garments. This indicates that the interaction design between electronic devices and clothing is a research hotspot.

By importing the data into Gephi 0.9.7 and utilizing Scimago Graphica software for timeline graphical analysis of literature keywords, we can correlate and analyze different clusters over time, illustrating the development and coherence of research content within each cluster. This approach clearly demonstrates trends in intelligent textile and clothing research.

A time zone map provides a view that represents the evolution of knowledge in the time dimension, which can intuitively show the changes and mutual influences of research hotspots [52]. In the time zone diagram of WOS keywords, as depicted in Figure 8, the literature clustering of smart textiles, carbon nanotubes, sensors, and wearable devices continue to be the present throughout the time span and have good time continuity. The pivotal node of smart textiles emerged in 2012. The most frequently cited literature on this topic within the experimental database [53] reveals that researchers have developed a flexible and stretchable electronic circuit technology that integrates electronic systems into elastomeric materials to produce complex functional, stretchable, and flexible electronic modules. The primary focus in 2013 was on biological materials. Since 2015, yarn supercapacitors have become a prominent topic in the field and are expected to remain so until 2022. In summary, research on smart textiles and carbon nanotubes began in 2012. Carbon nanotubes can optimize the cooperative loss mechanism of multiple components and the absorbing performance by combining with magnetic metals and metal compounds. This is an effective ways to achieve thinness, a light weight, a wide frequency band, and strong absorption of the absorbing materials. For example, using cellulose fibers as raw materials can realize the preparation of carbon nanotube absorbing materials. Using carbon nanotubes and metal materials as functional particles, textile absorbing materials can be prepared through electrospinning and finishing [54]. The timeline of smart materials research shows that in 2013, there was a notable focus on this area. Starting in 2014, there has been a focus on wearable electronics and energy storage within the realm of smart textile and clothing research. The research focus in 2015 was mainly on posture pressure, while from 2016 to 2018, the emphasis shifted towards wearable strain sensors and 3D printing, both of which are within the scope of intelligent textile clothing research. From 2019 to 2022, the main directions for intelligent textile clothing research will include shape memory and electronic skin development, as well as wearable strain sensing. The term "emergent keyword" refers to a word that experiences a sudden increase in frequency within a specific time period, with the growth rate of this word intensifying. This intensity can serve as an indicator of research hotspots and trends during the aforementioned period. By utilizing CiteSpace 6.2.R4 software to track emerging keywords in the literature from the WOS, we can gain insights into the evolutionary dynamics of research hotspots in intelligent textiles and clothing, ultimately enabling us to predict future development trends.

Figure 9 displays the emerging foreign research terms in the field of smart textiles and clothing from 2012 to 2022. By conducting a keyword emergence analysis of the WOS literature, we identified a total of 15 keywords with the highest emergence intensity. These can be roughly divided into two phases based on time: (1) From 2012 to 2018, fifteen emerging research hotspots were identified, including "circuit", "smart fabric", "yarn supercapacitor", and "flexible supercapacitor". Among these hotspots, the research on "yarn" has been continuously pursued for three to four years. Currently, foreign scholars are focusing their attention on smart fabrics and electronic components. Smart textile garments use yarn supercapacitors as energy storage devices. Large, high-stretch yarn electrodes are manufactured using CNI impregnation and PPy electrodeposition processes [55]. Flexible supercapacitors have unique advantages in terms of flexibility, shape, and weight due to the development of carbon-based materials, composite materials, and flexible micro-supercapacitors [56,57]. In addition, research on smart fabrics has make a significant contribution to the development of smart clothing. For example, superhydrophobic-coated fabrics have facilitated the creation of smart oil and water separators, microfluidic valves, and chip experimental devices [58]. (2) In the years 2019–2022, foreign scholars primarily focused on the development of pressure sensors in wearable devices and conductive textile research. For instance, thin-film flexible wireless pressure sensors can provide a wireless monitoring platform [59]. Conductive textiles, including graphene-based textiles, offer technical advantages in wearable products such as improved conductivity, ultra-flexibility, and machine washability [60]. The current focus of scholarly research is on wearable devices, which require a multidisciplinary approach to information collection, processing, storage, battery technology, intelligent operating systems, and human–computer interaction design. The integration of data processing, software, and haptic technologies enables the achievement of specific intelligent functions [61,62].

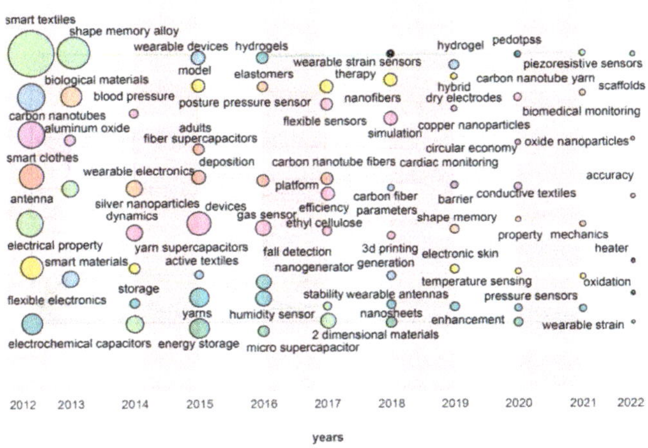

Figure 8. WOS keyword time zone map.

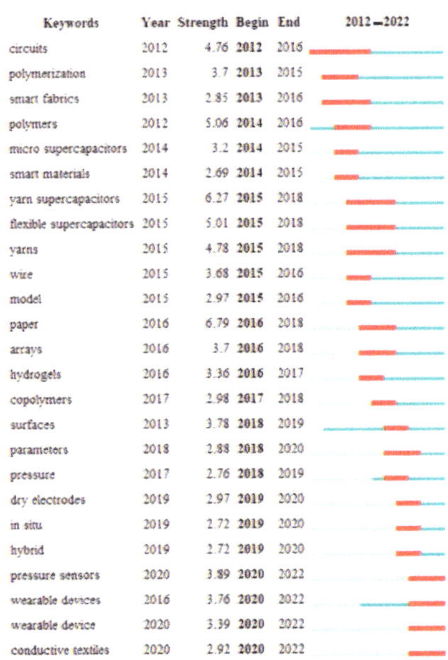

Keywords	Year	Strength	Begin	End	2012–2022
circuits	2012	4.76	**2012**	2016	
polymerization	2013	3.7	**2013**	2015	
smart fabrics	2013	2.85	**2013**	2016	
polymers	2012	5.06	**2014**	2016	
micro supercapacitors	2014	3.2	**2014**	2015	
smart materials	2014	2.69	**2014**	2015	
yarn supercapacitors	2015	6.27	**2015**	2018	
flexible supercapacitors	2015	5.01	**2015**	2018	
yarns	2015	4.78	**2015**	2018	
wire	2015	3.68	**2015**	2016	
model	2015	2.97	**2015**	2016	
paper	2016	6.79	**2016**	2018	
arrays	2016	3.7	**2016**	2018	
hydrogels	2016	3.36	**2016**	2017	
copolymers	2017	2.98	**2017**	2018	
surfaces	2013	3.78	**2018**	2019	
parameters	2018	2.88	**2018**	2020	
pressure	2017	2.76	**2018**	2019	
dry electrodes	2019	2.97	**2019**	2020	
in situ	2019	2.72	**2019**	2020	
hybrid	2019	2.72	**2019**	2020	
pressure sensors	2020	3.89	**2020**	2022	
wearable devices	2016	3.76	**2020**	2022	
wearable device	2020	3.39	**2020**	2022	
conductive textiles	2020	2.92	**2020**	2022	

Figure 9. Prominent words in WOS intelligent textile and apparel research, the timeline is shown in blue, and the interval at the time of the burst is shown in red, indicating the start year, the end year, and the duration of the burst.

4. Limitations

The present study attempted to conduct a comprehensive bibliometric analysis of papers related to intelligent textile clothing from 2012 to 2022. Although this paper presents some meaningful findings, there are some limitations. Due to the formatting requirements of the CiteSpace 6.2.R4 software, all the data were retrieved and downloaded from the Web of Science database, excluding other databases such as CNKI and Wanfang. Additionally, we have restricted the indexing, article type, and language type of the search, which may result in the omission of some high-quality articles. However, we believe that the volume of data retrieved is large enough to adequately reflect the current state of the research.

5. Conclusions

A combination of bibliometric and knowledge mapping analysis methods was used to reveal the multidimensional structure and network relationships of relevant research in the field of smart textiles and garments based on the WOS database. The main conclusions are as follows:

(1) In terms of the number of published papers and countries represented, the WOS has seen a steady increase in publications since 2015, with particularly rapid growth observed after 2017 in terms of the number of published papers and countries represented. The United States has maintained close collaborations with other countries and regions, followed by China, which has also emerged as a leading contributor to global research output in recent years.

(2) From the perspectives of the authors and institutions, the WOS database research institutions are very close, with institutions such as the University of Boras, Sweden, Sungkyunkwan University, Korea, and the University of Manchester, UK, forming the central collaborative forces. Among them, scholars such as Li, Yi, Qu, Lijun, Jin, and Lu form the core group of authors, whose research concentration and intensity of cooperation are greatest.

(3) Based on the journals and citations, SCIENCE and ADV MATER are the authoritative journals in the field in the WOS database, and together with journals such as ACS APPL MATER INTER, NANO LETT, and ADV FUNCT MATER, they constitute the top journals in the field. The most cited paper was written by the Stoppa M. group, which had a significant influence. It can be seen from the citation explosion that research in this field focuses on flexible, sensitive strain sensors, supercapacitors, integrated microelectronic systems, and functional textiles.

(4) Based on the analysis of keyword co-occurrence, the keywords in the WOS database have a high frequency of occurrence and diversified research topics. The overall intensity of keywords for smart textiles and clothing in the WOS database is large, and there is a significant difference in intensity among the emergent words. The literature in the WOS database focuses more on the manufacturing of electronic components and smart fibers and polymers in smart textiles and clothing. Therefore, future research on electronic component manufacturing and smart fibers and fabrics in the field of smart textiles and apparel is expected to be the focus of attention.

(5) From the point of view of research hotspots and stage characteristics, the research hotspots in the WOS in recent years have been more diverse, involving yarn supercapacitors, biomaterials, carbon nanotubes, wearable strains, and electronic skin. Future research is expected to pay more attention to the intelligence of fibers and fabrics and the accurate detectability of electronic components. This work will strengthen the development of intelligent core technology, which will also be a key focus in the future.

Author Contributions: T.W. and C.L. designed the study. J.Z. and A.W. collected the data. C.L., T.W. and A.W. re-examined the data. T.W., C.L. and A.W. analyzed the data. T.W. and C.L. wrote the manuscript. A.W. and J.Z. reviewed and revised the manuscript. All authors have read and agreed to the published version of the manuscript.

Funding: This research was funded by the Shaanxi Provincial Key Research and Development Program 2023 (2023-YBSF-203).

Data Availability Statement: Further inquiries can be directed to the corresponding authors.

Conflicts of Interest: The authors declare no conflict of interest.

References

1. Shen, L.; Sang, P.; Xie, Z. Development of intelligent cycling clothing based on outdoor visual warning functions. *Knitt. Ind.* **2019**, *1*, 57–61.
2. Tian, M.; Li, J. The design model and development trend of intelligent clothing. *J. Text.* **2014**, *35*, 109–115.
3. Van Langenhove, L.; Hertleer, C. Smart clothing: A new life. *Int. J. Cloth. Sci. Technol.* **2004**, *16*, 63–72. [CrossRef]
4. Tian, M.; Zhang, G.; Qu, L.; Wang, B.; Shi, Y. The application of conductive fibers and their flexible sensors in the field of wearable intelligent textiles. *Basic Sci. J. Text. Univ.* **2021**, *34*, 51–59.
5. Ahsan, M.; Teay, S.H.; Sayem, A.S.M.; Albarbar, A. Smart Clothing Framework for Health Monitoring Applications. *Signals* **2022**, *3*, 113–145. [CrossRef]
6. Su, Y.G.; Zhao, X. Modern garment design from the perspective of smart textiles. *Woolen Technol.* **2020**, *48*, 102–106.
7. Sun, Y.; Fan, J.; Wang, L.; Liu, Y. Research Progress on the Application of Wearable Technology in Textile and Apparel. *J. Text.* **2018**, *39*, 132–138.
8. Wang, X.; Yu, M.; Zhang, C.; Zheng, G.; Qin, R.; Liu, J. Based on flexible sensor application and progress of smart clothing. *Basic Sci. J. Text. Univ./Fangzhi Gaoxiao Jichu Kexue Xuebao* **2019**, *32*, 133–137.
9. He, S.; Jin, H.; Hong, W.; Liu, Z. Design and application research of knitted flexible intelligent patient garments. *Knitt. Ind.* **2020**, *7*, 81–84.
10. Yan, N.; Zhang, H.; Deng, Y. Research status and development trend of wearable medical monitoring clothing. *J. Text.* **2015**, *36*, 163–167.
11. Xie, H.Y.; Tang, H.; Gu, L.Y.; Sun, J.; Gao, Q.; Zhang, S.B. Research on Intelligent Fire Underwear Based on Temperature and Humidity Monitoring Function. *Knitt. Ind.* **2019**, *5*, 58–62.
12. Wang, Y.; Deng, Y.; Yang, X. Research status and development trend of breast monitoring intelligent underwear. *Knitt. Ind.* **2018**, *5*, 62–65.
13. Tang, Q.; Zhang, B.; Zheng, X. Infant wearable intelligent monitoring clothing design. *Acta Textile Sinica* **2021**, *42*, 156–160.
14. Sheng, Q.; Zheng, J.; Liu, J.; Shi, W.; Li, H. Research progress and trends in CiteSpace-based inner surface defect detection. *Spectrosc. Spectr. Anal.* **2023**, *43*, 9–15.
15. Zhang, X.; Wang, J.; Hu, J.R.; Wang, Y.; Lai, J.; Zhou, L.Y.; Zhu, Y.C. Visual analysis of surimi research based on CiteSpace and the bibliometric analysis platform. *Food Sci.* **2023**, *44*, 362–370.
16. Peng, Z.; Wu, Q.; Chen, H.; Zheng, Y.; Wang, S. A review of bibliometric-based research on machine vision defect detection. *Comput. Eng. Appl.* **2021**, *57*, 28–34.
17. Liu, Z.; Zheng, Y.; Jin, L.; Chen, K.; Zhai, H.; Huang, Q.; Chen, Z.; Yi, Y.; Umar, M.; Xu, L.; et al. Highly Breathable and Stretchable Strain Sensors with Insensitive Response to Pressure and Bending. *Adv. Funct. Mater.* **2021**, *31*, 2007622. [CrossRef]
18. Wang, Y.; Zhang, S.; Zhi, J.; Huang, M.; Pei, F. A bibliometric analysis: Current status and frontier trends of Schwann cells in neurosciences. *Front. Mol. Neurosci.* **2023**, *15*, 1087550. [CrossRef]
19. Zhou, Q.; Wu, F.; Zhao, M.; Yang, M. Bibliometric Evaluation of 2012–2020 Publications on Ferroptosis in Cancer Treatment. *Front. Cell Dev. Biol.* **2022**, *9*, 793347. [CrossRef]
20. Stoppa, M.; Chiolerio, A. Wearable Electronics and Smart Textiles: A Critical Review. *Sensors* **2014**, *14*, 11957–11992. [CrossRef] [PubMed]
21. Zeng, W.; Shu, L.; Li, Q.; Chen, S.; Wang, F.; Tao, X.M. Fiber-based wearable electronics: A review of materials, fabrication, devices, and applications. *Adv. Mater.* **2014**, *26*, 5310–5336. [CrossRef] [PubMed]
22. Heo, J.S.; Eom, J.; Kim, Y.H.; Park, S.K. Recent progress of textile-based wearable electronics: A comprehensive review of materials, devices, and applications. *Small* **2018**, *14*, 1703034. [CrossRef] [PubMed]
23. Amjadi, M.; Kyung, K.U.; Park, I.; Sitti, M. Stretchable, skin-mountable, and wearable strain sensors and their potential applications: A review. *Adv. Funct. Mater.* **2016**, *26*, 1678–1698. [CrossRef]
24. Dong, K.; Peng, X.; Wang, Z.L. Fiber/fabric-based piezoelectric and triboelectric nanogenerators for flexible/stretchable and wearable electronics and artificial intelligence. *Adv. Mater.* **2019**, *32*, 1902549. [CrossRef] [PubMed]
25. Weng, W.; Chen, P.N.; He, S.S.; Sun, X.M.; Peng, H.S. Smart Electronic Textiles. *Angew. Chem. Int. Ed.* **2016**, *55*, 6140–6169. [CrossRef]
26. Shi, J.; Liu, S.; Zhang, L.; Yang, B.; Shu, L.; Yang, Y.; Ren, M.; Wang, Y.; Chen, J.; Chen, W.; et al. Smart textile-integrated microelectronic systems for wearable applications. *Adv. Mater.* **2020**, *32*, 1901958. [CrossRef]
27. Pu, X.; Li, L.; Liu, M.; Jiang, C.; Du, C.; Zhao, Z.; Hu, W.; Wang, Z.L. Wearable self-charging power textile based on flexible yarn supercapacitors and fabric nanogenerators. *Adv. Mater.* **2016**, *28*, 98–105. [CrossRef]
28. Lee, J.; Kwon, H.; Seo, J.; Shin, S.; Koo, J.H.; Pang, C.; Son, S.; Kim, J.H.; Jang, Y.H.; Kim, D.E.; et al. Conductive fiber-based ultrasensitive textile pressure sensor for wearable electronics. *Adv. Mater.* **2015**, *27*, 2433–2439. [CrossRef]

29. Chen, J.; Huang, Y.; Zhang, N.; Zou, H.; Liu, R.; Tao, C.; Fan, X.; Wang, Z.L. Micro-cable structured textile for simultaneously harvesting solar and mechanical energy. *Nat. Energy* **2016**, *1*, 16138. [CrossRef]
30. Wen, Z.; Yeh, M.H.; Guo, H.; Wang, J.; Zi, Y.; Xu, W.; Deng, J.; Zhu, L.; Wang, X.; Hu, C.; et al. Self-powered textile for wearable electronics by hybridizing fiber-shaped nanogenerators, solar cells, and supercapacitors. *Sci. Adv.* **2016**, *2*, e1600097. [CrossRef]
31. Ren, J.; Wang, C.; Zhang, X.; Carey, T.; Chen, K.; Yin, Y.; Torrisi, F. Environmentally-friendly conductive cotton fabric as flexible strain sensor based on hot press reduced graphene oxide. *Carbon* **2017**, *111*, 622–630. [CrossRef]
32. Kou, L.; Huang, T.; Zheng, B.; Han, Y.; Zhao, X.; Gopalsamy, K.; Sun, H.; Gao, C. Coaxial wet-spun yarn supercapacitors for high-energy density and safe wearable electronics. *Nat. Commun.* **2014**, *5*, 3754. [CrossRef] [PubMed]
33. Hsu, P.C.; Song, A.Y.; Catrysse, P.B.; Liu, C.; Peng, Y.; Xie, J.; Fan, S.; Cui, Y. Radiative human body cooling by nanoporous polyethylene textile. *Science* **2016**, *353*, 1019–1023. [CrossRef] [PubMed]
34. Zhao, Z.; Yan, C.; Liu, Z.; Fu, X.; Peng, L.M.; Hu, Y.; Zheng, Z. Machine-washable textile triboelectric nanogenerators for effective human respiratory monitoring through loom weaving of metallic yarns. *Adv. Mater.* **2016**, *28*, 10267–10274. [CrossRef] [PubMed]
35. Hu, L.; Pasta, M.; La Mantia, F.; Cui, L.; Jeong, S.; Deshazer, H.D.; Choi, J.W.; Han, S.M.; Cui, Y. Stretchable, porous, and conductive energy textiles. *Nano Lett.* **2010**, *10*, 708–714. [CrossRef]
36. Cherenack, K.; Van Pieterson, L. Smart textiles: Challenges and opportunities? *J. Appl. Phys.* **2012**, *112*, 091301. [CrossRef]
37. Meng, Y.; Zhao, Y.; Hu, C.; Cheng, H.; Hu, Y.; Zhang, Z.; Shi, G.; Qu, L. All-Graphene Core-Sheath Microfibers for All-Solid-State, Stretchable Fibriform Supercapacitors and Wearable Electronic Textiles. *Adv. Mater.* **2013**, *25*, 2326–2331. [CrossRef]
38. Lee, Y.H.; Kim, J.S.; Noh, J.; Lee, I.; Kim, H.J.; Choi, S.; Seo, J.; Jeon, S.; Kim, T.S.; Lee, J.Y.; et al. Wearable textile battery rechargeable by solar energy. *Nano Lett.* **2013**, *13*, 5753–5761. [CrossRef]
39. Castano, L.M.; Flatau, A.B. Smart fabric sensors and e-textile technologies: A review. *Smart Mater. Struct.* **2014**, *23*, 053001. [CrossRef]
40. Lee, J.A.; Shin, M.K.; Kim, S.H.; Cho, H.U.; Spinks, G.M.; Wallace, G.G.; Lima, M.D.; Lepró, X.; Kozlov, M.E.; Baughman, R.H.; et al. Ultrafast charge and discharge biscrolled yarn supercapacitors for textiles and microdevices. *Nat. Commun.* **2013**, *4*, 1970. [CrossRef]
41. Wang, K.; Meng, Q.; Zhang, Y.; Wei, Z.; Miao, M. High-performance two-ply yarn supercapacitors based on carbon nanotubes and polyaniline nanowire arrays. *Adv. Mater.* **2013**, *25*, 1494–1498. [CrossRef] [PubMed]
42. Fu, Y.; Cai, X.; Wu, H.; Lv, Z.; Hou, S.; Peng, M.; Yu, X.; Zou, D. Fiber supercapacitors utilizing pen ink for flexible/wearable energy storage. *Adv. Mater.* **2012**, *24*, 5713–5718. [CrossRef]
43. Zhong, J.; Zhang, Y.; Zhong, Q.; Hu, Q.; Hu, B.; Wang, Z.L.; Zhou, J. Fiber-based generator for wearable electronics and mobile medication. *ACS Nano* **2014**, *8*, 6273–6280. [CrossRef] [PubMed]
44. Amjadi, M.; Pichitpajongkit, A.; Lee, S.; Ryu, S.; Park, I. Highly stretchable and sensitive strain sensor based on silver nanowire–elastomer nanocomposite. *ACS Nano* **2014**, *8*, 5154–5163. [CrossRef] [PubMed]
45. Cheng, Y.; Wang, R.; Sun, J.; Gao, L. A stretchable and highly sensitive graphene-based fiber for sensing tensile strain, bending, and torsion. *Adv. Mater.* **2015**, *27*, 7365–7371. [CrossRef]
46. Seyedin, S.; Razal, J.M.; Innis, P.C.; Jeiranikhameneh, A.; Beirne, S.; Wallace, G.G. Knitted strain sensor textiles of highly conductive all-polymeric fibers. *ACS Appl. Mater. Interfaces* **2015**, *7*, 21150–21158. [CrossRef]
47. Liu, M.; Pu, X.; Jiang, C.; Liu, T.; Huang, X.; Chen, L.; Du, C.; Sun, J.; Hu, W.; Wang, Z.L. Large-area all-textile pressure sensors for monitoring human motion and physiological signals. *Adv. Mater.* **2017**, *29*, 1703700. [CrossRef]
48. Ryu, S.; Lee, P.; Chou, J.B.; Xu, R.; Zhao, R.; Hart, A.J.; Kim, S.G. Extremely elastic wearable carbon nanotube fiber strain sensor for monitoring of human motion. *ACS Nano* **2015**, *9*, 5929–5936. [CrossRef]
49. Shen, L.; Fang, D.; Tang, Y.; Tong, X. Research status and development trend of intelligent clothing materials. *Shanghai Text. Sci. Technol.* **2016**, *44*, 36.
50. Peng, Y.; Gao, W.; Sun, F. Design of wool yarn artificial muscles of double-helix structure and application of moisture-sensing smart textiles. *J. Xi'an Polytech. Univ.* **2022**, *36*, 8–13.
51. Zhou, H.; Zhu, X.; Zhang, X. Design of intelligent and comfortable-shaping orthopedic garments. *Shanghai Text. Sci. Technol.* **2022**, *50*, 49–52.
52. Wang, J.; Deng, H.; Liu, B.; Hu, A.; Liang, J.; Fan, L.; Zheng, X.; Wang, T.; Lei, J. Systematic Evaluation of Research Progress on Natural Language. *J. Med. Internet Res.* **2020**, *22*, e16816. [CrossRef]
53. Vervust, T.; Buyle, G.; Bossuyt, F.; Vanfleteren, J. Integration of stretchable and washable electronic modules for smart textile applications. *J. Text. Inst.* **2012**, *103*, 1127–1138. [CrossRef]
54. Yu, Z.; Zhang, B.; Sun, J. Application of functional textile material innovation to intelligent clothing development. *J. Xi'an Polytech. Univ.* **2019**, *33*, 129–135.
55. Sun, J.; Huang, Y.; Fu, C.; Wang, Z.; Huang, Y.; Zhu, M.; Zhi, C.; Hu, H. High-performance stretchable yarn supercapacitor based on PPy@CNTs@urethane elastic fiber core spun yarn. *Nano Energy* **2016**, *27*, 230–237. [CrossRef]
56. Wang, X.; Lu, X.; Liu, B.; Chen, D.; Tong, Y.; Shen, G. Flexible Energy-Storage Devices: Design Consideration and Recent Progress. *Adv. Mater.* **2014**, *26*, 4763–4782. [CrossRef]
57. Zhao, Y.; Wu, C.; Wang, Y.; Lan, J.; Chen, S. Research progress of MXene-modified textiles in the field of flexible strain sensing. *Basic Sci. J. Text. Univ.* **2022**, *35*, 48–60.

58. Xu, Z.; Zhao, Y.; Wang, H.; Zhou, H.; Qin, C.; Wang, X.; Lin, T. Fluorine-Free Superhydrophobic Coatings with pH-Induced Wettability Transition for Controllable Oil-Water Separation. *ACS Appl. Mater. Interfaces* **2016**, *8*, 5661–5667. [CrossRef]
59. Farooq, M.; Iqbal, T.; Vazquez, P.; Farid, N.; Thampi, S.; Wijns, W.; Shahzad, A. Thin-Film Flexible Wireless Pressure Sensor for Continuous Pressure Monitoring in Medical Applications. *Sensors* **2020**, *20*, 6653. [CrossRef]
60. Afroj, S.; Tan, S.; Abdelkader, A.M.; Novoselov, K.S.; Karim, N. Highly Conductive, Scalable, and Machine Washable Graphene-Based E-Textiles for Multifunctional Wearable Electronic Applications. *Adv. Funct. Mater.* **2020**, *30*, 2000293. [CrossRef]
61. Qi, H.; Lu, J.; Lu, Z. Status and development trend of women's health smart wearable devices. *Knitt. Ind.* **2020**, *1*, 55–58.
62. Wang, A.; Xu, J.; Ma, D. Equipment sportswear: New category development and product design. *Knitt. Ind.* **2021**, *9*, 61–65.

Disclaimer/Publisher's Note: The statements, opinions and data contained in all publications are solely those of the individual author(s) and contributor(s) and not of MDPI and/or the editor(s). MDPI and/or the editor(s) disclaim responsibility for any injury to people or property resulting from any ideas, methods, instructions or products referred to in the content.

processes

MDPI

Article

Assessment of Wearable Cooling and Dehumidifying System Used under Personal Protective Clothing through Human Subject Testing

Yiying Zhou [1], Lun Lou [2] and Jintu Fan [1,3,4,*]

1 School of Fashion and Textiles, The Hong Kong Polytechnic University, Hong Kong; 21052594r@connect.polyu.hk
2 Energy Sector, Nano and Advanced Materials Institute, Hong Kong Science Park, Hong Kong
3 Research Center of Textiles for Future Fashion, The Hong Kong Polytechnic University, Hong Kong
4 Research Institute of Sports Science and Technology, The Hong Kong Polytechnic University, Hong Kong
* Correspondence: jin-tu.fan@polyu.edu.hk

Abstract: Healthcare professionals wearing personal protective equipment (PPE) during outbreaks often experience heat strain and discomfort, which can negatively impact their work performance and well-being. This study aimed to evaluate the physiological and psychological effects of a newly designed wearable cooling and dehumidifying system (WCDS) on healthcare workers wearing PPE via a 60 min treadmill walking test. Core temperature, mean skin temperature, heart rate, and subjective assessments of thermal sensation, wetness sensation, and thermal comfort were measured throughout the test. Additionally, ratings of wearing comfort and movement comfort were recorded during a wearing trial. The results showed that the WCDS significantly reduced core temperature, improved thermal sensation, and reduced wetness sensation compared to the non-cooling condition. The microclimatic temperature within the PPE was significantly lower in the cooling condition, indicating the WCDS's ability to reduce heat buildup. The wearing trial results demonstrated general satisfaction with the wearability and comfort of the WCDS across various postures. These findings contribute to the development of enhanced PPE designs and the improvement in working conditions for healthcare professionals on the frontlines during outbreaks.

Keywords: personal protective equipment; healthcare workers; wearable cooling and dehumidifying system; thermal comfort; wearing comfort

Citation: Zhou, Y.; Lou, L.; Fan, J. Assessment of Wearable Cooling and Dehumidifying System Used under Personal Protective Clothing through Human Subject Testing. *Processes* **2024**, *12*, 1126. https://doi.org/10.3390/pr12061126

Academic Editor: Kian Jon Chua

Received: 12 April 2024
Revised: 9 May 2024
Accepted: 28 May 2024
Published: 30 May 2024

Copyright: © 2024 by the authors. Licensee MDPI, Basel, Switzerland. This article is an open access article distributed under the terms and conditions of the Creative Commons Attribution (CC BY) license (https://creativecommons.org/licenses/by/4.0/).

1. Introduction

Healthcare professionals frequently work in environments with a high risk of viral transmission. Personal protective equipment (PPE) is an essential component of isolation precautions designed to protect these individuals from potential exposure to infectious agents [1,2]. However, the prolonged use of PPE can lead to additional physical and mental stress, such as heat strain, fatigue, and a potential decrease in work performance [3–6]. A study revealed that healthcare professionals experienced heat strain symptoms approximately 25 times more frequently when working with PPE compared to working without it [7].

To alleviate heat stress caused by PPE, personal cooling garments were developed to enhance heat loss and improve thermal comfort by regulating body temperature and reducing physiological stress. Several studies have evaluated the effects of existing personal cooling garments used with PPE, including PCM/ice vests [8–10], liquid cooling garments [11–14], and ventilation cooling systems [15,16], all of which demonstrated significant improvements in thermal comfort. However, PCM/ice vests and liquid cooling garments have limitations in operating time [17,18], and both of these cooling methods can cause skin wetness, which affects skin wetness comfort [19]. Additionally, existing air-cooling systems are impractical in environments with highly contagious infections as they require air intake from the surroundings. Considering the limitations of phase-change cooling vests, liquid cooling garments, and air-cooling systems, Lou et al. [20] developed a novel and lightweight wearable cooling and dehumidifying system (WCDS) for use in hazardous environments to reduce skin wetness and improve thermal comfort.

When examining the effects of personal cooling garments on healthcare professionals, it is crucial to differentiate between thermal sensation and thermal comfort. Thermal sensation refers to an individual's immediate perception of coldness or warmth, which is subjective and can vary among individuals. In contrast, thermal comfort is a state of mind that indicates contentment with the surrounding thermal environment, taking into account both physiological and psychological aspects. Although thermal sensation is an essential aspect of thermal comfort, it is not the only determining factor. For example, a healthcare worker may experience a neutral thermal sensation while wearing PPE but still feel uncomfortable due to skin wetness caused by sweating. Consequently, when assessing the effectiveness of personal cooling garments, it is necessary to consider both thermal sensation and thermal comfort.

Previous studies have not considered the specific challenges posed by the use of PPE during a pneumonia outbreak. Furthermore, the precise psychological and physiological effects of wearing WCDS under personal cooling garments in real working conditions have not been adequately investigated. To address this gap, additional research is necessary to gain a comprehensive understanding of the psychological and physiological impact of WCDS in the working environment, which will facilitate optimal design and improved performance. Therefore, this study was conducted to evaluate the physiological and psychological impact of a newly designed cooling suit on healthcare workers. The study analyzed the impact of WCDS on the thermal comfort of healthcare workers by monitoring skin temperature, core temperature, and heart rate. Additionally, a subjective evaluation questionnaire and a wearing trial, which included assessments of wearing comfort and movement comfort, were administered. The aim was to obtain a comprehensive assessment of the effectiveness of the WCDS application by examining the thermal perception, physiological responses, and wearing comfort of healthcare workers. This research provides valuable insights for the design of future PPE and contributes to enhancing the occupational environment for those on the healthcare frontlines.

2. Methods

2.1. Participants

Eight healthy male college students voluntarily participated in the study. The participants had a mean age of 27 ± 2.1 years, height of 178.3 ± 2.2 cm, weight of 69.6 ± 3.6 kg, BMI of 21.8 ± 1.3 kg/m^2, and body surface area of 1.8 ± 0.1 m^2. Before participation, each participant was comprehensively informed about the experimental procedure and potential risks and provided written informed consent. None of the participants had a history of heat-related illness, cardiovascular, metabolic, or respiratory disease. Additionally, they were instructed to refrain from consuming tea, coffee, or alcohol for at least 24 h prior to each test. The Human Subjects Ethics Sub-Committee of the Hong Kong Polytechnic University approved the experimental procedures under protocol number [HSEARS20210514003].

2.2. Clothing

The wearable cooling and dehumidifying system (WCDS) (Figure 1), developed by Lou et al. [20], is intended for use in hazardous and infectious environments without compromising the protective capabilities of personal protective equipment (PPE). The system is designed as a knitted vest with an embedded tube that provides cooling air. During each test, participants wore a standard PPE ensemble recommended for use during the COVID-19 pandemic, which included a surgical mask, eye protection, a Dupont Tyvek 600 (Tyvek 600 Plus, Dupont, Wilmington, DE, USA) plus coverall as a protective gown, and gloves. In addition to the PPE, subjects wore basic undergarments consisting of a t-shirt, trousers, underwear, and socks [21]. The WCDS was worn underneath the protective clothing. Prior to the human trials, all garments, including the cooling ones, were conditioned in a climatic chamber for 24 h.

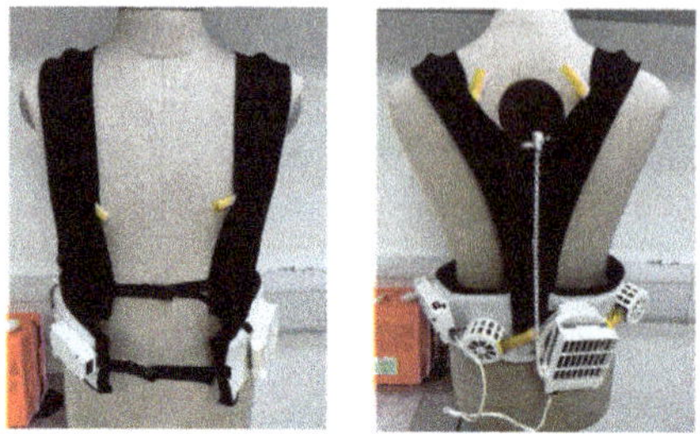

Figure 1. Wearable cooling and dehumidifying system (WCDS).

2.3. Protocol

The study took place in a climatic chamber set to 23 °C, 50% relative humidity, and an air velocity of 0.1 m/s, simulating the conditions of a clean workroom in a Hong Kong hospital. Participants underwent two 120 min experimental trials: one with a cooling garment (cooling) and a control trial (without cooling). Initially, subjects were given a 30 min acclimatization period in the chamber, during which they were briefed on the test procedures and the interpretation of perceptual rating scales. Following this, they were instructed to change into the provided attire and were equipped with monitoring instruments. For 60 min, the participants walked on a treadmill at a speed of 3.0 km/h, representing a typical work intensity level [22] (Figure 2). After a 30 min recovery period, they removed their clothing and equipment and exited the chamber. To ensure safety, the test was immediately terminated if any of the following conditions were met: (1) the subject's core temperature exceeded 38 °C, (2) the heart rate surpassed 95% of the average maximum heart rate, or (3) the subject expressed a desire to stop due to volitional fatigue. In such cases, subjects were scheduled to retake the same test during their next visit. It is important to note that participants had the option to stop the test at any time for any reason.

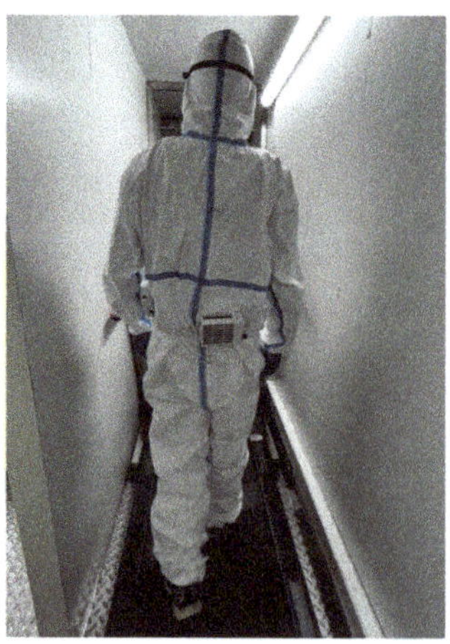

Figure 2. A subject wearing the WCDS walking on a treadmill.

2.4. Measurements

2.4.1. Physiological Measurement

Core body temperature measurements were recorded at one-minute intervals using a wearable sensor device (CORE, greenTEG). The microclimatic temperature was monitored every minute using digital thermometers (DS1923, iButton) attached inside the personal protective garment on the upper back area. Skin temperature sensors (DS1923, iButton) were placed at eight locations (forehead, left upper chest, right scapula, left and right arm, left hand and calf, and right anterior thigh; see Figure 3) using waterproof adhesive tape (PVC, 3M). These sensors consistently recorded local skin temperatures every minute throughout the entire test period. To calculate the mean skin temperature, an eight-point weighting scheme, as specified in the ISO 9886 standard, was employed (Table 1).

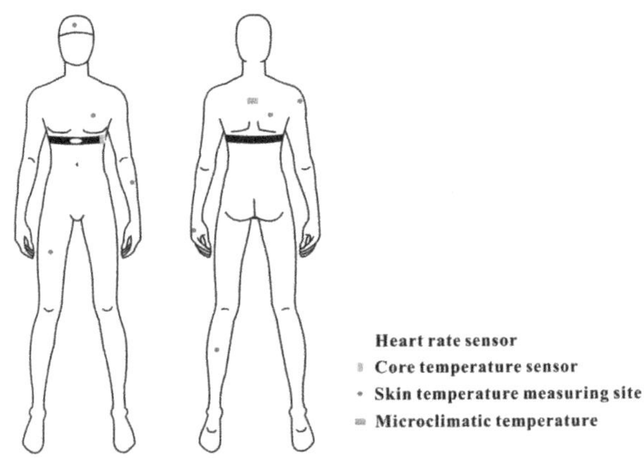

Figure 3. Measuring sites of heart rate, core temperature, skin temperature, and microclimatic temperature.

Table 1. Measuring sites and weighting coefficients.

Measuring sites	Forehead	Right scapula	Left Upper chest	Right arm	Left arm	Left hand	Right anterior	Left calf
Weighting coefficients	0.07	0.175	0.175	0.07	0.07	0.05	0.19	0.2

The mean skin temperature is obtained via the following formula:

$$t_{sk} = \sum k_i \, t_{ski}$$

2.4.2. Subjective Evaluation

Thermal sensation, wetness sensation, and thermal comfort sensation were assessed initially and then every 10 min throughout the entirety of the trials. The mean value was calculated for each parameter. Thermal sensation for both the whole body and the upper body was rated using a nine-point scale (Figure 4a), ranging from −4 (very cold) to +4 (very hot), with 0 indicating a neutral thermal sensation. Wetness sensation for the whole body and the upper body was measured using a seven-point scale (Figure 4b), where −3 corresponds to a very wet sensation, +3 to a very dry sensation, and 0 represents a neutral wetness sensation. Thermal comfort for the whole body and the upper body was assessed using a five-point scale (Figure 4c), ranging from −4 (extremely uncomfortable) to 0 (comfortable).

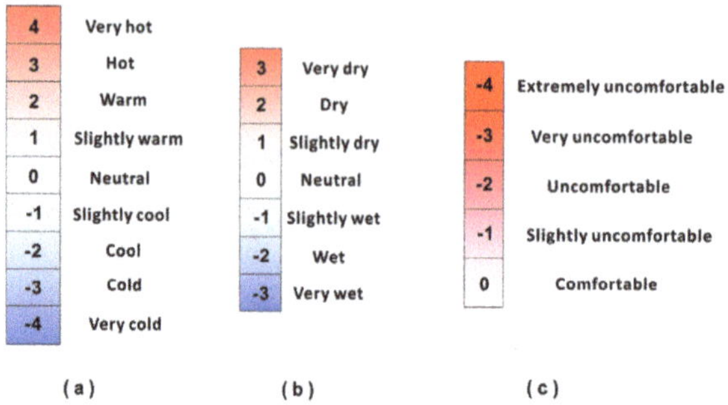

Figure 4. Subjective measurement scales: (**a**) thermal sensation; (**b**) wetness sensation; (**c**) thermal comfort.

2.4.3. Wearing Trial

Post-experimental assessments of wearing comfort were conducted using a seven-point Likert scale, ranging from −3 (very unsatisfied) to 3 (very satisfied) (Figure 5a). The assessed criteria included ease of wearing, ease of removal, fit, flexibility, and safety. Additionally, evaluations of body movement comfort were carried out in the same climatic chamber. Participants, dressed in identical clothing, simulated six distinct postures: transitioning from standing to sitting, sitting to standing, walking, bending forward and then straightening up, squatting and then standing upright, and twisting (refer to Figure 6). After each posture, participants provided subjective ratings of their comfort and freedom of movement using a seven-point scale, ranging from −3 (very uncomfortable) to 3 (very comfortable) (Figure 5b).

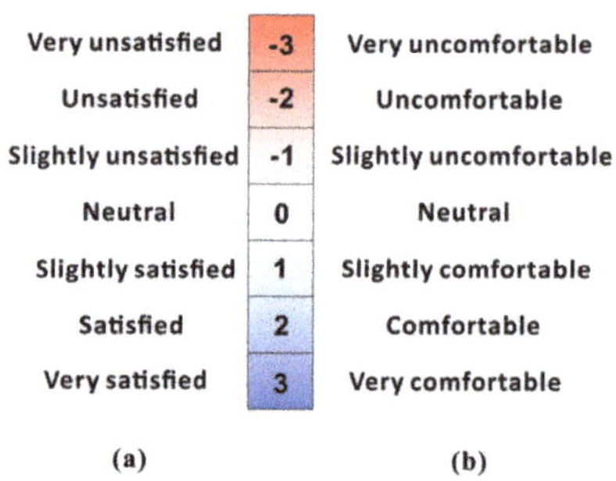

Figure 5. Wearing trial scales: (**a**) wearing comfort; (**b**) body movement comfort.

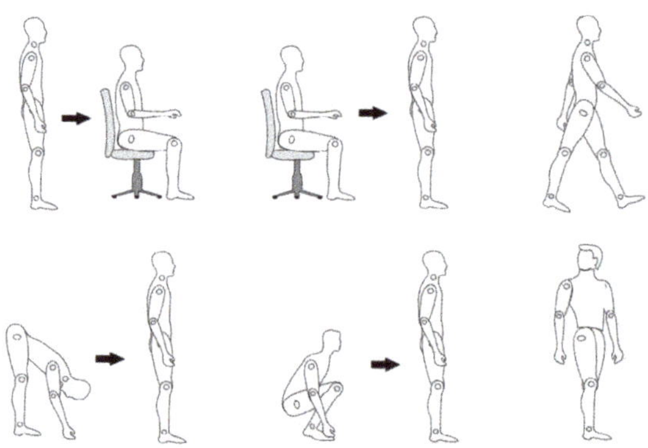

Figure 6. Six different postures for body movement comfort evaluation.

2.5. Statistical Analysis

Statistical data analysis was performed using IBM SPSS Statistics (Version 19.0, IBM Corporation). Unless otherwise specified, all values are presented as mean ± standard deviation. Paired *t*-tests were employed to assess both physiological and perceptual results over time and between treatments (with and without cooling). A 95% confidence level was set for all statistical analyses.

3. Results

3.1. Physiological Parameters

Core temperature, mean skin temperature and heart rate measurements were recorded at one-minute intervals. A significant difference ($p < 0.05$) in core temperature between the control and cooling conditions (Figure 7a) was observed from the 42nd minute onwards, with a maximum difference of 0.22 °C between the two conditions. Although both groups experienced an increase in core temperature throughout the exercise duration, the increase was less pronounced in the cooling group. The cooling intervention effectively maintained the core temperature at a lower level compared to the non-cooling condition, demonstrating the potential benefits of cooling in maintaining a stable core temperature and preventing excessive heat accumulation. Statistical analysis revealed no significant difference ($p > 0.05$) in average skin temperature between the control and cooling conditions (Figure 7b). The cooling group's mean skin temperature was slightly higher than that of the control group from the 29th minute of the testing period. This result could be attributed to the lower core temperature of human subjects. It is known that the perspiration rate is closely related to the change in core temperature. The cooling effect of the WCDS resulted in less perspiration and lower skin wetness and therefore reduced the evaporative heat loss rate from the skin. The thermal conditions of the experiment were carefully selected to avoid exposing participants to severe thermal stress, simulating real hospital and working conditions while ensuring participant safety under protective clothing. Regardless of the test variant, with or without cooling, the heart rate did not exceed 101 bpm, and there was no significant difference ($p > 0.05$) in the results (Figure 7c).

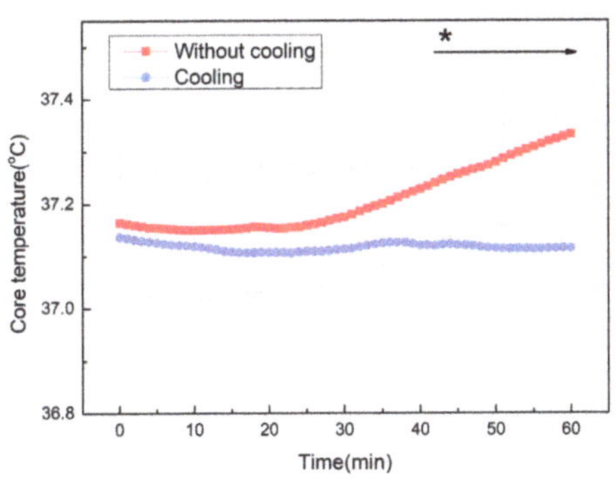

(a)

Figure 7. *Cont.*

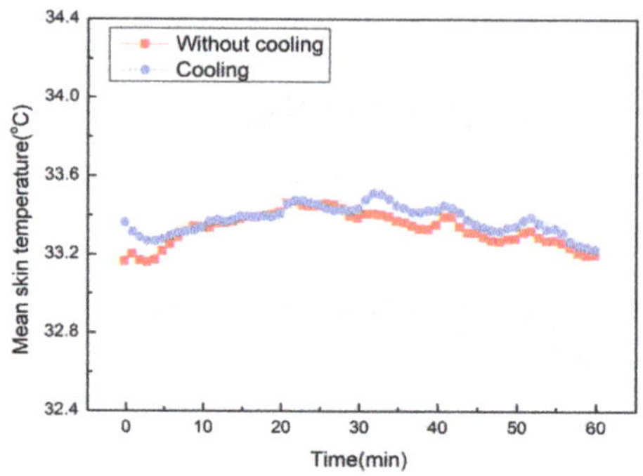

(b)

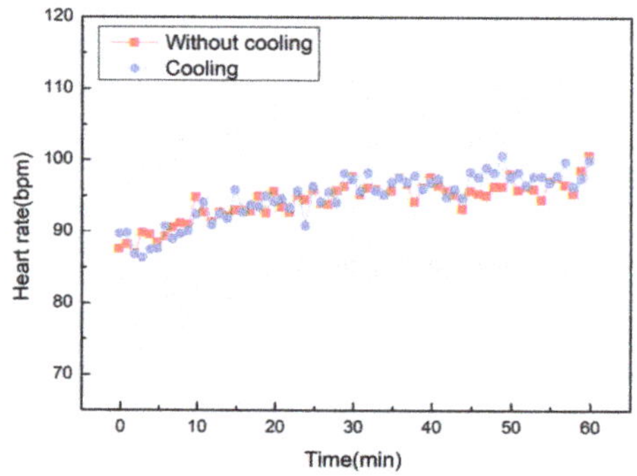

(c)

Figure 7. Comparison of changes in the core temperature (**a**), mean skin temperature (**b**), and heart rate (**c**) between the control and cooling groups; *: $p < 0.05$.

3.2. Microclimatic Temperature

A significant difference in the microclimatic temperature was observed from the 2nd minute onwards ($p < 0.05$), suggesting that the WCDS has the ability to reduce the temperature within personal protective clothing (Figure 8). The maximum temperature difference (5.3 °C) of the microclimate under the protective garment was recorded at the 28th minute of the experiment.

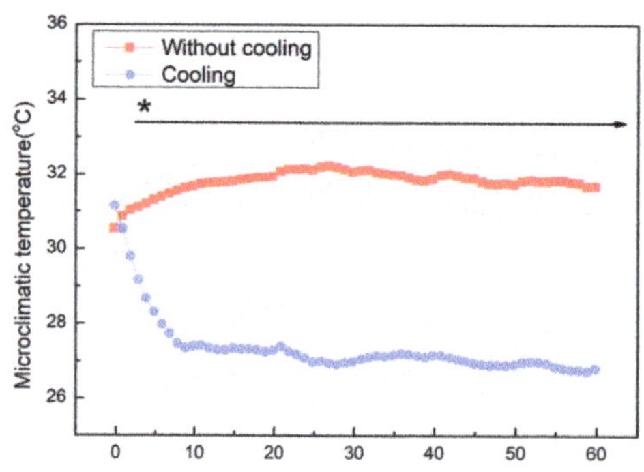

Figure 8. Changes in microclimatic temperature in the upper back area during the whole test period. *: $p < 0.05$.

3.3. Subjective Assessments

As depicted in Figure 9a,b, the overall and upper body thermal sensations under the cooling condition were consistently lower than those under the non-cooling condition throughout the testing period. A significant difference ($p < 0.05$), ranging from 0.8 to 1.7, was observed from the 10th minute in the overall body thermal sensation between the two conditions, suggesting that the WCDS enhanced the overall body thermal sensation. A significant difference ($p < 0.05$) was also noted in the upper body thermal sensation, with a disparity ranging from 0.7 to 1.8 between the two conditions.

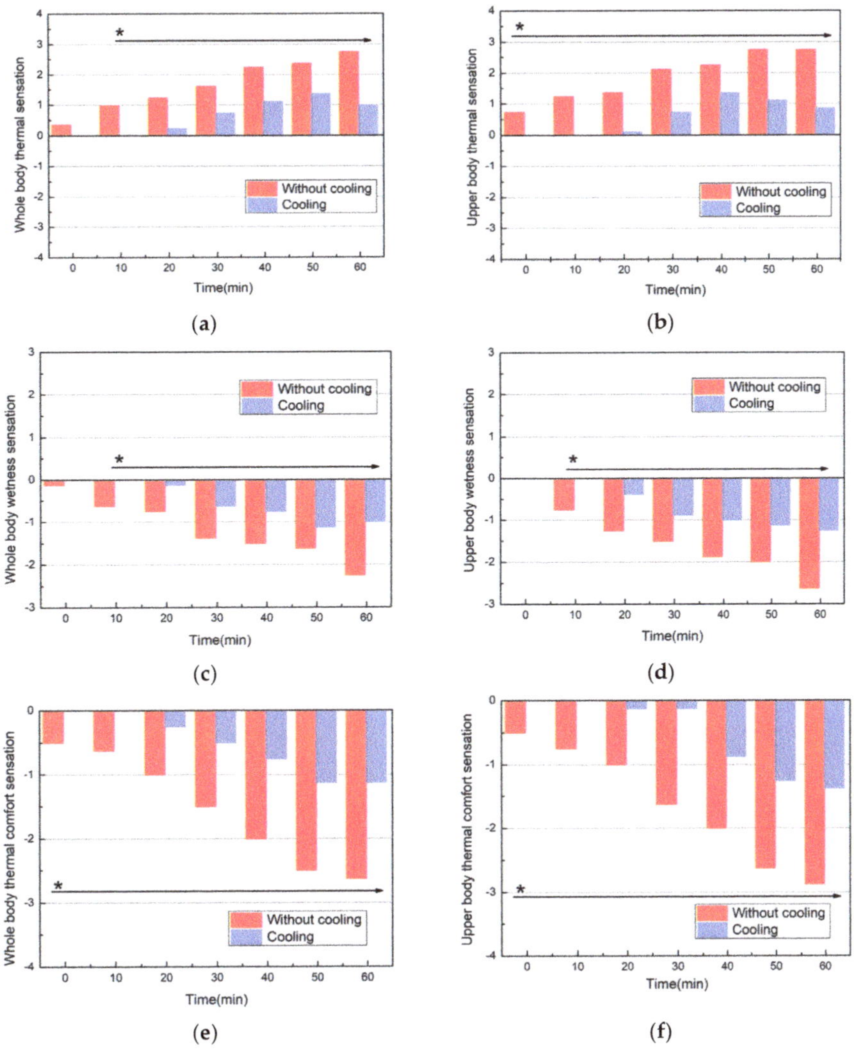

Figure 9. Time changes in the whole body-, upper body-thermal sensations (**a**,**b**), wetness sensation (**c**,**d**), and thermal comfort sensation (**e**,**f**) in the control and cooling groups. *: $p < 0.05$.

Figure 9c,d illustrate that the overall and upper body wetness sensations under the cooling condition were lower than those under the non-cooling condition throughout the testing period. A significant difference ($p < 0.05$) in the overall body wetness sensation was observed from the 10th minute, with a maximum difference of 1 between the two conditions. A significant difference ($p < 0.05$) in the upper body wetness sensation was also observed from the 10th minute, with a maximum difference of 1.3 between the two conditions. This result proves that the cooling effect of WCDS reduced the perspiration of human subjects.

The whole-body thermal comfort sensation and upper-body thermal comfort sensation were presented in Figure 9e and 9f, respectively. Throughout the testing period, participants reported a higher level of comfort under the cooling condition compared to the non-cooling condition, suggesting that the WCDS is an effective method to improve thermal comfort. A significant difference ($p < 0.05$) was observed in both the overall body thermal comfort sensation and the upper body thermal comfort sensation between the two conditions, with a range of 0.5 to 1.5.

3.4. Wearing Trial Results

The wearing comfort of WCDS is shown in Figure 10a. The data indicate a general satisfaction with wearability. Figure 10b illustrates the comfort levels associated with body movement in six different postures, as well as the evaluation of freedom of movement. The participants provided relatively positive comfort ratings for the WCDS across the various postures.

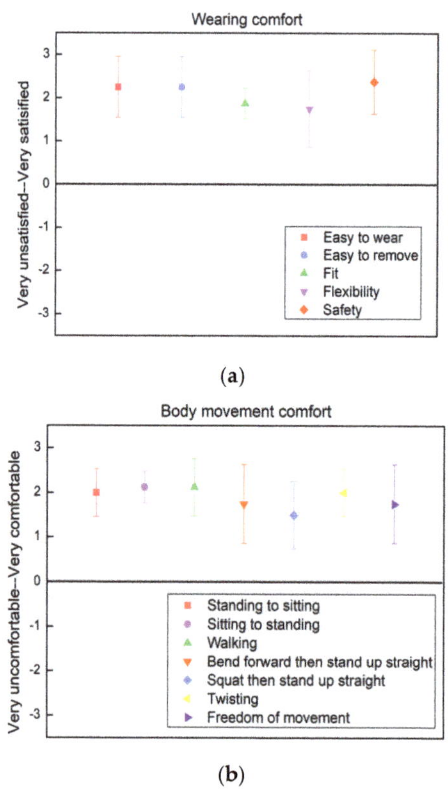

Figure 10. Wearing (**a**) and movement (**b**) comfort votes for subjects wearing WCDS.

4. Discussion

This study aimed to investigate the effectiveness of a wearable cooling and dehumidifying system (WCDS) in mitigating thermal stress and enhancing comfort under protective clothing. The findings provide valuable insights into the physiological and subjective responses to the WCDS, as well as its impact on the microclimate within protective clothing.

The physiological data revealed that the cooling intervention effectively maintained a lower core temperature throughout the exercise duration, indicating its potential to prevent excessive heat accumulation. Importantly, the heart rate remained below 101 bpm in both conditions, which may be associated with the exercise intensity [23], indicating that the WCDS did not impose additional cardiovascular strain. Mean skin temperature was more likely to be affected by skin wetness and sweat evaporation. The air-cooling system tended to reduce the sweating rate and enhanced the ventilative heat exchange. But in this experiment, the skin was covered by undergarments instead of being exposed to the cooling air, so the evaporative heat loss from the skin and mean skin temperatures were not significantly different for the cooling and non-cooling groups.

The WCDS also significantly altered the microclimatic temperature within the protective clothing. This is a crucial finding, as the microclimate is a key determinant of thermal comfort [24]. Compared to a traditional air-cooling system, this system can effectively lower the microclimatic temperature and reduce the risk of infectious diseases caused by air cooling, which cools via air circulation from the surroundings. The significant reduction in microclimatic temperature demonstrates the WCDS's potential to enhance comfort and reduce thermal stress in real-world settings.

The subjective assessments provide compelling evidence for the psychological benefits of the WCDS. The significant improvements in overall and upper body thermal sensations, along with reduced wetness sensations, highlight the WCDS's role in enhancing wearer comfort. This is particularly relevant for healthcare workers who must wear PPE for prolonged periods, as discomfort can lead to distraction, decreased work performance, and reduced compliance with necessary protective measures.

The wearing trial results further support the practicality of the WCDS, with participants reporting general satisfaction with its wearability and movement comfort. This suggests that the WCDS can be integrated into protective clothing without significantly compromising mobility, which is critical for occupational settings.

It is important to recognize certain limitations of this laboratory study. To ensure participant safety, variables such as environmental conditions and exercise intensity were strictly controlled. A field study may provide a more realistic assessment of the cooling effect. Furthermore, the study's participants were exclusively college students, not healthcare workers. The influence of gender should also be considered in future research.

5. Conclusions

This study provides valuable insights into the effectiveness of a wearable cooling device system (WCDS) in mitigating thermal stress and enhancing comfort under protective clothing. The WCDS significantly reduced core temperature, improved thermal sensation, and reduced wetness sensation compared to the non-cooling condition without imposing additional cardiovascular strain or compromising mobility. The wearing trial results indicated general satisfaction with the wearability and movement comfort of the WCDS. This suggests that the WCDS can be integrated into protective clothing without significantly compromising mobility, which is critical for occupational settings.

In conclusion, the WCDS appears to be a promising strategy for mitigating thermal stress and enhancing comfort under protective clothing. Future research should explore the long-term effects of the WCDS and its applicability in various occupational and clinical settings. The findings of this study have important implications for the design of protective clothing and the development of strategies to enhance comfort and safety for personal protective clothing.

Author Contributions: Conceptualization, J.F., L.L. and Y.Z.; methodology, L.L. and Y.Z.; software, Y.Z.; validation, Y.Z.; formal analysis, Y.Z.; investigation, Y.Z.; resources, Y.Z.; data curation, Y.Z.; writing—original draft preparation, Y.Z.; writing—review and editing, J.F., L.L. and Y.Z.; visualization,

Y.Z.; supervision, J.F.; project administration, J.F.; funding acquisition, J.F. All authors have read and agreed to the published version of the manuscript.

Funding: The authors acknowledge the financial support from the Hong Kong Innovation and Technology Commission (ITP/035/20TP). This work is also supported by the Hong Kong Research Institute of Textiles and Apparel Limited (HKRITA), and EPRO Advanced Technology Limited and Standard International Group (HK).

Data Availability Statement: Data will be made available upon request.

Conflicts of Interest: The authors declare no conflicts of interest.

References

1. Park, S.H. Personal protective equipment for healthcare workers during the COVID-19 pandemic. *Infect. Chemother.* **2020**, *52*, 165. [CrossRef] [PubMed]
2. Herron, J.; Hay-David, A.G.C.; Gilliam, A.D.; Brennan, P.A. *Personal Protective Equipment and COVID 19-a Risk to Healthcare Staff?* Elsevier: Amsterdam, The Netherlands, 2020; pp. 500–502.
3. Duan, X.; Sun, H.; He, Y.; Yang, J.; Li, X.; Taparia, K.; Zheng, B. Personal protective equipment in COVID-19: Impacts on health performance, work-related injuries, and measures for prevention. *J. Occup. Environ. Med.* **2021**, *63*, 221. [CrossRef] [PubMed]
4. Yánez Benítez, C.; Güemes, A.; Aranda, J.; Ribeiro, M.; Ottolino, P.; Di Saverio, S.; Alexandrino, H.; Ponchietti, L.; Blas, J.L.; International Cooperation Group on PPE and Emergency Surgery; et al. Impact of personal protective equipment on surgical performance during the COVID-19 pandemic. *World J. Surg.* **2020**, *44*, 2842–2847. [CrossRef] [PubMed]
5. Messeri, A.; Bonafede, M.; Pietrafesa, E.; Pinto, I.; de'Donato, F.; Crisci, A.; Lee, J.K.W.; Marinaccio, A.; Levi, M.; Morabito, M. A web survey to evaluate the thermal stress associated with personal protective equipment among healthcare workers during the COVID-19 pandemic in Italy. *Int. J. Environ. Res. Public Health* **2021**, *18*, 3861. [CrossRef] [PubMed]
6. Hunt, A.; Ting, J.; Schweitzer, D.; Laakso, E.L.; Stewart, I. Personal protective equipment for COVID-19 among healthcare workers in an emergency department: An exploratory survey of workload, thermal discomfort and symptoms of heat strain. *Emerg. Med. Australas.* **2023**, *35*, 483–488. [CrossRef] [PubMed]
7. Bongers, C.C.; de Korte, J.Q.; Zwartkruis, M.; Levels, K.; Kingma, B.R.M.; Eijsvogels, T.M.H. Heat Strain and Use of Heat Mitigation Strategies among COVID-19 Healthcare Workers Wearing Personal Protective Equipment—A Retrospective Study. *Int. J. Environ. Res. Public Health* **2022**, *19*, 1905. [CrossRef] [PubMed]
8. Bach, A.J.; Maley, M.J.; Minett, G.M.; Zietek, S.A.; Stewart, K.L.; Stewart, I.B. An evaluation of personal cooling systems for reducing thermal strain whilst working in chemical/biological protective clothing. *Front. Physiol.* **2019**, *10*, 437942. [CrossRef] [PubMed]
9. Li, Z.; Pan, B.; Yang, B.; Zhou, B.; Wang, F. Heat stress mitigation with ice cooling vests in PPE-clad medical workers: Effects of cooling area and gender differences. *Build. Environ.* **2023**, *245*, 110943. [CrossRef]
10. Smolander, J.; Kuklane, K.; Gavhed, D.; Nilsson, H.; Holmér, I. Effectiveness of a light-weight ice-vest for body cooling while wearing fire fighter's protective clothing in the heat. *Int. J. Occup. Saf. Ergon.* **2004**, *10*, 111–117. [CrossRef]
11. Aljaroudi, A.M.; Bhattacharya, A.; Yorio, P.; Strauch, A.L.; Quinn, T.D.; Williams, W.J. Probability of hyperthermia in a hot environment while wearing a liquid cooling garment underneath firefighters' protective clothing. *J. Occup. Environ. Hyg.* **2021**, *18*, 203–211. [CrossRef]
12. Yang, J.; Zhang, Y.; Huang, Y.; Chen, W. Effects of liquid cooling garment on physiological and psychological strain of firefighter in hot and warm environments. *J. Therm. Biol.* **2023**, *112*, 103487. [CrossRef] [PubMed]
13. Aljaroudi, A.M.; Kadis, D.S.; Bhattacharya, A.; Strauch, A.; Quinn, T.D.; Williams, W.J. Effect of continuous cooling on inhibition and attention while wearing firefighter's PPE in a hot environment. *J. Occup. Environ. Hyg.* **2020**, *17*, 243–252. [CrossRef]
14. Kim, J.-H.; Coca, A.; Williams, W.J.; Roberge, R.J. Subjective perceptions and ergonomics evaluation of a liquid cooled garment worn under protective ensemble during an intermittent treadmill exercise. *Ergonomics* **2011**, *54*, 626–635. [PubMed]
15. Zhao, Y.; Su, M.; Meng, X.; Liu, J.; Wang, F. Thermophysiological and Perceptual Responses of Amateur Healthcare Workers: Impacts of Ambient Condition, Inner-Garment Insulation and Personal Cooling Strategy. *Int. J. Environ. Res. Public Health* **2022**, *20*, 612. [CrossRef] [PubMed]
16. Glitz, K.; Seibel, U.; Rohde, U.; Gorges, W.; Witzki, A.; Piekarski, C.; Leyk, D. Reducing heat stress under thermal insulation in protective clothing: Microclimate cooling by a 'physiological' method. *Ergonomics* **2015**, *58*, 1461–1469. [CrossRef] [PubMed]
17. Gao, C.; Kuklane, K.; Holmér, I. Cooling vests with phase change material packs: The effects of temperature gradient, mass and covering area. *Ergonomics* **2010**, *53*, 716–723. [CrossRef] [PubMed]
18. Kayacan, Ö.; Kurbak, A. Effect of garment design on liquid cooling garments. *Text. Res. J.* **2010**, *80*, 1442–1455. [CrossRef]
19. Lou, L.; Wu, Y.S.; Shou, D.; Fan, J. Thermoregulatory clothing for personal thermal management. *Annu. Rev. Heat Transf.* **2018**, *21*, 205–244. [CrossRef]
20. Lou, L.; Zhou, Y.; Yan, Y.; Hong, Y. Wearable cooling and dehumidifying system for personal protective equipment (PPE). *Energy Build.* **2022**, *276*, 112510. [CrossRef]

21. Lembo, M.; Vedetta, C.; Moscato, U.; Del Gaudio, M. Thermal discomfort in healthcare workers during the COVID-19 pandemic. *La Med. Del Lav.* **2021**, *112*, 123.
22. Roskoden, F.C.; Krüger, J.; Vogt, L.J.; Gärtner, S.; Hannich, H.J.; Steveling, A.; Lerch, M.M.; Aghdassi, A.A. Physical activity, energy expenditure, nutritional habits, quality of sleep and stress levels in shift-working health care personnel. *PLoS ONE* **2017**, *12*, e0169983. [CrossRef] [PubMed]
23. Crandall, C.; Gonzalez-Alonso, J. Cardiovascular function in the heat-stressed human. *Acta Physiol.* **2010**, *199*, 407–423. [CrossRef] [PubMed]
24. Havenith, G. Heat balance when wearing protective clothing. *Ann. Occup. Hyg.* **1999**, *43*, 289–296. [CrossRef] [PubMed]

Disclaimer/Publisher's Note: The statements, opinions and data contained in all publications are solely those of the individual author(s) and contributor(s) and not of MDPI and/or the editor(s). MDPI and/or the editor(s) disclaim responsibility for any injury to people or property resulting from any ideas, methods, instructions or products referred to in the content.

processes

Article

Evaluation and Prediction of the Effect of Fabric Wetting on Coolness

Zijiang Wu †, Yunlong Shi *,†, Xiaoming Qian * and Haiyang Lei

School of Textile Science and Engineering, Tiangong University, Tianjin 300387, China; wuzijiang@tiangong.edu.cn (Z.W.); 2231010162@tiangong.edu.cn (H.L.)
* Correspondence: shiyunlong@tiangong.edu.cn (Y.S.); qxmtjpu@163.com (X.Q.)
† These authors contributed equally to this work.

Abstract: As an important parameter of garment comfort, the thermal sensation of fabrics changes with factors such as sweat-induced humidity, making it a crucial area of research. To explore the coolness sensation of fabrics under different humidities, we tested heat transfer between fabrics and skin for 20 different fabrics with varying thermal absorption rates using fuzzy comprehensive evaluation to objectively assess their coolness levels. Subjective evaluation was then obtained by having subjects touch the fabrics and provide feedback, resulting in a subjective evaluation of their coolness levels. We compared the objective and subjective evaluations and found them to be highly consistent (R^2 = 0.909), indicating accurate objective classification of fabric coolness levels. Currently, random forest regression models are widely used in the textile industry for classification, identification, and performance predictions. These models enable the prediction of fabric coolness levels by simultaneously considering the impact of all fabric parameters. We established a random forest regression model for predicting the coolness of wet fabrics, obtaining a high accuracy between predicted and tested thermal absorption coefficients (R^2 = 0.872, RMSE = 0.305). Therefore, our random forest regression model can successfully predict the coolness of wet fabrics.

Keywords: wet state; coolness sensation; sensory evaluation; thermal absorptivity; random forest regression model

Citation: Wu, Z.; Shi, Y.; Qian, X.; Lei, H. Evaluation and Prediction of the Effect of Fabric Wetting on Coolness. *Processes* **2023**, *11*, 2298. https://doi.org/10.3390/pr11082298

Academic Editors: Zhanxiao Kang and Qing Chen

Received: 22 June 2023
Revised: 27 July 2023
Accepted: 28 July 2023
Published: 31 July 2023

Copyright: © 2023 by the authors. Licensee MDPI, Basel, Switzerland. This article is an open access article distributed under the terms and conditions of the Creative Commons Attribution (CC BY) license (https://creativecommons.org/licenses/by/4.0/).

1. Introduction

Clothing is an important barrier to maintaining thermal and moisture balance in the human body. Choosing the right clothing can help people maintain a state of psychological and physiological comfort during work and activities, which is why consumers pay more attention to the thermal and moisture comfort of clothing [1]. When the skin comes into contact with fabric, heat on the skin surface will quickly diffuse to the cooler fabric surface, causing a brief decrease in skin temperature, stimulating skin temperature receptors, and forming a cool or warm sensorial judgment in the human brain [2]. The thermal sensation of fabrics is influenced by factors such as fabric type, structure, thickness, density, and surface treatment, and is also closely associated with fabric humidity [3]. The skin surface of the human body is constantly perspiring, including visible and invisible perspirations that affect the humidity of the fabric. In summer, high temperatures accelerate human sweat production to regulate body temperature, with an average sweating rate of up to 1.5 L/h [4]. As a result, the fabric can become soaked with sweat, which significantly changes the thermal sensation of the fabric and affects the comfort of the clothing. Therefore, accurately assessing the strength of the thermal sensation of fabrics is of great significance in the research and development of clothing products.

A fabric can be regarded as a mixture of fibers, air, and water [5] and the factors affecting the coolness of contact are numerous and complex, making it a hot research topic. Many studies have investigated the factors influencing the coolness of fabrics. Atalie et al. measured the thermal parameters of fabrics before and after composite finishing with different counts, twists, variation of mass (CVm), thickness, and strength, and found that the thermal absorption coefficient of the fabric changed significantly [6]. Akcagun et al. measured the thermal sensation of wool/PET-knitted fabrics with different water contents, and the results showed that wool causes a slow decrease in thermal sensation as humidity changes, giving it good thermal comfort [7]. Qian et al. studied knitted fabrics and explored the effects of fabric structure and porosity on thermal absorption coefficient and other parameters [8]. Mansoor et al. used the Alambeta method to test the thermal absorption coefficient of socks before and after water absorption, and the results showed that the thermal absorption coefficient of the fabric significantly increased after moisture absorption [9]. In summary, objective evaluation of the coolness of the fabric based on a single physical index measured using experimental instruments has been the common approach in research on the thermal sensation of fabrics. Unfortunately, there is little consideration of the subjective perception of the fabric's contact with human skin, and subjective evaluation of the coolness of fabrics often requires a lot of time and high labor costs. Improving the connection between subjective and objective evaluation and proposing an effective prediction model is an important way of solving this problem.

In recent years, many researchers have studied prediction models of fabric thermal performance based on fiber, yarn, and fabric structural parameters. Dias and Delkumburewatte proposed a prediction model that can effectively predict the thermal conductivity of fabrics [10]. Mangat et al. proposed a thermal resistance prediction model based on the combination of air, water, and fiber polymer in series and parallel in the wet state [11]. Bhattacharjee and Kothari proposed a model for predicting heat conduction through woven fabrics and radiation through yarns [12]. Kanat et al. used an Artificial Neural Network (ANN) model to predict thermal resistance at different moisture levels [13]. To more accurately classify the coolness levels of fabrics in the wet state and establish a fabric prediction model for evaluating the coolness contact of fabrics with different water contents, this study tested the thermal absorption coefficients of 20 types of fabrics with different water content levels. The fuzzy comprehensive evaluation method was then used to study the level of cooling sensation offered by the fabric. Finally, a comparison with the subjective evaluation level of the subjects was conducted to determine the level of cooling sensation that is in line with the actual user experience. A Random forest regression model was established to effectively predict the coolness of fabrics based on objective evaluation of parameter characteristics and human sensory evaluation results.

2. Methods

2.1. Objective Evaluation Method

The objective evaluation method divides the thermal sensation characteristics of the fabric by measuring related parameters. The main evaluation method currently used is the thermal absorption coefficient method. The thermal absorption coefficient method characterizes the heat absorption capacity of the fabric based on the heat value absorbed by the test sample during transient heat transfer following skin contact based on the size of the thermal absorption coefficient. The thermal absorption coefficient is also commonly referred to as the heat storage coefficient or heat dissipation coefficient and was first proposed by Hes and Dolezal [14] to describe the instantaneous sensation felt when contact occurs between the skin surface and the fabric surface, reflecting the fabric's ability to absorb heat from the skin during contact with the human body. The thermal absorption coefficient is calculated based on the thermal conductivity of the sample. The thermal conductivity of the fabric represents the amount of heat passing through 1 m^2 of material at a distance of 1 m in 1 s and is calculated using the Fan formula [15]:

$$\lambda = \frac{Qh}{tS\Delta T} \quad (1)$$

$$b = \sqrt{\lambda \rho c} \quad (2)$$

where λ is the thermal conductivity of the fabric, W/(m·°C); Q is the heat passing through the body, J; h is the measured thickness of the fabric, m; S is the area through which the heat passes, m^2; t is the flow time, s; ΔT is the temperature difference in the direction of heat conduction, °C; b is the thermal absorption coefficient of the fabric, Ws$^{1/2}$/(m^2·°C); c is the specific heat capacity of the fabric, J/(kg·°C); and ρ is the volume density of the fabric, kg/m^3.

As shown in Equation (2), the thermal absorption coefficient of the fabric is determined using three parameters: thermal conductivity, volume mass, and specific heat capacity. The larger the thermal absorption coefficient of the fabric, the stronger its heat absorption capacity, the more heat is lost when it comes into contact with the skin, and the better the coolness of the fabric. Conversely, if the thermal absorption coefficient of the fabric is small, the warming sensation of the fabric will be better. The actual thermal absorption coefficient of dry textiles usually ranges from 30 to 300 Ws$^{1/2}$/(m^2·°C) while for wet fabrics, the thermal absorption coefficient can even exceed 500 Ws$^{1/2}$/(m^2·°C) due to the high thermal conductivity and specific heat capacity of water [9]. The thermal absorption coefficient can only reflect the thermal transfer performance of the fabric in a steady-state environment and cannot reflect the dynamic heat transfer process when the fabric comes into contact with the skin.

2.2. Subjective Evaluation Methods

Objective evaluation methods have the advantages of strong repeatability and ease of operation. However, the coolness of fabrics is fundamentally a subjective sensation of human comfort, which can vary among individuals due to factors such as sensitivity to temperature and physical health conditions. Therefore, there is currently no accurate physical quantity that can directly measure the coolness of fabrics. Subjective evaluation is the most commonly used method for evaluating fabric comfort and mainly involves collecting subjects' feelings about fabric contact through questionnaire surveys and rating them based on subjective evaluation levels [16]. This can specifically express subjects' sensory perceptions of fabrics. Subjective evaluation methods mainly involve measuring the coolness sensation of a fabric when it comes into contact with the skin by directly touching the inner forearm of the subjects. Previous research has shown that the perception of heat sensation is similar between the forearm and the back, making it a more convenient option than directly measuring the coolness sensation by wearing the fabric. To accurately express the coolness of fabrics, a psychometric scale can be adopted to divide the human subjective sensation into levels and assign corresponding values, thereby quantitatively evaluating such qualitative issues [17]. In this study, the coolness levels of fabrics were divided into five levels based on a five-level psychometric scale. Level A indicated no coolness (x_1 = 1), level B indicated a slight coolness (x_2 = 2), level C indicated a general coolness (x_3 = 3), level D indicated an obvious coolness (x_4 = 4), and level E indicated a strong coolness (x_5 = 5). Subjects rated the fabric based on their feelings following contact between the fabric and the inner forearm. The number of ratings for each level was used as the weight for each level, which was denoted as ω_1, ω_2, ω_3, ω_4, and ω_5, respectively. The formula for the weighted average of the coolness of the fabric is:

$$x_i = \frac{x_1\omega_1+x_2\omega_2+x_3\omega_3+x_4\omega_4+x_5\omega_5}{\omega_1+\omega_2+\omega_3+\omega_4+\omega_5} \quad (i = 1, 2, \cdots, 20) \tag{3}$$

2.3. Coolness Sensation Level Classification

Fuzzy mathematics is based on fuzzy set theory [18], which is used to describe sets that are not clearly defined. The fuzzy comprehensive evaluation method is adopted to classify thermal absorption coefficient values of fabrics into membership degree levels. First, establish the evaluation factor object set $U = | u_1, u_2, u_3, \ldots, u_i |$ (i = 1, 2, ..., n), where the evaluation object u_i is the thermal absorption coefficient b_i. Second, establish the evaluation level set $V = | v_1, v_2, v_3, \ldots, v_j |$ (j = 1, 2, ..., n), where the evaluation level, v_i, is the coolness level of the fabric. Finally, by performing the membership degree function operation, the membership degree of the evaluation factors to the evaluation level is classified, making the fuzzy mathematical concepts more intuitive.

This study uses the trapezoidal distribution of fuzzy distribution. The mathematical model of the membership degree function is as follows:

$$\text{Level A}: \quad V_1(u_i) = \begin{cases} 1, \quad u_i < x_1 \\ \frac{1}{2}\left(1 + \frac{u_i - x_1}{x_2 - x_1}\right), & x_1 < u_i < x_2 \\ 0, \quad u_i > x_2 \end{cases} \tag{4}$$

$$\text{Level B}: \quad V_2(u_i) = \begin{cases} 0, \quad u_i < x_1 \\ \frac{1}{2}\left(1 + \frac{u_i - x_1}{x_2 - x_1}\right), & x_1 < u_i < x_2 \\ \frac{1}{2}\left(1 + \frac{u_i - x_2}{x_3 - x_2}\right), & x_2 < u_i < x_3 \\ \frac{1}{2}\left(1 - \frac{u_i - x_3}{x_4 - x_3}\right), & x_3 < u_i < x_4 \\ 0 \quad u_i > x_4 \end{cases} \tag{5}$$

$$\text{Level C}: \quad V_3(u_i) = \begin{cases} 0, \quad u_i < x_2 \\ \frac{1}{2}\left(1 + \frac{u_i - x_3}{x_3 - x_2}\right), & x_2 < u_i < x_3 \\ \frac{1}{2}\left(1 + \frac{u_i - x_3}{x_4 - x_3}\right), & x_3 < u_i < x_4 \\ \frac{1}{2}\left(1 - \frac{u_i - x_5}{x_5 - x_4}\right), & x_4 < u_i < x_5 \\ \frac{1}{2}\left(1 - \frac{u_i - x_5}{x_6 - x_5}\right), & x_5 < u_i < x_6 \\ 0 \quad u_i > x_6 \end{cases} \tag{6}$$

$$\text{Level D}: \quad V_4(u_i) = \begin{cases} 0, \quad u_i < x_4 \\ \frac{1}{2}\left(1 + \frac{u_i - x_5}{x_5 - x_4}\right), & x_4 < u_i < x_5 \\ \frac{1}{2}\left(1 + \frac{u_i - x_6}{x_6 - x_5}\right), & x_5 < u_i < x_6 \\ \frac{1}{2}\left(1 - \frac{u_i - x_7}{x_7 - x_6}\right), & x_6 < u_i < x_7 \\ 0 \quad u_i > x_7 \end{cases} \tag{7}$$

$$\text{Level E}: \quad V_5(u_i) = \begin{cases} 0, \quad u_i < x_6 \\ \frac{1}{2}\left(1 + \frac{u_i - x_6}{x_7 - x_6}\right), & x_6 < u_i < x_7 \\ 1, \quad u_i > x_7 \end{cases} \tag{8}$$

2.4. Random Forest Regression Model

Random Forest is a machine learning classification algorithm based on statistical theory, first proposed by Leo Breiman [19]. It combines Breiman's "Bootstrap aggregating" idea [20] and Ho's "random subspace" method [21] and can be used for both classification and regression. The Random Forest algorithm is a predictive tool that has high classification accuracy and is a new research hotspot in the field of data analysis. The principle is to generate a strong learning model by combining multiple weak learning models to improve the accuracy of the prediction model, a process called classifier combination. The Random Forest regression model can establish a complex nonlinear relationship between multiple independent variables and the dependent variable. Multiple sub-training sample sets are formed by using Bootstrap sampling to randomly extract data samples from the original training dataset, and each sub-training sample set has the same number of samples as the original training dataset. A decision tree model is then constructed. Due to randomness, hundreds or even thousands of decision trees can be generated, and the tree with the highest repeatability is selected as the final result. The prediction result of the Random Forest regression model is the average of all decision tree results. The specific process of constructing a Random Forest is as follows: first, randomly select m samples with replacement from the original sample set of n samples to construct m sub-sample sets. By training m sub-sample sets, m regression trees are constructed and the remaining samples are used as the test set to verify the prediction effect of the model. Then, set the regression model parameter "m_{try}". Assuming that there are k evaluation features of the fabric sample, randomly select a subset of m_{try} features (m_{try} < k) as the splitting node and determine the best splitting based on the optimal branching criterion. Finally, each regression tree stops growing based on the number of trees (n_{tree}), and after training, m decision trees are generated to form a Random Forest regression model. The arithmetic mean of the training results of the m decision trees is taken as the final model prediction result. The Random Forest prediction model is shown in Figure 1. The prediction effect of the model is evaluated using the root mean square error (*RMSE*) and the coefficient of determination (R^2) of the test set as shown in the following formula:

$$RMSE = \sum_{n=1}^{i} (y_i - y_i')^2/n \tag{9}$$

$$R^2 = 1 - RMSE/Y^2 \tag{10}$$

where y_i represents the true value of the test sample in the test set, y_i' represents the predicted value of the regression model, Y^2 represents the variance of the predicted value, and n is the number of samples in the test set.

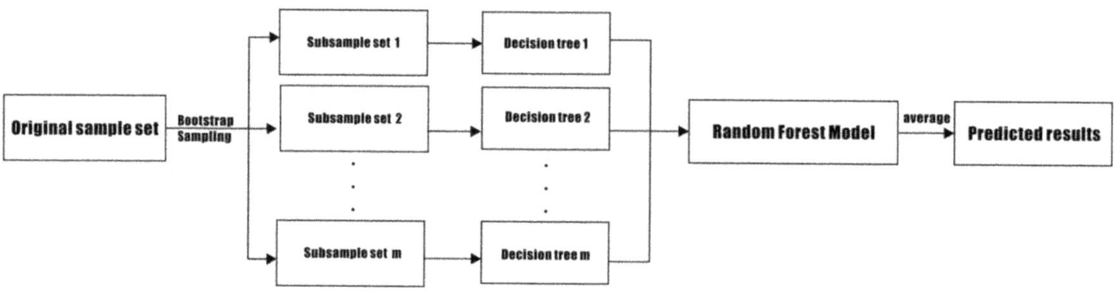

Figure 1. Random Forest regression model.

3. Experimental Section

3.1. Materials

This study involves analyzing the thermal contact coolness of multiple types of wet fabrics. All samples were selected from commonly used clothing fabrics in the market, and included fabrics of different materials and types. The parameters of all the fabrics are shown in Table 1. Before the experiment, the fabrics were subjected to low-temperature ironing to ensure a smooth surface. Then, the samples were cut into pieces with a length and width of 0.3 m and stored for later use after being placed in a room with a temperature of 24 °C and relative humidity of 65% for 24 h. Ionized water made in the laboratory was used to wet the fabrics.

The fabric thickness and surface density can be directly measured. There are many methods for measuring fabric porosity. In this study, fabric surface density, fabric thickness, and fabric fiber density were used to calculate the porosity [22] (see Equation (11)).

$$\varepsilon = 1 - \frac{m}{h \cdot \rho_{fib}} \tag{11}$$

where ε is the porosity of the fabric, %; ρ_{fib} is the fiber density at the stipulated regain, g/m^3; and m is the measured surface density of the fabric, g/m^3.

Table 1. Fabric specifications.

Symbol	Composition	Structure	Thickness (mm)	Weight (g/m²)	Fiber Density (kg/m³)	Porosity
#1	60S Cotton 100%	Plain	0.78	121.87	1540	0.90
#2	60S Cotton 100%	2/1 Twill	1.08	226.86	1540	0.86
#3	24S Jute 100%	2/1 Twill	0.93	249.25	1500	0.82
#4	21S Ramie 100%	3/1 Twill	1.13	208.55	1510	0.86
#5	60S Silk 100%	Plain	0.63	72.94	1360	0.91
#6	100S Wool 100%	2/2 Twill	1.02	175.39	1310	0.87
#7	40S Polyester 100%	Plain	0.66	90.05	1380	0.90
#8	21S Nylon 100%	Plain	0.72	161.06	1140	0.80
#9	120S Viscose 80% + Polyester 20%	Plain	0.90	130.44	1500	0.90
#10	60S Polyester 90% + Elastane 10%	Plain	0.89	155.37	1370	0.87
#11	60S Polyester 65% + Cotton 35%	Plain	1.41	227.26	1270	0.89
#12	40S Acrylic 70% + Viscose 30%	Plain	0.91	200.62	1142	0.80
#13	60S Nylon 85% + Elastane 15%	Warp knit	0.77	136.69	1419	0.87
#14	45S Polyester 98% + Elastane 2%	Warp knit	0.95	125.70	1378	0.90
#15	12S Nylon 70% + Polyester 30%	Warp knit	0.91	115.34	1212	0.89
#16	80S Polypropylene 65% + Polyester 35%	Weft knit	0.82	108.92	1068	0.87
#17	40S Cotton 80% + Polyester 20%	Weft knit	0.95	263.15	1508	0.81
#18	60S Polyester 58% + Cotton 42%	Weft knit	1.15	215.06	1447	0.87
#19	21S Wool 75% + Polyester 25%	Weft knit	1.16	133.69	1328	0.91
#20	32S Acrylic 80% + Polyester 20%	Weft knit	1.20	166.89	1196	0.88

When the fabric is made of two types of fibers blended together, the blended fiber density cannot be obtained directly but can be estimated using the following equation proposed by Militky [23].

$$\rho_{ab} = r\rho_a + (1 - r)\rho_b \tag{12}$$

In the formula, a and b represent the types of fibers; ρ_a and ρ_b represent their fiber densities; r represents the proportion of fiber a; and ρ_{ab} represents the density of the blended fiber. All tests were conducted under laboratory conditions at a temperature of 20 ± 1 °C, relative humidity of 50 ± 5%, and wind speed of less than 0.4 m/s.

3.2. Experimental Preprocessing

The study required measuring the maximum water absorption of each sample and dividing the fabric water content level based on the maximum water absorption. The experiment was conducted according to the method described by Tang et al. [24] and the pre-treated samples were ironed at low temperature and then dried in a 105 °C oven for 30 min before measurement to remove excess moisture inside the fabric and achieve the "ultra dry state" mentioned by Naka et al. [25]. The fabric samples were then soaked in deionized water for 3 min until completely wet, and then hung vertically to dry until there was no more liquid dripping for 30 consecutive seconds, indicating that the fabric was in a fully wet state. The weight difference between the fabric before and after soaking represents the maximum water content. The amount of humidity in each sample was determined based on the maximum water absorption. The study used fabric water content as the parameter for characterizing fabric moisture level, with the water content as the percentage of the sample's humidity level to the maximum water absorption reflecting the saturation level of the water content in the sample. The specific steps for adding humidity to the samples were as follows: the ultra-dry fabric was laid flat on a tabletop and a certain amount of deionized water was weighed using an electronic scale and poured into a humidifying spray bottle. The spray bottle was suspended about 2 cm above the center of the sample and the deionized water sprayed evenly over the fabric surface. The test was conducted after waiting for 5 min for the fabric to be fully wet. The study was conducted under conditions of ultra-dry state and water content levels of 20%, 40%, 60%, 80%, and 100%.

3.3. Objective Measurement Experiment

The experiment referred to the ISO 22007-2 [26] and ASTM D7984 [27] standards and was conducted at a constant temperature and humidity of 20 ± 0.5 °C and 60 ± 5%, respectively. A TPS 2500S thermal constants analyzer (Hot Disk AB., Co., Ltd., Göteborg, Sweden) was used to measure thermal conductivity and specific heat capacity of 20 fabrics with different water contents. Thermal absorption coefficient, which characterizes the thermal sensitivity performance of the fabric, was then calculated. The fabrics' thermal sensitivity levels were objectively evaluated based on the evaluation criteria. The instrument was turned on in advance and preheated for 30 min. The probe temperature was set to 35 °C, the same as the temperature of human skin. The resistance of the probe changed after contact with the sample. The thermal conductivity and thermal absorption coefficient of the fabric can be accurately calculated based on the change in probe resistance. The sample had to be in the same thermal state before each test and consecutive measurements were not allowed. Adequate time was required to ensure that the previously measured samples returned to their initial thermal states.

3.4. Subjective Measurement Experiment

Subjective evaluation was carried out in a laboratory at constant temperature and humidity (temperature setting of 20 ± 0.5 °C, humidity of 60 ± 5%, and wind speed less than 0.4 m/s) to ensure that the subjects' physiological functions were at their best state and their senses were more sensitive. At the same time, the psychological state was adjusted to ensure the smooth progression of the experiment. Subjects evaluated the coolness sensation of the fabric using the inside of their forearms [2]. Twelve college students—six males and six females—with an age of 21.8 ± 1.3 years (range 20–23), a height of 1.65 ± 0.04 m (range 1.52–1.82), and a weight of 55.6 ± 10.4 kg (range 43.3–82.4) volunteered to participate in this study. The subjects received relevant experimental training before the test, including an explanation of the definition of the thermal sensation from the fabric, the method for touching the samples during the test, and the rating method in the questionnaire. To avoid different scales, the same researcher explained the method for rating thermal sensation before each experiment to ensure smooth progression of the experiment and reliability and accuracy of the experimental results. Before touching the test samples, the subjects needed to sit quietly in the laboratory for 10 min to adapt to the surrounding environment and adjust their mental state. During the test, the subject extended their forearm forward with the inside of their forearm facing upwards, and the researcher placed the fabric on the forearm and pressed it appropriately to simulate the pressure exerted by clothing on the skin. As heat is conducted to the fabric when it comes into contact with human skin, to avoid errors, samples from the previous test were tested at an interval of 2 min to allow the fabric to return to its initial thermal state before the next test. The two forearms were covered alternately, and the test was completed when the two sensory evaluations given were consistent. If any inconsistency occurred, the process was repeated until all samples gave consistent evaluation results. The subjects filled out the questionnaires based on the warm/cool sensation felt by the forearm when it came into contact with the fabric. The coolness level was divided based on the individual's subjective feeling.

3.5. Statistical Analysis

The SPSS software (Version 27.0, IBM Co., Ltd., Armonk, NY, USA) was used to analyze the correlation between factors affecting the coolness sensation of wet-state fabrics using one-way analysis of variance and Pearson correlation analysis, with the significance level set at $p < 0.05$. Origin software (Version 2023, OriginLab Co., Ltd., Northampton, MA, USA) was used to compare the objective and subjective measurements by fitting the analysis and determining the linear relationship between the two.

4. Results and Discussion

4.1. Coolness via Objective Measurement

A study was conducted on 20 commonly used clothing fabrics as experimental samples. Before the experiment, all fabrics were pretested to ensure a smooth surface. The fabrics were then analyzed using a fabric thermal conductivity analyzer to measure the thermal absorption coefficient of different water contents. The thermal absorption coefficient of each sample is shown in Table 2. The coolness of the fabric was characterized based on the measured thermal absorption coefficient. A higher thermal absorption coefficient indicates a more significant coolness sensation upon contact with the fabric. Statistical analyses were conducted to identify factors that may affect the coolness of the fabric. The results showed that there were no significant differences in the coolness of different samples under dry conditions. A one-way ANOVA test showed that the water content of the fabric is the key factor affecting its coolness ($p < 0.05$). Pearson coefficients were used to characterize the relationships between the factors. The results showed that the type of fiber in the fabric (*Pearson* = 0.556), the thickness (*Pearson* = 0.157), the surface density (*Pearson* = 0.433), and the porosity (*Pearson* = −0.349) all affect the coolness of the fabric. Of these factors, water content was found to be the critical factor determining the thermal absorption coefficient of the sample. This is because the thermal conductivity of fabric fibers generally ranges from 0.03 to 0.10 W/(m·°C), while the thermal conductivity of still water at 20 °C is 0.57 W/(m·°C) and the thermal conductivity of water is generally 2–6 times higher than that of fiber polymers. At 37 °C, the volumetric heat capacity of water is 4200 J/(kg·°C), while the specific heat capacity of fiber polymers is approximately 1400–3300 J/(kg·°C). The volumetric heat capacity of water is generally 1.3–3 times higher than that of fiber polymers. Therefore, the thermal absorption coefficient of the fabric increases significantly when soaked in water [28,29]. The study also measured the thermal absorption coefficient of fabrics at moisture levels of 20%, 40%, 60%, 80%, and 100% (see Table 2). For each sample, each wet condition was analyzed in triplicate and outliers were removed to ensure that the coefficient of variation was less than 3%. The average result was rounded to two significant figures.

Table 2. Effect of water content on thermal absorptivity ($Ws^{1/2}/(m^2 \cdot °C)$).

Symbol	Ultra-Dry State	20%	40%	60%	80%	100%
#1	100.43	141.45	174.68	248.78	325.51	441.08
#2	112.72	161.58	184.66	253.77	330.31	455.97
#3	95.23	126.32	178.51	242.83	277.65	354.97
#4	70.33	111.54	157.80	267.90	302.14	326.57
#5	91.98	125.30	168.83	220.80	230.07	344.47
#6	95.30	136.83	155.74	232.62	293.51	402.60
#7	83.82	112.74	141.64	213.88	252.48	328.22
#8	109.49	138.81	164.54	224.52	253.28	321.52
#9	96.86	137.18	164.96	248.94	273.78	373.60
#10	111.40	139.66	165.65	222.35	260.57	317.49
#11	124.06	152.41	180.66	198.59	212.30	321.74
#12	118.97	148.77	192.06	232.60	251.46	358.26
#13	75.03	125.35	177.46	222.54	346.82	377.57
#14	89.56	97.29	137.43	190.10	243.79	284.75
#15	87.60	127.40	131.02	192.90	273.35	293.35
#16	85.89	101.75	106.39	194.50	220.74	331.11
#17	92.07	118.58	148.53	260.84	294.85	309.52
#18	83.01	149.91	207.67	312.59	331.89	344.66
#19	99.11	124.50	135.91	184.37	294.63	371.40
#20	91.08	130.89	117.35	179.77	303.61	374.26

4.2. Coolness Level Classification Results

The fabrics' thermal absorption coefficient values obtained during experimental testing were classified into different levels based on the fuzzy comprehensive evaluation method, giving a more scientifically reasonable range of coolness levels for the fabric. Fabrics were classified into different coolness levels based on the fuzzy mathematical model, as shown in Figure 2. Fabrics with thermal absorption coefficients of $\leq$100 $Ws^{1/2}/(m^2 \cdot °C)$ were defined as having no coolness (Level A), fabrics with thermal absorption coefficients ranging from 100 to 200 $Ws^{1/2}/(m^2 \cdot °C)$ were defined as having general coolness (Level B), fabrics with thermal absorption coefficients ranging from 200 to 300 $Ws^{1/2}/(m^2 \cdot °C)$ were defined as having slight coolness (Level C), fabrics with thermal absorption coefficients ranging from 300 to 340 $Ws^{1/2}/(m^2 \cdot °C)$ were defined as having obvious coolness (Level D), and fabrics with thermal absorption coefficients of $\geq$ 340 $Ws^{1/2}/(m^2 \cdot °C)$ were defined as having strong coolness (level E). Level A fabrics mainly contain 100% water, while Level B fabrics contain 80–100% water, Level C fabrics contain 60% water, and Level D fabrics contain 40% water. Level E fabrics mainly include ultra-dry fabrics and fabrics with a water content of 20%. As most dry fabrics had a thermal absorption coefficient of no more than 100 $Ws^{1/2}/(m^2 \cdot °C)$ (14/20), it can be concluded that fabrics conduct non-coolness only in the dry state.

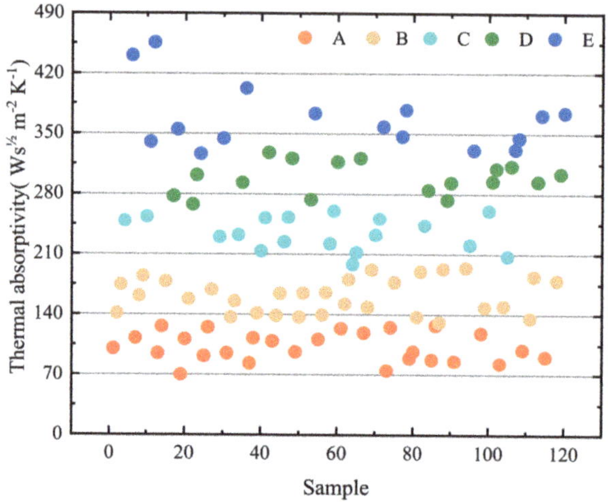

Figure 2. Thermal absorption coefficient class classification results.

By comparing the thermal absorption coefficient values of different types of fabrics, it was found that most fabrics have a thermal absorption coefficient of 100–300 $Ws^{1/2}/(m^2 \cdot °C)$. Sample #2 (100% Cotton) had the highest thermal absorption coefficient at different water contents, indicating the strongest coolness effect. Sample #16 (98% Polyester + 2% Elastane) had the lowest thermal absorption coefficient at different water contents, indicating the least obvious coolness effect. This may be due to the fact that pure cotton fabrics are hydrophilic materials, whereas polyester fibers are hydrophobic. Polyester fibers have a stronger drying sensation after water absorption compared with other fabrics; a similar phenomenon was observed in the study by Mansoor et al. [9]. Comparison of the thermal absorption coefficients of fabrics with different water contents showed that the thermal absorption coefficient of most fabrics increased by 200–300% from dry to completely wet, while Sample #15 (85% Nylon + 15% Elastane) had the highest increase of 403% and Sample #13 (65% Polyester + 35% Cotton) had the lowest increase of 159%, which may be associated with the degree of water absorption. Sample #12 (90% Polyester + 10% Elastane), which has a composition similar to Sample #13, also demonstrated a similar phenomenon, with only an 185% increase.

4.3. Coolness via Subjective Measurement

The subjective evaluation experiment was conducted in a laboratory under constant temperature and humidity, where 12 graduate students of different sexes evaluated 20 different types of fabrics for their coolness. The constant temperature and humidity environment ensured that the human sensory organs were in the best state and the psychological state was stable, which was conducive for the smooth progression of the experiment. The air velocity was set to not exceed 0.04 m/s to simulate a still environment, thereby increasing the accuracy and reliability of the subjective evaluation results and reducing experimental errors. The results showed that fabrics that were more compressible and more easily bent (had lower stiffness) tended to have higher coolness levels against the inner side of the forearm during subjective evaluations. When fabrics were wet, they tended to have a stronger tactile sensation due to the tighter contact with the skin, resulting in higher coolness levels and a feeling of stickiness that made the experience uncomfortable. To test the consistency of the subjects' evaluations, Spearman rank correlation coefficient was used to analyze the correlation between the coolness evaluations of the fabrics by the subjects. The results are shown in Table 3 and indicate that there was a significant positive correlation between the subjective evaluations of the coolness effects of the fabrics, with most significance levels having $p < 0.05$. This suggests that the subjects' evaluations were consistent and reliable and had a certain reference value. The evaluation levels were determined by taking the most common level for each sample among the subjects. The weighted average range for Level A was l–1.5, for Level B was 1.5–2.5, for Level C was 2.5–3.5, for Level D was 3.5–4.3, and for Level E was 4.3–5. The subjective evaluations were based on personal experiences and subjective judgments of the subjects, and due to individual differences, their experiences of coolness may differ.

Table 3. Spearman's rank correlation coefficients between coolness evaluations for each pair of subjects.

Subjects	S1	S2	S3	S4	S5	S6	S7	S8	S9	S10	S11	S12
Mean	0.951 *	0.852 *	0.854 *	0.765 *	0.865 *	0.758 *	0.876 *	0.875 *	0.876 *	0.858 *	0.855 *	0.890 *
S1		0.958 *	0.876 *	0.872 *	0.875 *	0.524	0.582	0.874 *	0.587	0.878 *	0.734 *	0.854 *
S2			0.912 *	0.654 *	0.756 *	0.675	0.847 *	0.784 *	0.875 *	0.914 *	0.821 *	0.587
S3				0.758 *	0.958 *	0.687	0.678	0.774 *	0.882 *	0.555	0.659	0.847 *
S4					0.707 *	0.879 *	0.911 *	0.768 *	0.745 *	0.576	0.879 *	0.861 *
S5						0.875 *	0.875 *	0.734 *	0.616	0.754 *	0.758 *	0.688
S6							0.758 *	0.548	0.702 *	0.662	0.725 *	0.651
S7								0.599	0.857 *	0.785 *	0.889 *	0.854 *
S8									0.798 *	0.714 *	0.854 *	0.741 *
S9										0.624	0.678	0.752 *
S10											0.758 *	0.732 *
S11												0.818 *

Note: * $p < 0.05$ (i.e., highly significant).

4.4. Consistency of Subjective and Objective Evaluations

Comparisons between the subjective evaluations and objective tests of the fabrics' coolness levels are shown in Table 4. Based on statistical analysis, there was a significant correlation between the subjective and objective evaluations of the coolness levels of the fabrics ($p < 0.05$). Additionally, the fitting curve in Figure 3 indicates good consistency between the subjective and objective evaluations of the fabrics' coolness levels ($R^2 = 0.909$; i.e., the closer R^2 is to 1, the higher the degree of fitting). The proportion of samples with consistent subjective and objective levels was 75.8%, validating the effectiveness of the thermal absorption coefficient method for evaluating the coolness performance of fabrics. It is worth noting that some samples showed differences between subjective and objective evaluations. For instance, samples #10, #11, and #12 in an ultra-dry state and samples #3 and #7, with a water content of 20%, had an objective evaluation level of B, while their subjective evaluation level was A. Sample #16, with a water content of 20%, and samples #1, #2, #11, and #14, with a water content of 40%, had an objective evaluation level of B, while their subjective evaluation level was upgraded to C. Samples #3, #5, and #9, with a water content of 60%, had an objective evaluation level of C, while their subjective evaluation level was B. Samples #1, #2, #5, and #14, with a water content of 60%, had an objective evaluation level of C, while their subjective evaluation level was B. Samples #12, #15, #17, and #19, with a water content of 80%, had an objective evaluation level of C, while their subjective evaluation level was D. Samples with different subjective and objective evaluation results can also be explained using their thermal absorption coefficients. Most of the measured results lie around the boundary of the objective classification, thus the differences in subjective and objective evaluations between different samples will not exceed one grade. In addition, when observing the results of subjective evaluation, it was found that more subjects tended to choose grades B or C. After questioning the subjects afterward, it was discovered that when the subjects did not feel the fabric's characteristics clearly or experienced sensory fatigue due to the experiment lasting too long, they would choose a more moderate grade to avoid making mistakes. Thus, grades A and E were rarely chosen, which could easily cause errors in the subjective evaluation experiment. However, the subjective and objective evaluations maintained a high degree of consistency for samples with obvious characteristics.

Table 4. Objective and subjective evaluations of fabric coolness levels.

Symbol	Ultra-Dry State		20%		40%		60%		80%		100%	
	Ob	Sub	Ob	Sub	Ob	Sub	Ob	Sub	Ob	Sub	Ob	Sub
#1	A	A	B	B	B	C	C	C	D	D	E	E
#2	B	B	B	B	B	C	C	C	D	D	E	E
#3	A	A	B	B	B	B	C	B	C	C	E	E
#4	A	A	B	A	B	B	C	C	D	C	D	E
#5	A	A	B	B	B	B	C	B	C	C	E	E
#6	A	A	B	B	B	B	C	C	C	C	E	E
#7	A	A	B	A	B	B	C	C	C	C	D	E
#8	B	B	B	B	B	B	C	C	C	C	E	E
#9	A	A	B	B	B	B	C	B	C	C	D	D
#10	B	A	B	B	B	B	C	D	C	C	D	D
#11	B	A	B	B	B	C	C	C	C	C	D	E
#12	B	A	B	B	B	B	B	C	C	D	E	E
#13	A	A	B	B	B	B	C	C	E	D	E	E
#14	A	A	A	B	B	C	B	C	C	C	C	E
#15	A	A	B	B	B	B	B	C	C	D	C	C
#16	A	A	B	C	B	B	B	B	C	C	D	D
#17	A	A	B	B	B	B	B	B	C	D	D	D
#18	A	A	B	B	B	B	D	C	D	D	E	E
#19	A	A	B	B	B	B	B	B	C	D	E	E
#20	A	A	B	B	B	B	B	B	D	D	E	E

Note: Ob and Sub represent objective and subjective evaluations, respectively.

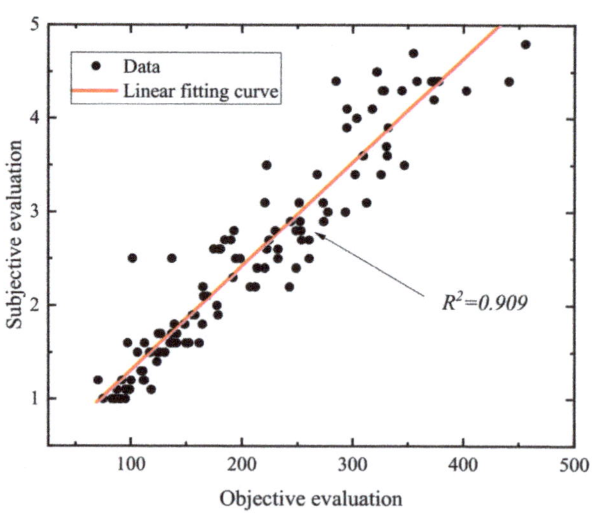

Figure 3. Fitting curves for subjective and objective evaluations.

4.5. Random Forest Model Predicts Coolness

A random forest regression model was used to predict the coolness of fabrics with different water contents. Five indicators, including the density of fiber bodies, the thickness of fabrics, the surface density, the porosity, and thermal conductivity in the dry state, were used as evaluation features in the random forest regression model. The original fabric sample set contained 100 samples, with 3/4 used as the training set and 1/4 used as the testing set. Combined with the subjective evaluation results, a random forest regression model was established, with the training set used for building the random forest algorithm and the testing set used for evaluating the remaining data. When the accuracy of the testing set is much higher than that of the training set, there is underfitting, while the opposite indicates overfitting. After analyzing the parameters of the random forest regression model, a random forest regression model for fabric coolness was established, and the model was used to predict the coolness level of fabrics with different water contents. The predicted results were compared with the measured values, as shown in Figure 4. The evaluation index of the model, R^2, was 0.872 and the RMSE was 0.305 (an RMSE of 0.2–0.5 indicates that the model can accurately predict the data and the smaller the value, the better the prediction effect). The low R^2 may be caused by a low sample size, although it still indicates that the model has good predictive performance and is effective at predicting and evaluating fabric coolness.

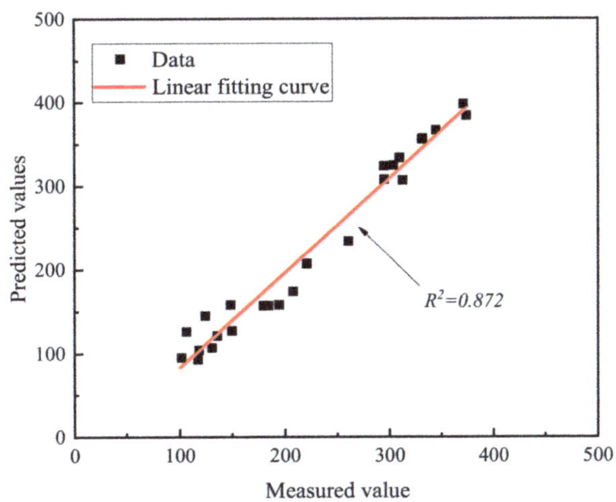

Figure 4. Comparison of the consistency between predicted and measured values.

4.6. Limitations

This study had its limitations. First, because the TPS 2500S thermal constant analyzer has different testing principles from Heat flux sensor-based instruments such as Alambeta, the thermal conductivity of the sample may be underestimated, which may also lead to the low thermal absorption coefficient of the sample. Second, the subject experiments are affected by steady-state heat conduction when the contact time between the skin and the sample exceeds 2 s. Even though emphasis was placed on the subjects to evaluate the instantaneous contact coolness of the fabric during the experiment, they are inevitably affected by steady-state heat conduction. We must admit that the above errors will occur during the experiment.

5. Conclusions

A thermal constant analyzer was used to test the thermal absorption coefficients of 20 commonly used clothing fabrics—including natural fibers, synthetic fibers, and blended fabrics—at different levels of water contents. The fabrics were objectively classified into five levels of coolness using fuzzy comprehensive evaluation and subjectively classified using participant evaluations. The results are as follows:

(1) The five levels of coolness classification provided by the fuzzy comprehensive evaluation method can give specific level indicators. For example, fabrics with a coolness level of A have a thermal absorption coefficient lower than 100 $Ws^{1/2}/(m^2{\cdot}°C)$ and the coolness upon contact with the fabric is defined as none. Fabrics with a coolness level of B have a thermal absorption coefficient of 100–200 $Ws^{1/2}/(m^2{\cdot}°C)$ and the coolness upon contact with the fabric is defined as general. Fabrics with a coolness level of C have a thermal absorption coefficient of 200–300 $Ws^{1/2}/(m^2{\cdot}°C)$ and the coolness upon contact with the fabric is defined as slight. Fabrics with a coolness level of D have a thermal absorption coefficient of 300–340 $Ws^{1/2}/(m^2{\cdot}°C)$ and the coolness upon contact with the fabric is defined as obvious. Fabrics with a coolness level of E have a thermal absorption coefficient greater than 340 $Ws^{1/2}/(m^2{\cdot}°C)$ and the coolness upon contact with the fabric is defined as strong.

(2) Analysis of the consistency between the subjective and objective coolness levels of the fabrics indicates that using the thermal absorption coefficient as the objective evaluation index for perceived coolness is reliable. A comprehensive evaluation of fabric coolness based on both subjective and objective aspects can accurately reflect the real perception of the fabric when in contact with the skin. This can provide reliable data support for consumers when purchasing related products in the future and can also serve as a reference for developing fabric coolness level standards.

(3) The thermal absorption coefficient of the fabric made of 100% cotton under wet conditions is high, ranging from 112.72 to 455.97 $Ws^{1/2}/(m^2{\cdot}°C)$, while the thermal absorption coefficient of the blended fabric made of 98% polyester + 2% elastane under wet conditions is low, ranging from 85.89 to 331.11 $Ws^{1/2}/(m^2{\cdot}°C)$. This is because the fabric made of 100% cotton has more water absorption than the 98% polyester + 2% elastane blend fabric, resulting in stronger contact coolness.

(4) The established random forest regression model can effectively predict the coolness of fabrics at different water content levels. The evaluation indicators for the training set prediction results show that the R^2 is 0.872 and the $RMSE$ is 0.305, indicating that the model has good predictive performance.

(5) Water content is the most important factor affecting the coolness of fabrics. As the water content of the fabric increases, the coolness of the fabric continuously improves. However, the corresponding humidity of the fabric also increases, potentially causing discomfort to the wearer. Therefore, when choosing summer clothing, it is important to consider fabric coolness upon contact under humid conditions and try to avoid the decrease in clothing comfort due to sweat-soaking.

Author Contributions: Methodology, Y.S.; Resources, X.Q.; Data curation, H.L.; Writing—original draft, Z.W. All authors have read and agreed to the published version of the manuscript.

Funding: This research was funded by [National Natural Science Foundation of China] grant number [No. U1933111] and [Tianjin Research Innovation Project for Postgraduate Students] grant number [No.2021YJSO2B06].

Data Availability Statement: Not applicable.

Conflicts of Interest: The authors declare no conflict of interest.

References

1. Wu, Z.; Shi, Y.; Yang, R.; Qian, X.; Fang, S. Modification and Validation of a Dynamic Thermal Resistance Model for Wet-State Fabrics. *Processes* **2023**, *11*, 1630. [CrossRef]
2. Park, J.; Yoo1, H.; Hong, K.; Kim, E. Knitted fabric properties influencing coolness to the touch and the relationship between subjective and objective coolness measurements. *Text. Res. J.* **2018**, *88*, 1931–1942. [CrossRef]
3. Shi, Y.; Wang, L.; Qian, X. Effect of non-uniform skin of "Walter" on the evaporative resistance and thermal insulation of clothing. *Int. J. Cloth. Sci. Technol.* **2017**, *29*, 686–695. [CrossRef]

4. Kaplan, S.; Okur, A. Determination of coolness and dampness sensations created by fabrics by forearm test and fabric measurements. *J. Sens. Stud.* **2009**, *24*, 479–497. [CrossRef]
5. Hes, L.; De Araujo, M. Simulation of the effect of air gaps between the skin and a wet fabric on resulting cooling flow. *Text. Res. J.* **2010**, *80*, 1488–1497. [CrossRef]
6. Atalie, D.; Gideon, R.; Melesse, G.; Ferede, E.; Getnet, F.; Nibret, A. Thermo-physiological comfort of half bleached woven fabrics made from different cotton yarns parameters. *J. Nat. Fibers* **2022**, *19*, 5034–5049. [CrossRef]
7. Akcagun, E.; Bogusławska-Baczek, M.; Hes, L. Thermal insulation and thermal contact properties of wool and wool/PES fabrics in wet state. *J. Nat. Fibers* **2019**, *16*, 199–208. [CrossRef]
8. Qian, J.; Xie, T.; Chen, L.; Li, Z.; Guo, N.; Fu, S.; Zhang, P. Effect of Knitting Structure and Polyethylene Content on Thermal-wet Comfort and Cooling Properties of Polyethylene/polyester Fabrics. *Fibers Polym.* **2022**, *23*, 3297–3308. [CrossRef]
9. Mansoor, T.; Hes, L.; Bajzik, V.; Noman, M.T. Novel method on thermal resistance prediction and thermo-physiological comfort of socks in a wet state. *Text. Res. J.* **2020**, *90*, 17–18. [CrossRef]
10. Dias, T.; Delkumburewatte, G.B. The influence of moisture content on the thermal conductivity of a knitted structure. *Meas. Sci. Technol.* **2007**, *18*, 1304–1314. [CrossRef]
11. Mangat, M.M.; Hes, L. Thermal resistance of denim fabric under dynamic moist conditions and its investigational confirmation. *Fibres Text. East Eur.* **2014**, *22*, 101–105.
12. Bhattacharjee, D.; Kothari, V.K. Heat transfer through woven textiles. *Int. J. Heat Mass Transf.* **2009**, *52*, 2155–2160. [CrossRef]
13. Kanat, Z.E.; Ozdil, N. Application of artificial neural network (ANN) for the prediction of thermal resistance of knitted fabrics at different moisture content. *J. Text. I.* **2018**, *109*, 1247–1253. [CrossRef]
14. Hes, L.; Dolezal, I. New method and equipment for measuring thermal properties of textiles. *J. Textile Mach. Soc. Jpn.* **1989**, *42*, 24–28. [CrossRef]
15. Fan, J. A Study of Heat Transfer through Clothing Assemblies. Ph.D. Thesis, Department of Textile Industries, The University of Leeds, Leeds, UK, 1998.
16. Tang, K.; Kan, C.; Fan, J. Assessing and predicting the subjective wetness sensation of textiles: Subjective and objective evaluation. *Text. Res. J.* **2014**, *85*, 838–849. [CrossRef]
17. Wu, Z.; Yang, R.; Qian, X.; Yang, L.; Lin, M. A multi-segmented human bioheat model under immersed conditions. *Int. J. Therm. Sci.* **2023**, *185*, 108029. [CrossRef]
18. Zhang, F.; Ignatius, J.; Lim, C.; Zhao, Y. A new method for deriving priority weights by extracting consistent numerical-valued matrices from interval-valued fuzzy judgement matrix. *Int. J. Fuzzy Syst.* **2017**, *19*, 27–46. [CrossRef]
19. Breiman, L. Random forests. *Mach. Learn.* **2001**, *45*, 5–32. [CrossRef]
20. Breiman, L. Bagging Predictors. *Mach. Learn.* **1996**, *24*, 123–140. [CrossRef]
21. Ho, T.K. The Random Subspace Method for Constructing Decision Forests. *IEEE Trans. Pattern Anal.* **1998**, *20*, 832–844.
22. Yang, R.; Wu, Z.; Qian, X.; Shi, Y. Development of thermal resistance prediction model and measurement of thermal resistance of clothing under fully wet conditions. *Text. Res. J.* **2023**, *93*, 911–924. [CrossRef]
23. Militky, J. Prediction of textile fabrics thermal conductivity. In *Thermal Manikins and Modelling*; Fan, J., Ed.; The Hong Kong Polytechnic University: Hongkong, China, 2006.
24. Tang, M.; Chau, K.; Kan, C.; Fan, J.T. Magnitude estimation approach for assessing stickiness sensation perceived in wet fabrics. *Fibers Polym.* **2018**, *19*, 2418–2430.
25. Naka, S.; Kamata, Y. Thermal conductivity of wet fabrics. *J. Text. Mach. Soc. Jpn.* **1977**, *23*, 114–119. [CrossRef]
26. *ISO 22007-2-2015*; Plastics—Determination of Thermal Conductivity and Thermal Diffusivity—Part 2—Transient Plane Heat Source (Hot Disc) Method. ISO: Geneva, Switzerland, 2015.
27. *ASTM D7984-2016*; Standard Test Method for Measurement of Thermal Effusivity of Fabrics Using a Modified Transient Plane Source (MTPS) Instrument. ISO: Geneva, Switzerland, 2016.
28. Yang, R.; Wang, L.; Zou, C.; Li, S.; Geng, D. Life preservers: Concepts, progress, and challenges. *Int. J. Aerosp. Psychol.* **2020**, *30*, 77–88. [CrossRef]
29. Transportation Safety Board (TSB). Loss of Control and Collision with Water Cochrane Air Service de Havilland DHC-2 Mk.1, C-FGBF Lillabelle Lake, Ontario, 25 May 2012. In *Aviation Investigation Report*; Report No. A12O0071; Transportation Safety Board (TSB): Ottawa, ON, Canada, 2012.

Disclaimer/Publisher's Note: The statements, opinions and data contained in all publications are solely those of the individual author(s) and contributor(s) and not of MDPI and/or the editor(s). MDPI and/or the editor(s) disclaim responsibility for any injury to people or property resulting from any ideas, methods, instructions or products referred to in the content.

processes

Article

Modification and Validation of a Dynamic Thermal Resistance Model for Wet-State Fabrics

Zijiang Wu [1,†], Yunlong Shi [1,†], Ruiliang Yang [2,*], Xiaoming Qian [1,*] and Shuting Fang [1]

[1] School of Textile Science and Engineering, Tiangong University, Tianjin 300387, China; wuzijiang@tiangong.edu.cn (Z.W.); shiyunlong@tiangong.edu.cn (Y.S.); fangshutingfst@163.com (S.F.)
[2] School of Aeronautics and Astronautics, Tiangong University, Tianjin 300387, China
* Correspondence: yangruiliang2001@sina.com (R.Y.); qxmtjpu@163.com (X.Q.)
† These authors contributed equally to this work.

Abstract: To investigate the dynamic thermal resistance of woven fabrics in different wetting states, ten commonly used clothing fabrics were selected and tested for fabric thermal resistance under different levels of water saturation in accordance with Chinese national standards. Based on Mangat's eight thermal resistance prediction models, the study improved the models by replacing the original moisture content with water content saturation. The suitability of the eight models in predicting the thermal resistance of woven fabrics in wet states was compared using the sum of squared deviations (*SSD*), sum of absolute deviations (*SAD*), and correlation coefficient (R^2). The results showed that during the process from initial wetting to complete immersion, the measured thermal resistance values of the ten fabric samples were consistent with the predicted values from Model 5 in the theoretical model of thermal resistance ($R^2 > 0.955$). The characteristic of Model 5 is that the air thermal resistance and water thermal resistance are first connected in parallel and then connected in series with the fiber thermal resistance. The corrected predicted values from Model 5 were highly consistent with the experimental measurement values and can be used to approximate the thermal resistance of woven fabrics in wet states.

Keywords: wet state; ultradry state; thermal resistance; thermal comfort; empirical model

Citation: Wu, Z.; Shi, Y.; Yang, R.; Qian, X.; Fang, S. Modification and Validation of a Dynamic Thermal Resistance Model for Wet-State Fabrics. *Processes* **2023**, *11*, 1630. https://doi.org/10.3390/pr11061630

Academic Editor: Zhanxiao Kang and Qing Chen

Received: 10 May 2023
Revised: 24 May 2023
Accepted: 25 May 2023
Published: 26 May 2023

Copyright: © 2023 by the authors. Licensee MDPI, Basel, Switzerland. This article is an open access article distributed under the terms and conditions of the Creative Commons Attribution (CC BY) license (https://creativecommons.org/licenses/by/4.0/).

1. Introduction

As the "second skin" of the human body, clothing is the most important barrier for maintaining thermal stability [1]. Clothing should help maintain the body's thermal and moisture balance, enabling the body to be in a state of psychological, physiological, and sensory comfort during long periods of work and activity. When factors such as sweating, rainwater, and accidental immersion cause clothing to become wet, both the thermal resistance and moisture resistance of the clothing will change, affecting its comfort. Fabric thermal resistance is influenced by factors such as the fabric structure, density, humidity, and surface treatment and is closely related to fabric thickness, yarn density, fabric surface friction coefficient, and fabric type [2–4]. In wet conditions, the moisture absorption and release of the fabric will cause a constant change in the proportion of water and air in the fabric, and the thermal resistance of the wet fabric will also dynamically change [5]. The three key factors affecting fabric moisture absorption are thickness, porosity, and fiber type, and thicker and higher-porosity fabrics can absorb more water, with natural fibers having greater moisture absorption than synthetic fibers. In wet conditions, due to the participation of water, the structure and composition of the fabric will undergo small changes, making the thermal performance of clothing more complex than in dry conditions.

There have been numerous studies on fabric thermal resistance in wet conditions worldwide. Hes et al. [6] tested the thermal resistance of fabrics per unit thickness in dry and wet conditions, and the results showed that the thermal resistance of fabrics in wet conditions was significantly lower than in dry conditions. Oğlakcioğlu et al. [7] tested the thermal resistance of 10 kinds of knitted pure cotton fabrics in wet conditions; the conclusion was that increasing the moisture content of the fabric significantly increased its heat transfer capability. Wang et al. [8] tested the thermal resistance of seven different thicknesses of cotton and polyester fabrics in fully saturated conditions, confirming that fabric thickness and fiber material have a significant impact on thermal resistance. Akckgun et al. [9] tested the thermal resistance changes in wool and wool/polyester blend fabrics at different levels of moisture, and the results showed that fabric porosity also has a significant impact on thermal resistance. Yang et al. [10] tested the thermal resistance of three typical clothing items in saturated conditions and found that the thermal insulation performance was significantly reduced compared with dry conditions. Therefore, the fundamental factors affecting fabric thermal resistance in wet conditions are fabric fiber type, the water in the fabric, and the distribution of water in the fabric. Establishing a theoretical model for the thermal resistance of wet fabrics based on fabric composition and structure is scientifically significant for analyzing complex wet thermal resistance.

In the past few decades, many researchers have studied the prediction of thermal resistance models for fabrics, both theoretically and experimentally. Nake et al. [11] were the first to propose a three-parameter theoretical model of air, water, and fiber polymers, including series, parallel, and combined usage, some of which could be used for thermal resistance prediction; however, these models were very complex and limited to the dry state. Hes [12] assumed that the thermal resistance of the fabric was parallel to the thermal resistance of water in his proposed thermal resistance model, which had higher predictability. Mangat et al. [3] proposed a theoretical model for the combination of series and parallel air, water, and fiber polymer thermal resistance under wet conditions. They tested the thermal resistance of cotton knitted fabrics at different moisture levels, and the results showed that two sets of models had the best consistency with experimental data. In addition, the experimental conclusion also suggested that the model may be applicable to other types of fabrics; this conclusion needs to be verified by subsequent experiments. In addition to the above thermal resistance models, the literature also mentions six thermal resistance models [13]. Through the comparison of experimental data and models, it is believed that these models have poor correlation with the thermal resistance of wet fabrics. Although individual thermal resistance models involve the influence of moisture, they do not consider the dynamic thermal resistance changes of fabrics due to moisture increase. In summary, previous studies mostly involved the prediction of thermal resistance of knitted fabrics, lacking the prediction of the thermal resistance of woven fabrics as a large category. In order to more comprehensively analyze the thermal resistance changes in various types of fabrics under wet conditions, this study refers to national standards to test the thermal resistance values at each moisture level and analyzes the effect of fabric humidity on thermal resistance by studying the thermal resistance changes in different fabrics after wetting. Based on the test results, the theoretical model of fabric thermal resistance is modified and verified more accurately.

2. Thermal Resistance Theoretical Model

Fabric is typically composed of fiber polymers, stagnant air within the fabric, and absorbed water molecules [3]. The material that provides warmth is primarily the stagnant air, and the main role of the fiber polymers is to provide storage space for this air. Under natural conditions, the water content in fabrics is minimal and mainly achieved through the binding of hydrophilic groups within the fiber polymers to water vapor in the air or in the form of adsorption onto the fabric surface. The thermal resistance of fabrics is a critical parameter for measuring their ability to insulate heat transfer, is an important indicator of fabric thermal comfort, and primarily dependent on fabric thickness and thermal conductivity [14,15] (see Equation (1)),

$$R_t = \frac{h}{\lambda} \tag{1}$$

where R_t is the thermal resistance of the fabric in $m^2 \cdot °C/W$, h is the fabric thickness in m, and λ is the thermal conductivity of the fabric in $W/(m \cdot °C)$.

The thermal resistance of fabric is largely dependent on the amount of stagnant air within it. Therefore, it can be concluded that factors affecting the air content in fabrics determine their thermal resistance. These factors include fabric fiber type, thickness, surface density, organizational structure, and porosity, all of which are critical in determining clothing thermal resistance. Fabric thickness and surface density can be directly measured, and there are many methods for measuring fabric porosity. Studies have used fabric surface density, fabric thickness, and fabric fiber density to calculate porosity [3] (see Equation (2)).

$$\varepsilon = 1 - \frac{m}{h \cdot \rho_{fib}} \tag{2}$$

where ε is the porosity of the fabric in percentage, ρ_{fib} is the fiber density under standardized moisture regain in g/m^3, and m is the measured areal density of the fabric in g/m^2.

The dry fabric system is composed of fibers and air, with the air uniformly distributed in the voids of the entire fabric system, including the interstices between the fabric structure and yarns, the interstices between fibers, and the internal voids of the fibers. Assuming a constant capacity of the fabric system, when the external environment is humid, the dry fabric continuously absorbs moisture from the outside. Unlike air, the interaction between water molecules and fibers is more complex, which Hes [16] describes in four forms: after the fabric absorbs water, water molecules first enter the micro-pores of the fibers and quickly form strong hydrogen bonds with the hydrophilic groups in the fiber polymers, whereas the remaining water molecules quickly occupy all the voids in the entire fabric system outside the fiber polymers until almost all the air in the fabric system is driven out. Finally, some of the water molecules are adsorbed on the fabric surface and the fabric reaches the state of saturation absorption, with the maximum water content. Sugawara and Yoshizawa [17] proposed that the thermal conductivity of porous materials depends on the thermal conductivity of the fluid and the solid. For fully saturated porous fabrics, the fabric thermal resistance can be regarded as the combination of fiber thermal resistance and water thermal resistance. Since the thermal conductivity of water is 22 times higher than that of air (0.57/0.026) [18], as the proportion of water in the fabric increases, the overall thermal conductivity of the fabric will significantly increase, resulting in a decrease in the fabric's insulation ability.

Many scholars have constructed theoretical and empirical models of fabric thermal resistance under different humidity levels through practical experiments and theoretical analyses. Among them, Mangat et al. [3] proposed a theoretical model for the wet-state thermal resistance of single-layer fabrics based on previous theories, which macroscopically considers the wet-state thermal resistance of fabrics as consisting of fiber thermal resistance, air thermal resistance, and water thermal resistance. The following assumptions were made: (1) the voids are uniformly distributed throughout the fabric system, and the moisture content contained in the fabric is constant. When the fabric is soaked, the water entering the fabric replaces some of the air in the voids; (2) when the hygroscopic fabric expands after absorbing water, the fabric undergoes changes and the thickness and area of the fabric increase, whereas the voids between the fibers and yarns decrease. However, since the macroscopic changes are still relatively minor, the changes can be ignored; (3) when the fabric is completely immersed, the water in the voids can be regarded as countless water columns, and the height of the water columns in the voids can be regarded as the thickness of the fabric. If the fabric has not reached complete wetting, the height of the water columns and air columns cannot be measured; (4) this model is based solely on thermal conduction and does not consider the effects of convection, radiation, and evaporation.

This series of models defines wet-state fabrics as a mixed system of fibers, water, and air, and the total thermal resistance of the fabric is jointly determined by the thermal resistance of the fibers, air, and water. The thermal resistance calculation formulas for the three factors are [3] (see Equations (3)–(5)):

$$\text{Fiber thermal resistance}: \ R_f = \frac{h(1-\varepsilon)}{\lambda_f} \tag{3}$$

$$\text{Air thermal resistance}: \ R_a = \frac{h\varepsilon}{\lambda_a(1-\mu)} \tag{4}$$

$$\text{Water thermal resistance}: \ R_w = \frac{h\varepsilon}{\lambda_w\mu} \tag{5}$$

where R_f, R_a, and R_w represent the thermal resistance of fibers, air, and water, respectively, $m^2{\cdot}°C/W$. λ_f is the thermal conductivity of fiber polymers, $W/(m{\cdot}°C)$. λ_a is the thermal conductivity of air, in $W/(m{\cdot}°C)$, whereas λ_w is the thermal conductivity of water, $W/(m{\cdot}°C)$. μ is the moisture content of the fabric, %.

The thermal resistance of a fabric in a wet state can be predicted by combining the thermal resistance of its fiber, air, and water components in series, parallel, or a combination of both. Based on different serial and parallel configurations in practice, Mangat et al. [3] summarized eight thermal resistance models for wet fabrics (see Equations (6)–(13)). By comparing the predicted results of these models with actual measurements, the thermal resistance model with the best correlation was finally selected as the theoretical prediction model for thermal resistance. The eight thermal resistance models are as follows:

$$\text{Model 1}: \ R_t = R_f + R_a + R_w \tag{6}$$

$$\text{Model 2}: \ R_t = \left(R_f^{-1} + R_a^{-1} + R_w^{-1}\right)^{-1} \tag{7}$$

$$\text{Model 3}: \ R_t = \frac{R_a \cdot R_f}{R_a + R_f} + R_w \tag{8}$$

$$\text{Model 4}: \ R_t = \frac{R_f \cdot R_w}{R_f + R_w} + R_a \tag{9}$$

$$\text{Model 5}: \ R_t = \frac{R_a \cdot R_w}{R_a + R_w} + R_f \tag{10}$$

$$\text{Model 6}: \ R_t = \frac{R_w(R_a + R_f)}{R_a + R_w + R_f} \tag{11}$$

$$\text{Model 7}: \ R_t = \frac{R_a(R_w + R_f)}{R_a + R_w + R_f} \tag{12}$$

$$\text{Model 8}: \ R_t = \frac{R_f(R_w + R_a)}{R_a + R_w + R_f} \tag{13}$$

Among the eight models summarized by Mangat et al. [3], Models 1 and 2 represent the predicted thermal resistance values for fiber, air, and water in direct series and in parallel and therefore should be excluded from consideration when selecting the optimal thermal resistance model. In their subsequent experiments, Mangat et al. validated the accuracy of the above models in predicting the thermal resistance of a single-layer knitted fabric at different levels of moisture content. The results showed that Model 3 in their model group had a good correlation with a certain twill cotton fabric tested, whereas Models 5 and 7 exhibited the best consistency with all experimental data and were applicable to different types of knitted fabrics.

It should be noted that Mangat et al. used the moisture content μ and $1 - \mu$ to represent the ratio of water and air in the interstices of the fabric, respectively, in the calculation formulas for air and water thermal resistance. According to the formula for moisture content, it represents the proportion of water in the wet fabric. Therefore, using moisture content to represent the ratio of water and air in the fabric interstices is not rigorous. The thermal resistance prediction curve in Mangat et al.'s results often underestimated the actual measurement values, which may be caused by the definition of moisture content. In this study, we followed the thermal resistance prediction model framework of Mangat et al. and used water content saturation to represent the ratio of water and air in the fabric interstices, which is the maximum water content that the fabric can absorb. The water content saturation was calculated as the ratio of the weight of water measured during the test to the maximum water absorption weight of the fabric. The maximum water absorption of the fabric was calculated based on the weight of the sample before and after immersion, using the Formulas (14) and (15):

$$WAC = \frac{m_{water}}{SA} \times 100\% \tag{14}$$

$$\eta = \frac{m_{wet} - m_{dry}}{WAC} \times 0.09 \tag{15}$$

where WAC is the maximum water absorption of the fabric in g/m^2, SA is the fabric area in m^2, m_{water} is the maximum water weight that the fabric can absorb, g, m_{wet} is the weight of the wet fabric, g, m_{dry} is the weight of the ultradry fabric, g, and η is the water content saturation, %. The fabric samples are all large samples with a size of 0.3 m $\times$ 0.3 m; the total area of the samples is 0.09 m^2.

3. Experimental

3.1. Materials

Previous research on the thermal performance of wet fabrics has primarily focused on porous and highly moisture-absorbent knitted fabrics. In this study, we investigated the thermal performance of various types of woven fabrics commonly used in clothing, including natural fiber fabrics such as cotton, linen, silk, and wool, as well as synthetic fiber fabrics such as nylon and polyester. We also included blended fabrics such as polyester/cotton, nylon/spandex, polyester/viscose, and polyester/ammonia. To prevent moisture evaporation from the fabric during thermal resistance testing, we sealed the fabric samples in flat bags (0.4 m $\times$ 0.3 m). The fabric samples were wetted with laboratory-made deionized water to ensure consistency across all samples. See Table 1 for details on all fabric parameters.

Table 1. Fabric specifications.

Symbol	Composition	Structure	Thickness (mm)	Weight (g/m^2)	Porosity
CO	Cotton 100%	Plain	0.78	121.87	0.8985
JU	Jute 100%	Twill	0.93	249.25	0.8213
SI	Silk 100%	Plain	0.63	72.94	0.9148
WO	Wool 100%	Twill	1.02	175.39	0.8697
PO	Polyester 100%	Plain	0.66	90.05	0.9011
NY	Nylon 100%	Plain	0.72	161.06	0.8037
PE	Polyester 90% + Elastane 10%	Plain	0.89	155.37	0.8739
PC	Polyester 65% + Cotton 35%	Plain	1.41	227.26	0.8877
AV	Acrylic 70% + Viscose 30%	Plain	0.89	200.62	0.8047
NE	Nylon 85% + Elastane 15%	Plain	0.77	136.69	0.8748

3.2. Sample Preparation

Measuring the maximum water absorption of a fabric sample is the first step in evaluating a fabric's ability to absorb and store moisture. The experimental method followed that by Tang et al. [19]. To ensure that the fabric was completely dry before measurement, all test fabrics were dried in an oven at 105 °C for 30 min to remove any excess moisture and achieve an "ultra dry state" as proposed by Naka et al. [11]. Then, the fabric sample was soaked in deionized water for 3 min and hung vertically until there were no liquid droplets falling for 30 s, indicating complete wetting of the fabric.

Based on the maximum water absorption of the fabric, the fabric saturation level, and the area of the fabric sample, the amount of water added to the fabric sample is determined. The fabric saturation level is defined as the percentage of the added water to the maximum water absorption of the fabric, reflecting the degree of water content saturation in the fabric. Since the testing time of the thermal resistance tester is relatively long, the evaporation of water from the test fabric during the test will not only affect the temperature and humidity of the surrounding environment but also cause a decrease in the moisture content of the fabric. To address this issue, the method proposed by Raccuglia et al. [20] was adopted to seal the wet fabric. Specifically, the ultradry fabric was placed flat in a sealed bag and a certain amount of deionized water was measured and poured into a humidifying spray bottle. The spray bottle was then suspended about 2 cm above the center of the fabric sample and sprayed evenly with deionized water. After 5 min of standing, the bag is sealed once the fabric is fully wetted.

In addition, the thermal resistance tests were conducted on the ultradry fabric and on fabrics with saturation levels of 20%, 40%, 60%, 80%, and 100%. If the saturation level of the test sample exceeded 100%, i.e., there is excess water between the fabric and the simulated skin, the humidification procedure is performed in two steps to avoid insufficient water due to excess water adhering to the inner wall of the sealed bag. First, the added water is determined based on the maximum water absorption of the sample, followed by adding the excess water using a micro pipette to the center of the sample before the thermal resistance test is performed.

3.3. Equipment and Methods

The thermal conductivity, thermal resistance, and insulation properties of fabric samples were measured using a textile heat transfer performance tester (Ningbo Textile Instrument Co., Ltd., Ningbo, China, YG606E, see Figure 1) in accordance with GB/T 11048-2018 (Textiles—Physiological effects—Measurement of thermal and water-vapour resistance under steady-state conditions (sweating guarded-hotplate test)) [21]. The instrument provides rapid measurement of both steady-state and transient thermal performance. To some extent, this instrument simulates the heat flow density q (W/m^2) from human skin to fabric when the fabric is initially in contact with the skin in an environment without forced convection. The fabric thickness was determined using a digital fabric thickness gauge (Wenzhou Interco Testing Instruments Co., Ltd., Wenzhou, China, YG(B)141D) in accordance with GB/T 3820-1997 (Determination of thickness of textiles and textile products) [22], whereas the surface density was determined using an electronic balance with an accuracy of ±0.1 g in accordance with GB/T 4743-2009 (Textiles—Yarn from packages—Determination of linear density (mass per unit length) by the skein method) [23]. The fabric samples were dried in a ventilated oven (Wenzhou Baien Instrument Co., Ltd., Wenzhou, China, Y802K) and sealed using a hand-press sealing machine; a humidifier was used.

When two fibers are blended during the fabric weaving process, the density of the blended fibers cannot be directly obtained. However, it can be estimated using the following equation proposed by Militky [24].

$$\rho_{ab} = r\rho_a + (1-r)\rho_b \tag{16}$$

where a and b represent the types of fibers, ρ_a and ρ_b represent their fiber densities, r represents the proportion of fiber a, and ρ_{ab} represents the density of the blended fiber. All tests were conducted under laboratory conditions with a temperature of 20 ± 1 °C, relative humidity of 50 ± 5%, and wind speed < 0.4 m/s.

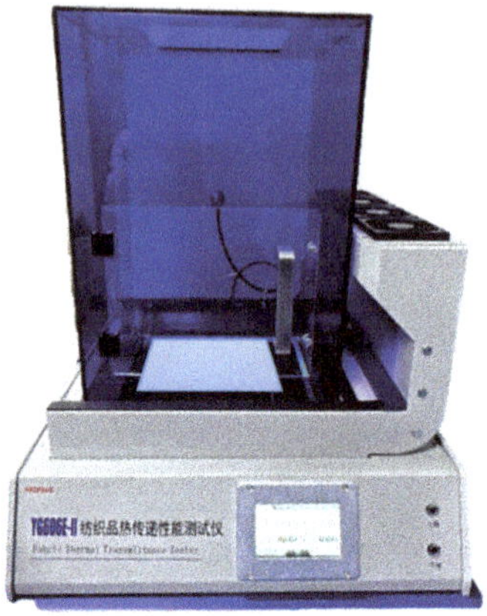

Figure 1. Textile heat transfer performance tester (test plate to do highlighting effect).

3.4. Thermal Properties Characterization

The thermal properties of textiles, such as the heat transfer coefficient (U), thermal resistance (R_t), and thermal insulation (Q), are influenced by factors such as the fabric type and environment. In this study, a textile heat transfer performance tester was used to measure these properties in a constant temperature and humidity room. The laboratory temperature was set at 25 ± 1 °C, with a humidity of 65 ± 5% and a wind speed less than 0.4 m/s. The temperature of the test plate was set at 35 °C, and square fabric samples measuring 0.3 m × 0.3 m were prepared for each fabric type. Before each day's experiment, a blank test was conducted, then the instrument chamber was opened and the sealed bag with the wetted fabric was laid flat onto the test plate of the thermal resistance tester. During the experiment, care was taken to maintain stable air inside the instrument as much as possible. After a period of testing, the thermal properties of the fabric could be read from the instrument panel. Each experiment was performed three times and the results were averaged, with a coefficient of variation between the test results being less than or equal to 3%.

The heat transfer coefficient of a fabric refers to the heat flux passing through a unit area of fabric when there is a surface temperature difference of 1 °C [25]. The formula for calculating the heat transfer coefficient is as follows:

$$U = \frac{U_{bp} \times U_1}{U_{bp} - U_1} \tag{17}$$

where U is the heat transfer coefficient of the fabric, W/(m^2·°C), U_{bp} is the heat transfer coefficient of the experimental plate without specimen, W/(m^2·°C), and U_1 is the heat transfer coefficient of the experimental plate with specimen, W/(m^2·°C).

The thermal resistance is the reciprocal of the heat transfer coefficient, and its conversion formula for thermal resistance is:

$$R_t = \frac{1}{U} \tag{18}$$

$$R_{tm} = R_{ct} - R_{ct0} \tag{19}$$

where R_{tm} is the actual measured thermal resistance of the fabric in m^2·°C/W; R_{ct} is the final thermal resistance test reading of the thermal resistance tester, which is the sum of the fabric thermal resistance and the sealing bag thermal resistance in m^2·°C/W; and R_{ct0} is the thermal resistance of the sealing bag in $m^2 \cdot °C/W$, which can be directly tested by the thermal resistance tester. The thermal resistance of the sealing bag R_{ct0} is 12.8×10^{-3} $m^2 \cdot °C/W$.

The thermal insulation rate is the percentage of the difference between the heat dissipation with and without the sample to the heat dissipation without the sample. It is related to the thermal conductivity and porosity of the fibers and reflects the ability of the fabric to prevent the loss of body heat. According to the latest textile thermal insulation testing standard GB/T 35762-2017 (Textiles—Test method for thermal transmittance—Flat plate test), the equation for calculating the thermal insulation rate is as follows [26]:

$$Q = \frac{W_2 - W_1}{W_1} \times 100\% \tag{20}$$

where W_1 is the heat dissipation of the blank test plate (W/°C) and W_2 is the heat dissipation of the sample test plate (W/°C).

4. Result and Discussion

4.1. Effect of Water Content Saturation on Fabric Thermal Resistance

In this study, the thermal performance changes of ten different fiber types of woven fabrics were measured at various humidity levels. The measurement results of thermal resistance and insulation rate in dry and wet states are shown in Figures 2 and 3; each sample was measured three times. The coefficient of variation (CV) for the measurement results was less than 3%. As shown in the figures, the fabrics deteriorated in thermal performance after wetting and the thermal resistance and insulation rate both decreased as the water content saturation level of the fabric increased. Compared with the thermal resistance of all fabrics in the ultradry state, when the water content saturation level reached 20%, the thermal resistance significantly decreased in the range of 8.3~12.1 $\times 10^{-3}$ $m^2 \cdot °C/W$, with an average decrease of 42.5%. Subsequently, when the water content saturation level reached 40%, 60%, 80%, and 100%, the average decreases in thermal resistance were 61.8%, 71.5%, 74.9%, and 77.8%, respectively. It was observed that the degree of decrease in thermal resistance gradually decreased as the water content saturation level of the fabric increased; this ultimately decreased to 22.2% of the thermal resistance in the ultradry state. This may be because a small amount of water in the fabric provides a shortcut for heat transfer when the water content saturation level of the fabric is low, significantly increasing the thermal conductivity of the fabric. As the water content saturation level increases, water molecules gradually enter the interior of the fabric until all the voids in the fabric system are occupied [27]. The thermal conductivity of fabric fibers generally ranges from 0.033 to 0.100 W/(m·°C) according to literature [17]. Air, as a poor conductor of heat, has a stable thermal conductivity of only 0.026 W/(m·°C) at 20 °C, whereas the thermal conductivity of still water at 20 °C is 0.57 W/(m·°C), which is 22 times higher than that of air. The specific heat capacity of water is 3431 times that of air at 37 °C [28]. When water occupies the entire fabric system, it not only destroys the loose structure of the fabric but also significantly reduces the thermal resistance of the fabric due to its higher thermal conductivity. When a large number of water molecules are adsorbed on the surface of the fabric, the fabric surface becomes moist [29]. As the proportion of adsorbed water in the fabric increases, the impact on thermal resistance gradually becomes less significant because the adhesion force and the action force of adsorbed water are weak [10]. For blended fabrics, due to the different blending proportions of hydrophilic and hydrophobic fibers, it can be found that the proportion of hydrophilic fibers is greater than hydrophobic fibers in materials such as NE (Nylon 85% + Elastane 15%), whose thermal resistance loss after complete wetting is 84.21% of the ultradry-state thermal resistance, whereas the proportion of hydrophobic fibers is greater than hydrophilic fibers in materials such as PC (Polyester 65% + Cotton 35%), whose thermal resistance loss was 68.70% of the ultradry-state thermal resistance. This finding may indicate that the thermal resistance loss after wetting is greater for blended fabrics with a higher percentage of hydrophilic fibers.

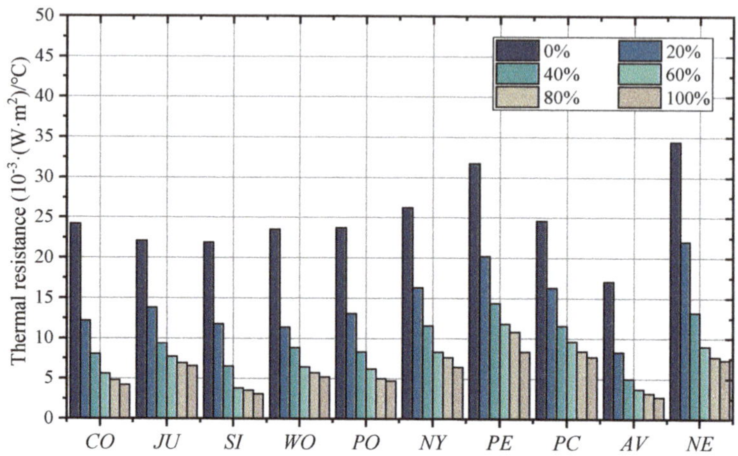

Figure 2. Thermal resistance of the fabrics at different water content saturation.

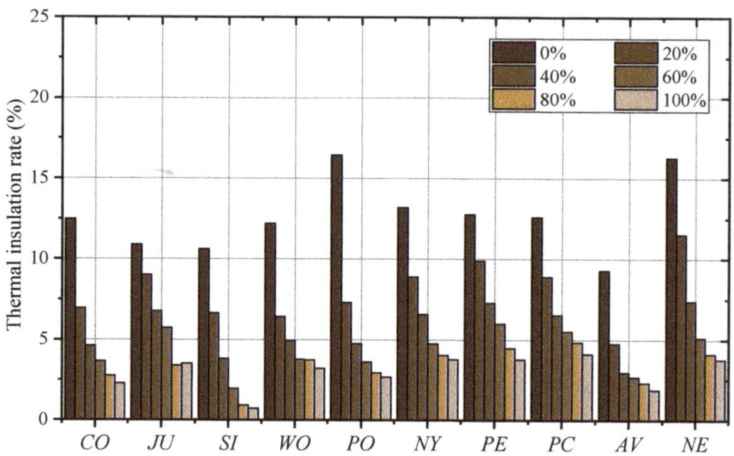

Figure 3. Thermal insulation rate of the fabrics at different water content saturation.

4.2. Selection of a Theoretical Model for Thermal Resistance

Exploring the accuracy of different models in predicting the thermal resistances of wet fabrics, we conducted statistical comparisons between the predicted values from each model and the experimental values. We used three methods: sum of squared deviations (*SSD*), sum of absolute deviations (*SAD*), and coefficient of determination (R^2), according to standard procedures, to analyze the accuracy of the model predictions. Tables 2–4 compare the predicted thermal resistance values and actual test values for Model 1 to Model 8, with the two best results highlighted in bold. The results show that all three evaluation methods gave consistent results and that Model 5 was the closest to the actual test values for all ten fabric samples ($R^2 \geq 0.955$). This model's characteristics are that the air resistance and water resistance are connected in parallel, followed by a series connection with the fiber resistance. Although Model 7 did not perform as well as Model 5 ($R^2 \geq 0.917$), it still demonstrated good predictive ability. This model's characteristics are that the water resistance and fiber resistance are connected in series, followed by a parallel connection with the air resistance. Therefore, we consider that the predictions of Model 5 and Model 7 for the thermal resistance of moist fabrics have some reference value. This finding is similar to the results of Mangat et al. [3], indicating that the use of woven fabric samples is also applicable to Model 5 in this study.

Table 2. Sum of squares of deviations of each model.

Symbol	Model 1	Model 2	Model 3	Model 4	Model 5	Model 6	Model 7	Model 8
CO	0.933199	0.000101	0.000904	0.928485	**0.000002**	0.000021	**0.000008**	0.000085
JU	1.219922	0.000149	0.000977	1.212564	**0.000007**	1.212564	**0.000019**	0.000048
SI	1.22135	0.001666	0.000076	0.000462	**0.000003**	0.000027	**0.000012**	0.000056
WO	0.859346	0.011524	0.005531	0.859995	**0.000009**	0.010149	**0.000023**	0.011331
PO	1.001217	0.000123	0.000901	0.998148	**0.000011**	0.000015	**0.000013**	0.000121
NY	0.774761	0.000105	0.000676	0.770626	**0.000012**	0.000019	**0.000021**	0.000097
PE	0.639571	0.000084	0.000565	0.637524	**0.000008**	0.000012	**0.000011**	0.000821
PC	0.605522	0.000072	0.000581	0.602423	**0.000005**	0.000017	**0.000009**	0.000061
AV	0.843322	0.005482	0.001254	0.598752	**0.000014**	0.000017	**0.000027**	0.000121
NE	0.755424	0.000382	0.000951	0.754122	**0.000022**	0.000017	**0.000034**	0.000201

Table 3. Sum of absolute deviations of each model.

Symbol	Model 1	Model 2	Model 3	Model 4	Model 5	Model 6	Model 7	Model 8
CO	1.101330	0.019659	0.033001	1.053953	**0.003696**	0.010501	**0.005818**	0.015354
JU	1.255706	0.025461	0.036638	1.202308	**0.004415**	1.202308	**0.008084**	0.015765
SI	1.263636	0.008987	0.018391	1.024102	**0.003267**	0.012435	**0.006348**	0.012336
WO	1.089659	0.154268	0.106831	1.120350	**0.005283**	0.145965	**0.009126**	0.146836
PO	1.139877	0.017560	0.034325	1.094537	**0.006141**	0.007583	**0.007124**	0.017053
NY	1.005387	0.016181	0.031567	0.961944	**0.007786**	0.008308	**0.009639**	0.016016
PE	0.909487	0.015569	0.027219	0.873232	**0.005527**	0.007134	**0.006164**	0.014718
PC	0.886296	0.016374	0.027933	0.847941	**0.004618**	0.009027	**0.006348**	0.013441
AV	0.756122	0.022356	0.032515	0.235551	**0.003155**	0.004655	**0.008759**	0.012581
NE	0.842523	0.015666	0.012111	0.616515	**0.004235**	0.023151	**0.007989**	0.014587

Table 4. Correlation between the measured and predicted values.

Symbol	Model 1	Model 2	Model 3	Model 4	Model 5	Model 6	Model 7	Model 8
CO	−0.365	0.880	0.958	−0.400	**0.985**	0.883	**0.979**	−0.038
JU	−0.382	0.838	0.914	−0.410	**0.999**	0.811	**0.956**	−0.046
SI	−0.349	0.841	0.951	−0.346	**0.987**	0.876	**0.961**	−0.059
WO	−0.292	0.816	0.894	−0.324	**0.976**	0.933	**0.939**	−0.066
PO	−0.310	0.839	0.921	−0.344	**0.958**	0.858	**0.961**	−0.068
NY	−0.310	0.850	0.920	−0.344	**0.984**	0.856	**0.959**	−0.070
PE	−0.384	0.860	0.937	−0.418	**0.972**	0.871	**0.971**	−0.094
PC	−0.441	0.850	0.907	−0.473	**0.997**	0.846	**0.939**	−0.182
AV	−0.522	0.872	0.891	−0.423	**0.965**	0.817	**0.921**	−0.132
NE	−0.221	0.897	0.905	−0.373	**0.955**	0.873	**0.916**	−0.061

The coefficient of determination (R^2), sum of squared deviations (SSD), and sum of absolute deviations (SAD) were calculated as follows:

$$R^2 = S_{xy}^2 / S_x^2 S_y^2 \tag{21}$$

$$SSD = \sum_{i=1}^{n} (R_{tm,i} - R_t)^2 \tag{22}$$

$$SAD = \sum_{i=1}^{n} |R_{tm,i} - R_t| \tag{23}$$

where $R_{tm,i}$ represents the actual thermal resistance of the fabric measured by the experimental instrument, in units of $(W{\cdot}m^2)/{}^{\circ}C$, i indicates the order of the actual thermal resistance test, and R_t is the predicted thermal resistance value of the model, in units of $(W{\cdot}m^2)/{}^{\circ}C$.

4.3. Modification of Thermal Resistance Model

According to the assumptions of the theoretical model (i.e., the voids in the fabric are uniform and constant), the model was revised by using the water content saturation instead of the moisture content to characterize the proportion of water in the fabric system voids. The study compared the fitting relationship between the prediction curve and experimental measurements before and after the revision of Model 5, as shown in Figure 4. From the figure, it can be clearly seen that the prediction curve is generally lower than the measured value, especially when using the original model. The revised model can obtain results closer to the measured value than the original model. Among the 10 sets of revised model prediction curves, 8 sets (80%) had R^2 values ranging from 0.95 to 0.99. Only one set of the revised model (NE group) had an R^2 value lower than 0.90 ($R^2 = 0.896$), which was higher than the original model ($R^2 = 0.824$) by 0.072. However, this was still higher than the acceptable range of R^2 values required by correlation analysis ($R^2 > 0.8$) [30]. It was also found that, especially at low water content saturation levels, the slope of the prediction curve of the revised model was smaller than that of the original model and the downward trend was more gentle, which was in line with the experimental test results. Comparison of the model prediction curve with the experimental results by Mangat et al. [3] also revealed that the original model always underestimated the thermal resistance compared with the actual measurement. Using water content saturation instead of moisture content can effectively improve the prediction accuracy of the model. The experimental results support the applicability of the model in predicting the wet-state thermal resistance of single-layer woven fabrics, indicating a great consistency between the measured and predicted values. Therefore, this study suggests that the revised Model 5 be used as the optimal model for predicting the wet-state thermal resistance of fabrics.

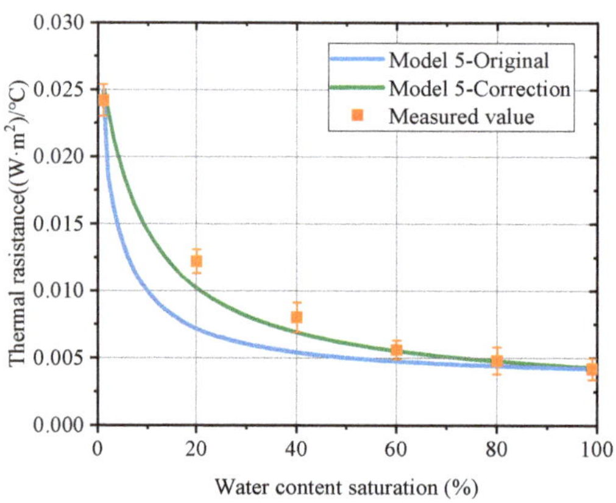

CO

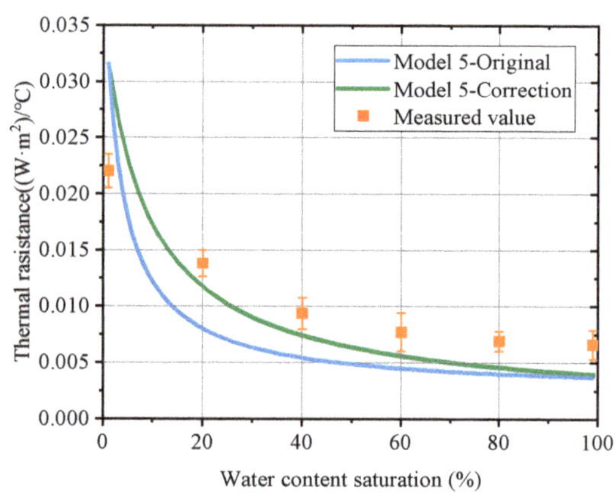

JU

Figure 4. *Cont.*

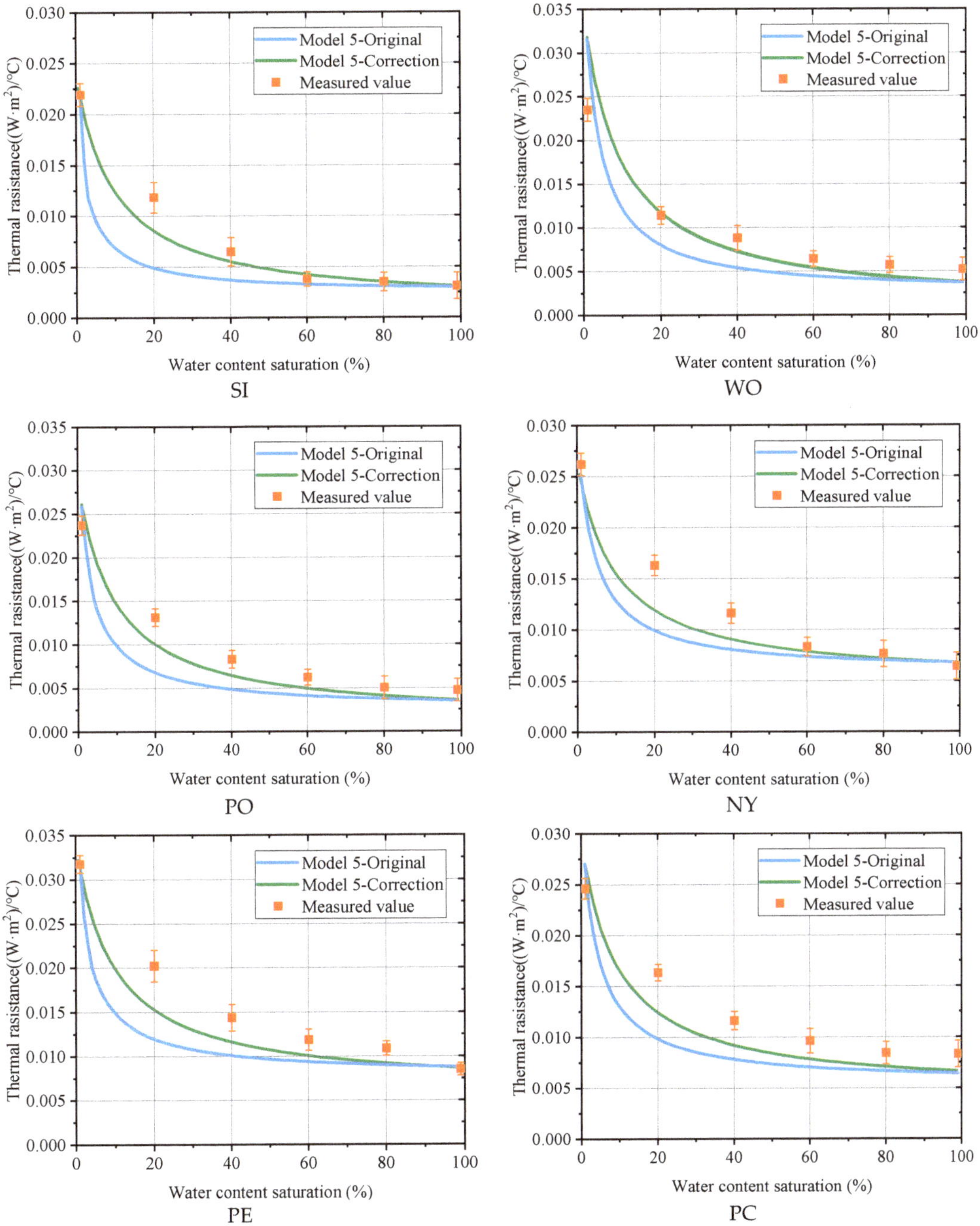

Figure 4. *Cont.*

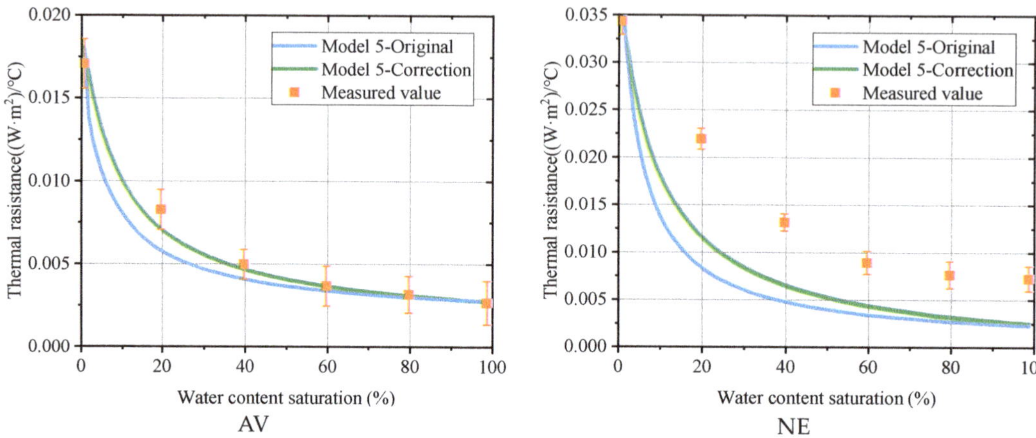

Figure 4. Simulated and measured values of the thermal resistance.

4.4. Relationship between the Theoretical Models and Test Results

Exploring the composition of the fabric thermal resistance and the effect of moisture on fabric thermal resistance, we explain the changes in thermal resistance observed in our experiments based on a theoretical model. As shown in Figure 4, the prediction curve of Model 5 more clearly illustrates the rate at which fabric thermal resistance deteriorates after absorbing moisture. According to the trend of Model 5's curve, we found that as the fabric's water saturation level increases, its thermal resistance significantly decreases. Moreover, we observed that the thermal resistance curve changes significantly around a water saturation level of 20%. Therefore, the fabric moisture absorption can be roughly divided into two processes: the initial stage of moisture absorption when the water saturation level is between 0~20% and the later stage of moisture absorption when the water saturation level is above 20%. The initial stage is characterized by a rapid decline in thermal resistance, with a decrease of approximately 51.92% to 64.28% compared with the ultradry state. The later stage is characterized by a slow decline in thermal resistance, with a decrease of only approximately 23.08% to 27.14% compared with the ultradry state.

According to the thermal resistance composition analysis of Model 5, the air thermal resistance and water thermal resistance are parallel components, whereas the fiber thermal resistance is a series component. The primary factor that affects fabric thermal resistance is the composition of the filler in the fabric's void spaces. The reason for the abrupt decline in the first half of the thermal resistance curve during the initial stage of moisture absorption is that the air content in the fabric's void spaces gradually decreases, whereas the water content gradually increases. Therefore, the factors that determine the amount of void space in the fabric, such as fabric thickness and porosity, become critical in affecting the thermal resistance of the wet fabric. When humidity changes, the fiber thermal resistance in the fabric's total thermal resistance does not change, but the composition of air thermal resistance and water thermal resistance does change. Therefore, the fiber's ability to adsorb water is also an important factor affecting the change in thermal resistance of wet fabrics. When the fiber becomes wet, its physical and chemical properties change, leading to significant microscopic effects on the system's internal energy, which manifests as a significant change in thermal conductivity. When a small amount of water enters the fabric, it first forms strong hydrogen bonds with hydrophilic fibers and then penetrates into the microporous structure of the porous fibers [31]. With an increase in moisture content, the water gradually fills the large pores formed by the fabric's tissue, forming numerous water columns that provide many transmission channels to accelerate heat transfer. After filling the void spaces of the fabric with water, excess water will accumulate on the fabric's surface due to its adsorption effect, resulting in a slow decline in the fabric's thermal resistance.

However, the effect of water absorption on thermal resistance is significantly smaller than that of moisture absorption, as indicated by the gentle slope of the curve in the latter stage.

4.5. Limitations and Applications

The theoretical model of fabric thermal resistance is not a perfect simulation of the real fabric structure due to the inherent complexity of the fabric itself. When the filling material in the fabric is only water or air, the thermal resistance of the fabric is composed of only two components: fiber thermal resistance and filling material thermal resistance, with the total fabric thermal resistance being the sum of the two. When the wetting is incomplete, the distribution of water and air inside the fabric becomes more complex. Water, which possesses surface tension and viscosity compared with air [32], has a more complicated contact situation with the fibers, and the number and position of the contact points have an impact on the overall thermal resistance of the fabric. This is difficult to measure or calculate; instruments can only measure the average thermal resistance of the fabric. As a part that is favorable for heat transfer is added to the filling material, the overall thermal resistance of the fabric decreases. However, as the proportion of one component continues to increase, the rate of change of the overall thermal resistance with respect to that component also increases, reflecting the heat conduction ability of that component [33]. In summary, the thermal resistance with respect to changes in water content cannot ignore the influence of the various components in the filling material. Such situations are difficult to predict and can only be derived from empirical models. The modified Model 5 can be used to approximate the thermal resistance of fabrics in the wet state and is a good alternative empirical model.

5. Conclusions

The thermal resistance of ten typical fabrics at different moisture levels was measured in this study to observe the changes in thermal resistance of woven fabrics as they are gradually humidified. The study found that as the humidity of the fabric increased, the thermal parameters including thermal resistance and thermal insulation rate deteriorated. The thermal resistance and thermal insulation rate of completely wetted fabrics were only 22.2% and 23.4% of that in the ultradry state. Based on Mangat et al.'s model, a model was selected and improved to predict the thermal resistance of woven fabrics at different moisture levels. After replacing the moisture content with water content saturation, the thermal resistance prediction model showed a more accurate prediction ability. The modified model had a significant improvement compared with the original model, as demonstrated by the correlation coefficient (R^2) test. The improved Model 5 fills the gap in predicting the thermal resistance of woven fabrics in a wet state. These results indicate that the modified Model 5 can be used to predict the thermal resistance of various types of woven fabrics under different moisture levels. This method provides an effective reference for quantifying the impact of moisture on the thermal resistance of woven fabrics and provides a theoretical basis for evaluating the thermal and moisture comfort of fabrics. Future research will consider using physiological saline to humidify fabrics to predict the thermal resistance of fabrics after being humidified by human sweat and to test other types of mathematical models using this method.

Author Contributions: Conceptualization, Z.W. and Y.S.; methodology, R.Y.; formal analysis, Z.W.; investigation, S.F.; resources, X.Q.; data curation, S.F.; writing—original draft preparation, Z.W.; writing—review and editing, Z.W.; visualization, Z.W.; supervision, X.Q.; project administration, Y.S.; funding acquisition, R.Y. All authors have read and agreed to the published version of the manuscript.

Funding: This research was funded by [National Natural Science Foundation of China] grant number [No. U1933111] and [Tianjin Research Innovation Project for Postgraduate Students] grant number [No.2021YJSO2B06].

Institutional Review Board Statement: Not applicable.

Informed Consent Statement: Not applicable.

Data Availability Statement: Not applicable.

Conflicts of Interest: The authors declare no conflict of interest.

References

1. Yang, R.; Wu, Z.; Qian, X.; Shi, Y. Analysis and evaluation of the thermal performance of the combinations of suits and life preservers in water. *Text. Res. J.* **2023**, *93*, 1043–1056. [CrossRef]
2. Wang, F.; Shi, W.; Lu, Y.; Song, G.; Rossi, R.M.; Anaheim, S. Effects of moisture content and clothing fit on clothing apparent 'wet' thermal insulation: A thermal manikin study. *Text. Res. J.* **2016**, *86*, 57–63. [CrossRef]
3. Mangat, M.M.; Hes, L. Thermal resistance of denim fabric under dynamic moist conditions and its investigational confirmation. *Fibres Text. East. Eur.* **2014**, *22*, 101–105.
4. Li, Y.; Zhu, Q.; Yeung, K.W. Influence of thickness and porosity on coupled heat and liquid moisture transfer in porous textiles. *Text. Res. J.* **2002**, *72*, 435–446. [CrossRef]
5. Mangat, M.M.; Hes, L.; Bajzık, V. Thermal resistance models of selected fabrics in wet state and their experimental verification. *Text. Res. J.* **2015**, *85*, 200–210. [CrossRef]
6. Hes, L.; Loghin, C. Heat, Moisture and Air Transfer Properties of Selected Woven Fabrics in Wet State. *J. Fibre Bioeng. Inform.* **2009**, *2*, 141–149.
7. Oğlakcioğlu, N.; Marmarali, A. Thermal comfort properties of some knitted structures. *Fibres Text. East. Eur.* **2007**, *15*, 64–65.
8. Wang, F.; Lai, D.; Shi, W.; Fu, M. Effects of fabric thickness and material on apparent 'wet' conductive thermal resistance of knitted fabric 'skin' on sweating manikins. *J. Therm. Biol.* **2017**, *70*, 69–76. [CrossRef]
9. Akcagun, E.; Bogusławska-Bączek, M.; Hes, L. Thermal insulation and thermal contact properties of wool and wool/PES fabrics in wet state. *J. Nat. Fibers* **2019**, *16*, 199–208. [CrossRef]
10. Yang, R.; Wu, Z.; Qian, X.; Shi, Y. Development of thermal resistance prediction model and measurement of thermal resistance of clothing under fully wet conditions. *Text. Res. J.* **2023**, *93*, 911–924. [CrossRef]
11. Naka, S.; Kamata, Y. Thermal conductivity of wet fabrics. *J. Text. Mach. Soc. Jpn.* **1977**, *23*, 114–119. [CrossRef]
12. Hes, L.; De Araujo, M. Simulation of the effect of air gaps between the skin and a wet fabric on resulting cooling flow. *Text. Res. J.* **2010**, *80*, 1488–1497. [CrossRef]
13. Mansoor, T.; Hes, L.; Bajzik, V.; Noman, M.T. Novel method on thermal resistance prediction and thermo-physiological comfort of socks in a wet state. *Text. Res. J.* **2020**, *90*, 17–18. [CrossRef]
14. Wang, F. Comments on "important of air spaces when comparing fabric thermal resistance". *Text. Res. J.* **2012**, *82*, 521. [CrossRef]
15. Boguslawska-Baczek, M.; Hes, L. Effective water vapour permeability of wet wool fabric and blended fabrics. *Fibres Text. East. Eur.* **2013**, *1*, 67–71.
16. Hes, L.; Dolezal, I. Precise measurement of water vapour permeability of wet fabrics. In Proceedings of the AUTEX International Textile Conference, Tampere, Finland, 26–28 June 2007.
17. Sugawara, A.; Yoshizawa, Y. An investigation on the thermal conductivity of porous materials and its application to porous rock. *Aust. J. Phys.* **1961**, *14*, 469. [CrossRef]
18. Yang, R.; Wang, L.; Zou, C.; Li, S.; Geng, D. Life preservers: Concepts, progress, and challenges. *Int. J. Aerosp. Psychol.* **2020**, *30*, 77–88. [CrossRef]
19. Tang, M.; Chau, K.; Kan, C.; Fan, J.T. Magnitude estimation approach for assessing stickiness sensation perceived in wet fabrics. *Fibers Polym.* **2018**, *19*, 2418–2430.
20. Raccuglia, M.; Pistak, K.; Heyde, C.; Qu, J.; Mao, N.; Hodder, S.; Havenith, G. Human wetness perception of fabrics under dynamic skin contact. *Text. Res. J.* **2017**, *88*, 2155–2168. [CrossRef]
21. *GB/T 11048–2018*; Textiles-Physiological Effects-Measurement of Thermal and Water-Vapour Resistance under Steady-State Conditions (Sweating Guarded-Hotplate Test). China Standard Press: Beijing, China, 2018.
22. *GB/T 3820–1997*; Determination of Thickness of Textiles and Textile Products. China Standard Press: Beijing, China, 1997.
23. *GB/T 4743–2009*; Textiles–Yarn from Packages–Determination of Linear Density (Mass Per Unit Length) by the Skein Method. China Standard Press: Beijing, China, 2009.
24. Militky, J. Prediction of textile fabrics thermal conductivity. In *Thermal Manikins and Modelling*; Fan, J., Ed.; The Hong Kong Polytechnic University: Hongkong, China, 2006.
25. Shi, Y.; Wang, L.; Qian, X. Effect of non-uniform skin of "Walter" on the evaporative resistance and thermal insulation of clothing. *Int. J. Cloth. Sci. Technol.* **2017**, *29*, 686–695. [CrossRef]
26. *GB/T 35762-2017*; Textiles-Test Method for Thermal Transmittance-Flat Plate Test. China Standard Press: Beijing, China, 2017.
27. Wei, J.; Xu, S.; Liu, H.; Zheng, L.; Qian, Y. Simplified model for predicting fabric thermal resistance according to its microstructural parameters. *Fibers Text. East. Eur.* **2015**, *23*, 57–60.
28. Transportation Safety Board (TSB). Loss of control and collision with water Cochrane Air Service de Havilland DHC-2 Mk.1, C-FGBF Lillabelle Lake, Ontario, 25 May 2012 (Report No. A12O0071). In *Aviation Investigation Report*; Transportation Safety Board (TSB): Ottawa, ON, Canada, 2012.
29. Mangat, M.M.; Militky, J.; Hes, L. Thermal resistance of cotton denim fabric under various moisture conditions. *Compar. Wet. Milling. Action. Fibrous. Solid. Mat.* **2012**, *16*, 35.

30. Wu, Z.; Yang, R.; Qian, X.; Yang, L.; Lin, M. A multi-segmented human bioheat model under immersed conditions. *Int. J. Therm. Sci.* **2023**, *185*, 108029. [CrossRef]
31. Hes, L. Fundamentals of design of fabrics and clothing with demanded thermophysiological comfort. In Proceedings of the International Round Table Clothing Comfort-Condition of Life Quality, Iasi University, Iasi, Romania, 29–31 October 2009.
32. Richards, M.G.M.; Rossi, R.; Meinander, H.; Broede, P.; Candas, V.; den Hartog, E.; Holmér, I.; Nocker, W.; Havenith, G. Dry and wet heat transfer through clothing dependent on the clothing properties under cold conditions. *Int. J. Occup. Saf. Ergon.* **2008**, *14*, 69–76. [CrossRef] [PubMed]
33. Du, N.; Fan, J.; Wu, H. Optimum porosity of fibrous porous materials for thermal insulation. *Fibers Polym.* **2008**, *9*, 27–33. [CrossRef]

Disclaimer/Publisher's Note: The statements, opinions and data contained in all publications are solely those of the individual author(s) and contributor(s) and not of MDPI and/or the editor(s). MDPI and/or the editor(s) disclaim responsibility for any injury to people or property resulting from any ideas, methods, instructions or products referred to in the content.

processes

MDPI

Article

Optimization of Deep Learning Models for Enhanced Respiratory Signal Estimation Using Wearable Sensors

Jiseon Kim [1] and Jooyong Kim [2,*]

[1] Department of Smart Wearables Engineering, Soongsil University, Dongjak-gu, Seoul 06978, Republic of Korea
[2] Department of Material Science and Engineering, Soongsil University, Dongjak-gu, Seoul 06978, Republic of Korea
* Correspondence: jykim@ssu.ac.kr; Tel.: +82-2-820-0631

Abstract: Measuring breathing changes during exercise is crucial for healthcare applications. This study used wearable capacitive sensors to capture abdominal motion and extract breathing patterns. Data preprocessing methods included filtering and normalization, followed by feature extraction for classification. Despite the growing interest in respiratory monitoring, research on a deep learning-based analysis of breathing data remains limited. To address this research gap, we optimized CNN and ResNet through systematic hyperparameter tuning, enhancing classification accuracy and robustness. The optimized ResNet outperformed the CNN in accuracy (0.96 vs. 0.87) and precision for Class 4 (0.8 vs. 0.6), demonstrating its capability to capture complex breathing patterns. These findings highlight the importance of hyperparameter optimization in respiratory monitoring and suggest ResNet as a promising tool for real-time assessment in medical applications.

Keywords: deep learning; respiratory signal; wearable sensor; convolutional neural network; residual networks

Academic Editors: Zhanxiao Kang and Qing Chen

Received: 22 January 2025
Revised: 17 February 2025
Accepted: 20 February 2025
Published: 4 March 2025

Citation: Kim, J.; Kim, J. Optimization of Deep Learning Models for Enhanced Respiratory Signal Estimation Using Wearable Sensors. *Processes* **2025**, *13*, 747. https://doi.org/10.3390/pr13030747

Copyright: © 2025 by the authors. Licensee MDPI, Basel, Switzerland. This article is an open access article distributed under the terms and conditions of the Creative Commons Attribution (CC BY) license (https://creativecommons.org/licenses/by/4.0/).

1. Introduction

Respiratory changes during exercise are critical indicators in healthcare fields, such as sports monitoring. Previous studies have suggested that fast inputs, including central command, primarily regulate respiratory signal during exercise [1]. This regulation may explain the strong association between respiratory signal and perceived exertion across various exercise conditions. Emerging evidence highlights that respiratory signal can effectively reflect the level of effort during cycling exercise [1–3]. In this way, respiratory signal serves as a prominent indicator of physical effort during exercise, surpassing the relevance of other commonly monitored physiological variables [4]. For example, respiration during exercise is primarily regulated by central command, along with input from muscle afferent fibers and metabolic signals [5]. Respiration is influenced by inputs that operate at distinct timings in response to sudden changes in exercise intensity. Rapid increases in respiration at the onset of exercise are driven by fast inputs, including central command and afferent feedback, while metabolic stimuli contribute to respiration with a delayed response [6]. Evidence indicates that respiratory signal more accurately represents physical effort compared to blood lactate, particularly in conditions such as post-exercise muscle damage [7], glycogen depletion [8], and in individuals with McArdle's disease [9]. Thus, respiratory signal provides corresponding data on physical effort during exercise.

Moreover, measuring respiratory signals is not an easy task, making wearable devices, data preprocessing, and analysis methods such as deep learning critical factors in the measurement process. Wearable devices come in a variety of forms, including clothing, bracelets, and belts, each with their own advantages and disadvantages. Wearable respiratory sensors have gained significant attention for their ability to non-invasively monitor breathing patterns. Recent advancements, including deep learning-assisted portable biosensors, have enabled real-time respiratory analysis using compact and cost-effective devices [10]. Among these, clothing-based wearables are particularly well suited for everyday use. By utilizing lightweight and flexible materials, they allow for unrestricted movement, enabling the convenient measurement of body activity data, such as respiratory monitoring. For instance, a study developed a resistive sensor that can measure changes in breathing caused by emotional states. This was achieved by detecting changes in waist circumference, showing that clothing-based sensors can effectively capture respiratory changes associated with bodily movements [11]. Another study investigated textile-based sensors that measure breathing through contact resistance changes between fibers, even under different postural conditions [12]. Research on respiratory monitoring using clothing-based sensors is on the rise. However, while the classification of breathing patterns and data processing methods are critical, these aspects remain relatively underexplored.

Typically, respiratory signal estimation is directly performed from previously detected respiratory signals using deep learning, particularly Convolutional Neural Networks (CNNs) [13,14]. A CNN is a computational model composed of multiple processing layers capable of learning data representations [15]. This model enables tasks such as image classification, image segmentation, and action recognition. Despite their high applicability, CNNs occasionally encounter issues where neurons "die" during the learning process and fail to recover [16,17]. Additionally, as the model introduces more parameters, overfitting may occur, or training errors may increase [18]. To overcome these challenges, various methods have been proposed, one of which involves using a Residual Network (ResNet) [19,20]. Deep learning is utilized not only in medical image analysis but also in the classification of bio signals using wearable sensors, making it particularly useful for processing time series such as breathing patterns [21]. The key distinction of ResNet lies in the inclusion of shortcut connections within convolutional layers. These connections help address common CNN issues, allowing the gradient to backpropagate more effectively, thereby accelerating the training process [22]. As demonstrated, different training methods have distinct characteristics, and selecting the appropriate deep learning model that aligns with the research objective is essential. Additionally, choosing suitable parameters based on the characteristics of the data can significantly impact the accuracy of the results. Therefore, the selection of an appropriate deep learning model and parameter configuration that fit the research objective is crucial.

In this study, respiratory data generated during physical activity were trained using deep learning models such as CNN and ResNet to identify the most suitable approach for improving prediction accuracy. To measure respiration during exercise, a wearable respiration sensor was developed. The sensor was designed to be positioned at the front and back of the torso, specifically near the xiphoid process, to measure changes in capacitance caused by thoracic expansion and contraction during breathing. The respiration changes were observed according to different exercise conditions. To accurately classify and detect these changes, a comparative analysis of CNN and ResNet models was conducted. The method that demonstrated higher accuracy was considered the most appropriate for the given dataset. These findings highlight the development of an optimized system for selecting deep learning models that best suit the characteristics and objectives of the dataset, ultimately enhancing prediction accuracy. The research process is illustrated in Figure 1, providing a clear visualization of the study's workflow. This diagram helps readers easily understand the research objectives.

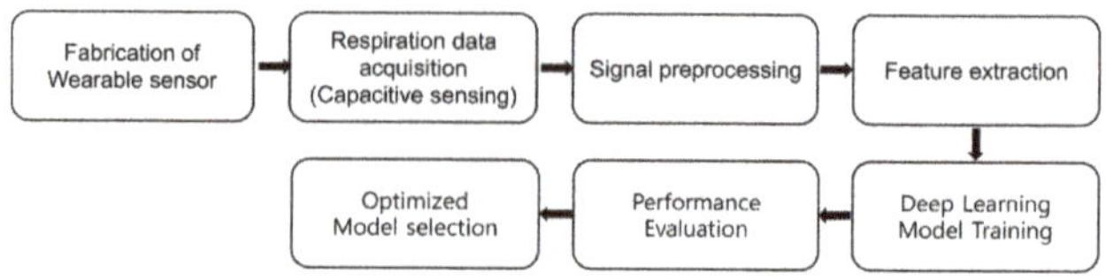

Figure 1. A schematic of the proposed work.

2. Materials and Methods

2.1. Respiration Measurement System

2.1.1. Breathing Mechanism

Understanding the mechanism of breathing is essential for accurate respiratory measurement. Equally important is the selection of a sensor measurement method that effectively corresponds to this mechanism. Various methods exist for measuring respiration, including monitoring airflow through the nose, capturing respiratory sounds, and tracking abdominal movement. Among these, this study focuses on measuring abdominal movement. This approach was chosen because it is the most suitable for wearable sensor applications and can be easily implemented in daily life.

In this method, respiration is measured using capacitive values (C) derived from abdominal movement (Equation (1)). The principle of measurement is as follows: when inhaling, air enters the lungs, causing the lungs to expand. This expansion leads to the spreading of the ribcage and the subsequent expansion of the thorax and abdomen. Conversely, when exhaling, air exits the lungs, the ribcage contracts, and the abdomen compresses. This process results in variations in abdominal volume, which correspond to changes in the distance between electrodes in a capacitive sensing system. By analyzing these differences in capacitive values, respiration can be assessed.

The two sensors were placed on the front and back of the body near the xiphoid process, where the abdominal movements caused by breathing can be most effectively measured. When the abdomen expands, the distance between the electrodes (d) is denoted as d_0, and during contraction, it is d_f. The difference between these distances, Δd, is used to analyze respiration (Equation (2)). The measurement principle and the breathing mechanism are illustrated in Figure 2 [23].

$$C = \frac{\varepsilon A}{d} \tag{1}$$

$$\Delta d = d_f - d_0 \tag{2}$$

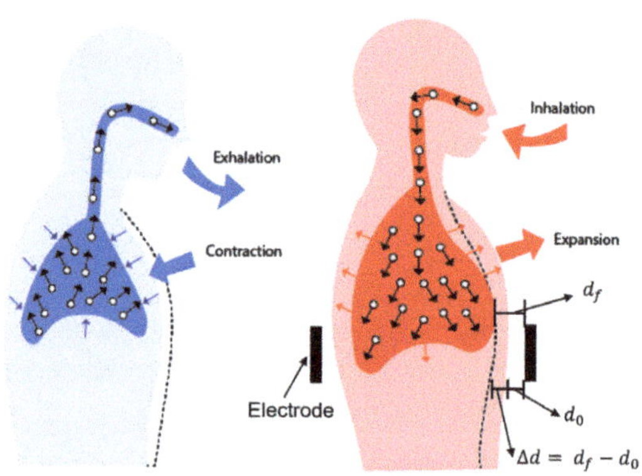

Figure 2. Abdominal movement in response to breathing [23].

2.1.2. Fabrication of Wearable Sensors

The electrodes were designed to have a rectangular shape with dimensions of 100 mm in width and 50 mm in height. The connectors for measurement were circular, measuring 20 mm × 20 mm. The stitch connecting the sensors to the connector was made in a zigzag pattern with a total length of 335 mm. The placement of the sensors is shown in Figure 3a, where the distances relative to reference points can also be observed. The area density of the embroidery (line/mm) was set to 6 for the sensor electrodes and 4.5 for the connectors. As the connection line consists of stitch lines, it does not have an embroidery area density. Regarding the stitch pattern, the electrodes and connectors were designed with a running stitch for the top thread and a fill stitch for the bobbin thread, while the connection line used a zigzag stitch.

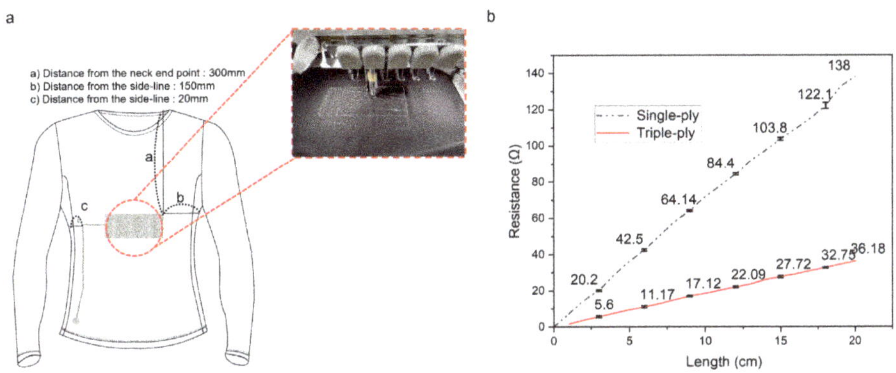

Figure 3. (**a**) Wearable sensors in the shape of a finished garment [23]; (**b**) the effect of single-ply and triple-ply thread length on the resistance value.

The stitch length for the electrodes and connectors was set to a width of 2 mm and a height of 4 mm, whereas the connection line had a width of 5 mm and a height of 1 mm. Finally, the total amount of thread used was 9526 stitches for the electrodes and 606 stitches for the connectors. This information is provided only for embroidered area measurements. Detailed specifications of the sensors are provided in Table 1 [24].

Table 1. Specifications of the wearable sensors.

Parameter	Electrodes	Connection Line	Connectors
Dimension (width × height)	100 × 50 (mm^2)	335 (mm^2)	20 × 20 (mm^2)
Density (line/mm)	6	-	4.5
Shape of stitch	Running, fill	Zigzag	Running, fill
Length of stitch (mm)	2, 4	5/1 (width/height)	2, 4
Number of stitches	9526	-	606

The electrodes of the sensors were directly embroidered onto fabric (87% polyester, 13% spandex) (Sumnfit, Seoul, Republic of Korea) using silver-coated thread (AMANN, Bönnigheim, Germany) and an embroidery machine (BROTHER, Bridgewater, NJ, USA). Silver-coated thread was used as the top thread, while 100% rayon embroidery thread was used as the bobbin thread. The top thread was placed on the inner side of the garment, which encounters the body, while the bobbin thread remained on the outer side. To protect the outer side, a polyurethane (PU) film was heat-bonded onto the surface.

As shown in Figure 3a [23], rectangular electrodes were directly embroidered onto the front and back of the garment near the xiphoid process. The electrode placement was chosen to align with the areas of the abdomen that exhibit the most significant volume changes during respiration. To prevent any obstruction to breathing during measurements, the connector was positioned at the bottom hem of the garment, allowing seamless connection to the measurement device. The connectors were symmetrically aligned on both the front and back of the garment. To connect the electrodes and the connectors, a zigzag stitch was used, creating a straight pathway from the rectangular electrodes to the circular connectors.

The silver-coated thread consists of 34 twisted nylon filaments. This thread is widely used in wearable applications due to its compatibility with embroidery [25] and sewing machines. It has also been evaluated as safe in cytotoxicity tests following the biological evaluation standard for medical devices (DIN EN ISO 10993-5) [26]. Therefore, it is suitable for placement on the inner side of garments where it comes into direct contact with the skin. In this regard, ply refers to the unit indicating the number of strands twisted together to form a thread. A single-ply thread consists of one twisted strand, while a triple-ply thread consists of three twisted strands. As shown in Figure 3b [27], the resistance of the thread decreases as the number of twists increases, demonstrating that the thread used in this study minimizes measurement errors caused by resistance.

2.2. Experimental Setup and Acquistion

The experiment was conducted to measure breathing under different intensity levels through cycling. The measurement process involved wearing a garment-based sensor and using an indoor cycling apparatus (XAMEN KANGTO FITNESS FOUPWENT CO, Xiamen, China). For precise measurements, an LCR meter (KEYSIGHT, Santa rosa, CA, USA) and the Keysight Benchvue program (KEYSIGHT, Santa Rosa, CA, USA) were employed instead of simple devices.

The intensity levels were divided into four categories: resting state (no activity), low intensity, moderate intensity, and high intensity. Low intensity corresponded to an average speed of 15–20 km/h, moderate intensity to 20–30 km/h, and high intensity to 30–40 km/h. Each intensity level was measured for 5 min, followed by a 1 min rest period before proceeding to the next measurement. The schematic diagram of the measurement setup is shown in Figure 4a [23], and the actual measurement process is illustrated in Figure 4b. The temperature was measured at room temperature (18 °C to 20 °C), and the experiment was conducted at approximately 60% humidity.

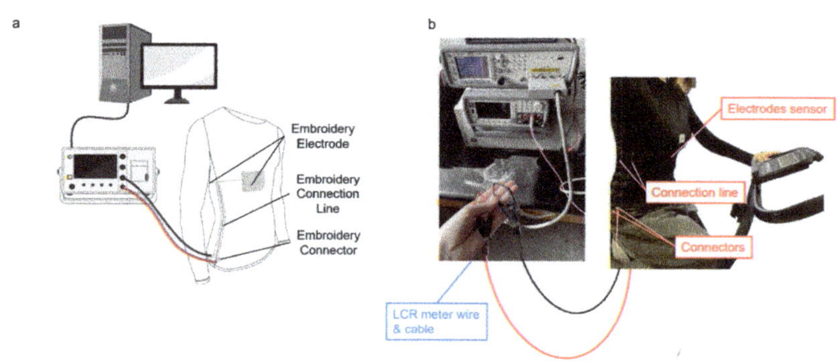

Figure 4. (**a**) Schematic of LCR meter [23]; (**b**) measurement of wearable sensors.

2.3. Signal Processing with Deep Learning

Signal processing was conducted using MATLAB R2023a. The data, representing the capacitive values of the sensor corresponding to respiration, were normalized to a range between −3 and 3 to enhance model generalization and improve training stability. Using the normalized data, classification results were analyzed with CNN and ResNet to determine the most suitable algorithm for the dataset used in this study. Figure 5 illustrates the hierarchical structures of the networks, where 5a represents the CNN architecture and 5b depicts the ResNet architecture.

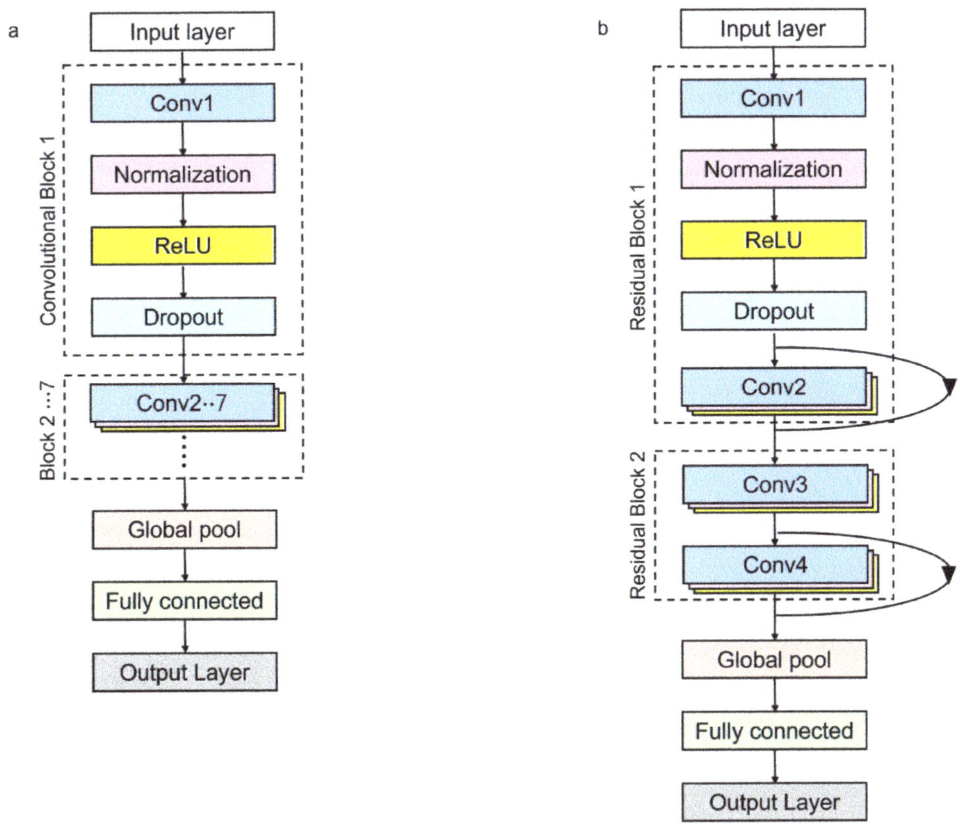

Figure 5. (**a**) CNN architecture; (**b**) ResNet architecture.

A CNN is an algorithm that processes multiple layers sequentially. The CNN used in this study consists of seven convolutional blocks, each containing convolution, normalization, ReLU, and dropout layers. After repeating these blocks seven times, the network concludes with a global pooling layer and a fully connected layer for final classification. Through the pooling layer, features can be directly extracted, and hierarchical learning makes it effective for capturing complex data. However, as mentioned in the introduction, issues such as neuron death and non-recovery occasionally occur during this process. To address these limitations, ResNet, which has a similar structure but overcomes such challenges, is one of the alternative algorithms.

The structure of ResNet consists of two residual blocks, each containing two sets of convolutional blocks. A distinguishing feature of ResNet is its ability to learn the residual function $F(x)$, which represents the difference between the input (x) and the output (y), unlike conventional neural networks. The mathematical representation of the residual function is provided in Equation (3), where $H(x)$ denotes the target function. Furthermore, as illustrated in Figure 5b, simple skip connections ensure that information is preserved during transmission, preventing performance degradation even as the network depth increases. The mathematical formulation of these skip connections can be found in Equation (4). After the residual blocks, the network concludes with global pooling and a fully connected layer, like CNN, to produce the final classification results.

$$F(x) = H(x) - x \tag{3}$$

$$y = F(x) + x \tag{4}$$

The total dataset consists of 157 samples, with 100 sets used for training, 24 sets for validation, and 23 sets for testing. Both algorithms classify four classes using a single input. The dropout rate is set at 0.5, with a filter size of 5 and 32 filters. The maximum number of epochs is configured as 200 for CNN and 1000 for ResNet. The initial learning rate is set to 0.0005 for CNN and 0.001 for ResNet.

Table 2 includes all materials and methods used in this work together with their purposes.

Table 2. Summary of analyzed materials and test devices used.

Component	Remark	Manufacture	Purpose
Electrodes	Silver-coated outer nylon cores covered by PU	AMANN, Bönnigheim, Germany	Creating sensing area of upper and lower plates
Embroidery machine	PR670E	BROTHER, Bridgewater, NJ, USA	Sewing electrodes, connection wires, and connectors
Indoor cycling apparatus	MKHB-01	XAMEN KANGTO FITNESS FOUPWENT CO, Xiamen, China	Device for measuring respiration based on exercise intensity
LCR meter	E4980AL	KEYSIGHT, Santa rosa, CA, USA	Measurement of capacitance values based on respiration
MATLAB	R2023a	MathWorks, Natick, MA, USA	Data classification and optimization using deep learning

3. Results and Discussion

Breathing data were collected under four conditions, resting, low intensity, moderate intensity, and high intensity, based on changes in cycling speed. Using this dataset, classification was performed using both CNN and ResNet algorithms, and their results were compared to determine the most suitable algorithm for this classification task. Figure 6 sequentially presents the capacitive data measured by the respiratory sensor for each condition: (a) resting, (b) low intensity, (c) moderate intensity, and (d) high intensity. Additionally, the results were derived by comparing accuracy, precision, recall, and the F1-score. Accuracy refers to the proportion of correctly predicted samples out of the total samples. Precision measures the proportion of true positive predictions among all positive predictions, while recall measures the proportion of true positive instances among all actual positive instances. The F1-score represents the harmonic meaning of precision and recall.

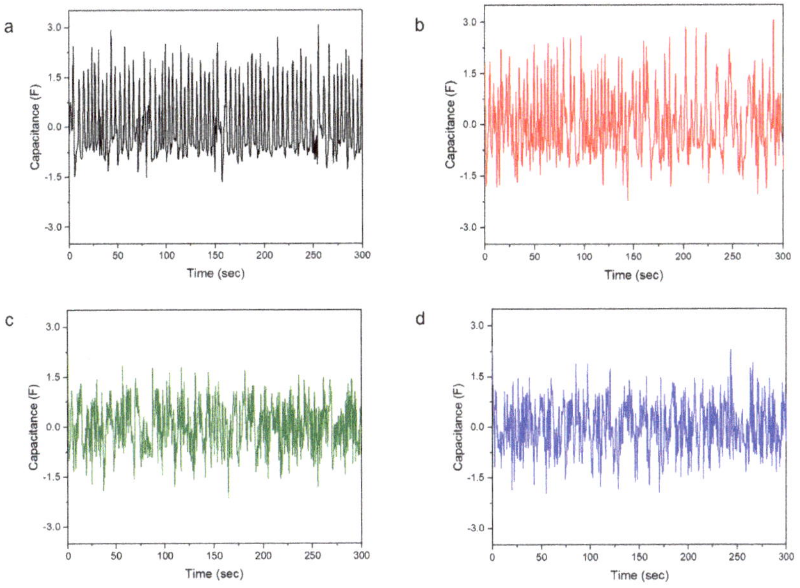

Figure 6. Breathing data under different conditions: (**a**) resting; (**b**) low intensity; (**c**) moderate intensity; and (**d**) high intensity.

3.1. Analysis of Results from CNN Algorithms

During the process of result derivation, the parameters and the number of blocks were adjusted to identify the most suitable configuration. In the initial model, CNN-1, the baseline number of blocks was set to seven, with the hypothesis that increasing the number of blocks would enhance training capacity and ultimately improve final accuracy. Deep learning training was conducted under this assumption.

Contrary to expectations, the training results failed to identify meaningful patterns, resulting in significantly low validation and test accuracy of 25%. Instead, all predictions were classified as resting (Class 1) breathing. These outcomes are visualized in Figure 7, where (a) displays the training results window, and (b) presents the confusion matrix.

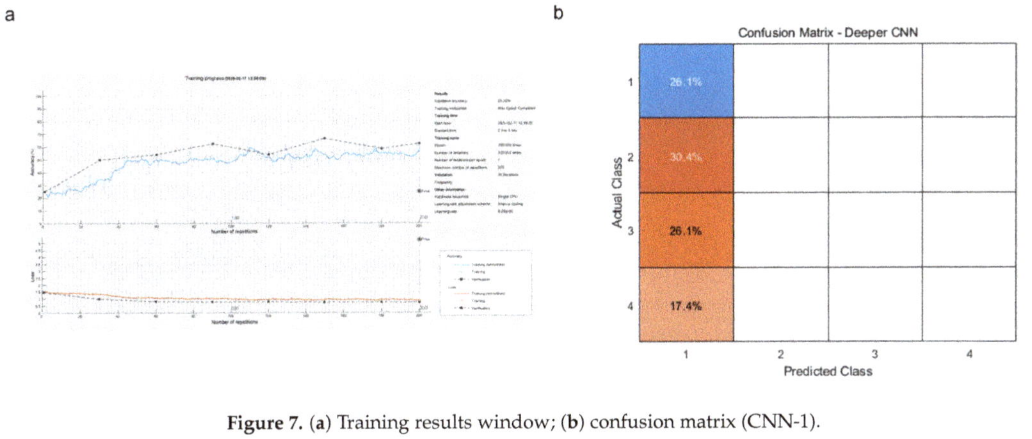

Figure 7. (**a**) Training results window; (**b**) confusion matrix (CNN-1).

This result suggests that increasing the number of blocks led to overfitting, thereby degrading generalization performance. Additionally, it may be attributed to information loss during propagation due to neuron deletion, which was previously identified as an issue.

In the second training session (O-CNN), improvements were made by fixing the number of blocks at seven and varying the number of epochs between 150 and 300 to identify the point where accuracy steadily increased without overfitting. Ultimately, the most stable gradient was achieved at 200 epochs. The final validation accuracy reached 79.16%, while the test accuracy was 83.92%. Additionally, the final loss value decreased to 0.89. Therefore, this model represents the optimized CNN (O-CNN) configuration, and the validation accuracy results along with the final loss value graph are shown in Figure 8a,b, respectively. These findings suggest that a deeper network than the CNN classification algorithm and an approach capable of preserving information throughout propagation are required to achieve the optimal outcome for this dataset.

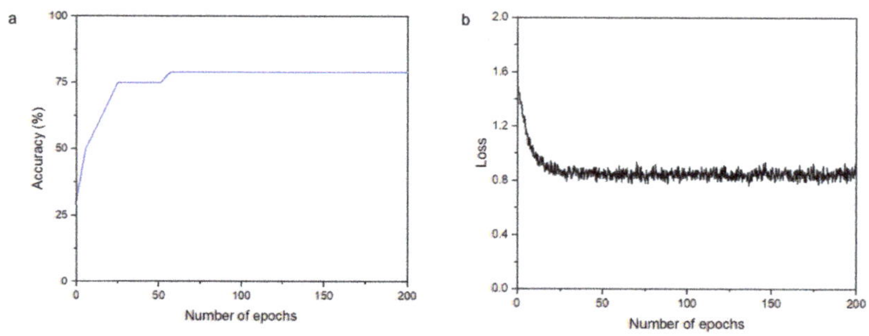

Figure 8. The results of O-CNN. (**a**) Validation accuracy; (**b**) training loss.

3.2. Analysis of Results from ResNet Algorithms

Training with ResNet also involved varying the number of epochs to test for accurate improvements. Compared to CNN, ResNet utilizes a deeper network, requiring a longer training duration to achieve higher accuracy. Accordingly, the training was conducted with epochs ranging from 1000 to 2000, with the initial training set to approximately 2000 epochs. Unlike CNN, ResNet demonstrated superior initial performance, achieving a validation accuracy of 87.5% and a notably low loss value of 0.4. Both the testing and validation curves showed similar trends with gradual gradients. The test accuracy reached an impressive 91%. These results indicate that ResNet effectively overcomes the limitations of CNN by leveraging skip connections to mitigate neuron loss, leading to higher accuracy. The results are illustrated in Figure 9a,b.

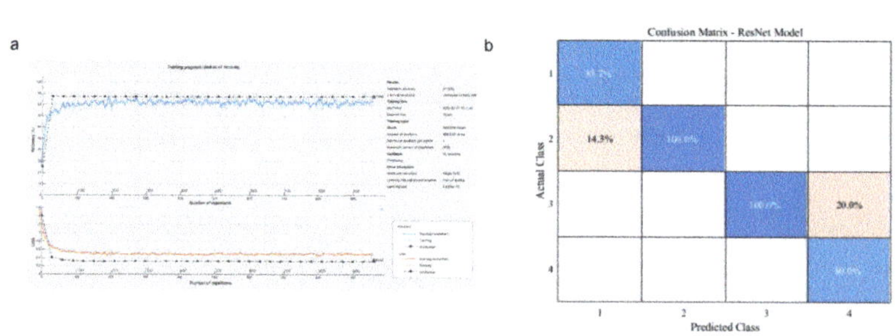

Figure 9. (**a**) Training results window; (**b**) confusion matrix (ResNet-1).

The test results were further improved when we used the optimized ResNet (O-ResNet), with the number of epochs reduced to 1000 for retraining. While the validation accuracy remained similar at 87%, the loss value decreased significantly to 0.003, and the test accuracy increased to 95%. These results suggest that an excessive number of epochs can negatively impact test performance and confirm that ResNet is more suitable for classifying this dataset compared to CNN. These results can be observed in Figure 10.

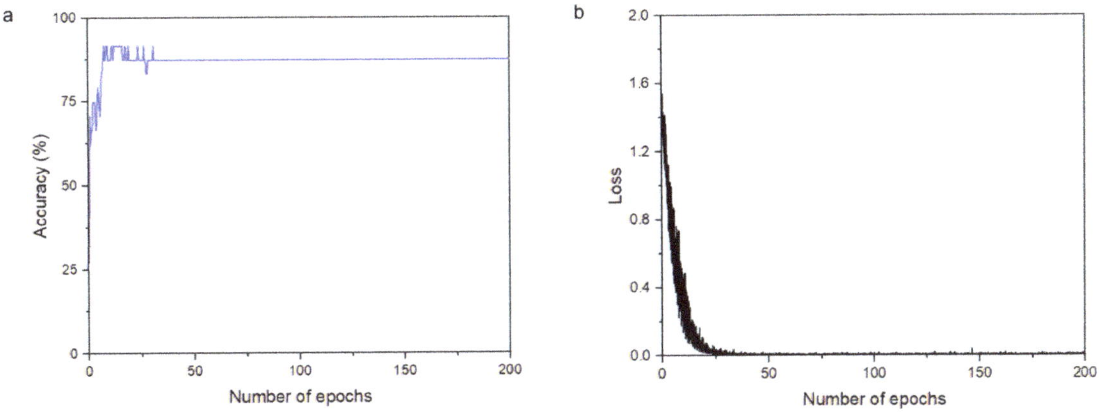

Figure 10. The results of O-Resnet. (**a**) Validation accuracy; (**b**) training loss.

The comparative analysis of CNN and ResNet in this study highlights the effectiveness of deep learning architectures in respiratory signal classification. While both models share similar layer structures, ResNet's skip connections play a crucial role in preserving important features during training, thereby enhancing classification performance. The ability of ResNet to maintain higher accuracy across different training conditions underscores the importance of selecting appropriate network structures and training parameters. These findings demonstrate that deep learning model selection and optimization are critical factors in achieving robust classification performance for physiological signal processing.

3.3. A Comparison to the State of the Art

The proposed models' performance is presented in comparison with existing deep learning-based studies on similar datasets, as shown in Table 3. The results indicate that the optimized CNN (O-CNN) model achieved an accuracy of 83.92%, while the optimized ResNet (O-ResNet) model attained an accuracy of 95%. Although our models did not surpass all previously published studies, they demonstrated performance comparable to the highest accuracy levels reported in the literature.

A review of related research reveals that many studies focus on combining models optimized for specific datasets or conducting comparative analyses with existing architectures. However, there remains a lack of studies exploring parameter tuning experiments and the application of deep learning to respiratory signal-based data analysis. These findings highlight the necessity of this research and underscore the importance of optimizing deep learning models for specific datasets.

While our models did not achieve the highest reported accuracy, they successfully reached the performance range of state-of-the-art studies. More importantly, this study contributes to the field by identifying dataset-specific optimal models through adjustments in the number of epochs and network blocks, demonstrating the significance of tailored optimization in deep learning applications.

Table 3. A comparison with state-of-the-art deep learning models.

Authors	Model	Accuracy (%)
[28]	Convolutional Neural Network—Mixture of Experts (CNN-MoE)	84.6
[29]	Deep Convolutional Autoencoder (CAE)	94.2
[30]	Convolutional Autoencoder for Multi-Person Activity Sensing (CAE-MAS)	97.13
[31]	Artificial Hydrocarbon Network (ANN)	98.24
[19]	ResNet-18, ResNet-34, ResNet-50, ResNet-101, ResNet-152	96.43
[32]	Convolutional Neural Network (CNN)	94.5
[33]	Residual Neural Network (ResNet)	99.99
[20]	ResNet-RS (Improved ResNet Scaling Strategies)	86.2
	Optimized Convolutional Neural Network (O-CNN)	83.92
	Optimized Residual Neural Network (O-ResNet)	95

3.4. Classification Analysis Across Different Breathing Intensities

The classification of the four types of breathing was analyzed based on accuracy, precision, recall, and F1-score to evaluate how well each breathing type was classified. The results of the classification using O-CNN are presented in a confusion matrix, as shown in Figure 11. The four classes are as follows: Class 1 represents resting, Class 2 represents low-intensity breathing, Class 3 represents moderate-intensity breathing, and Class 4 represents high-intensity breathing.

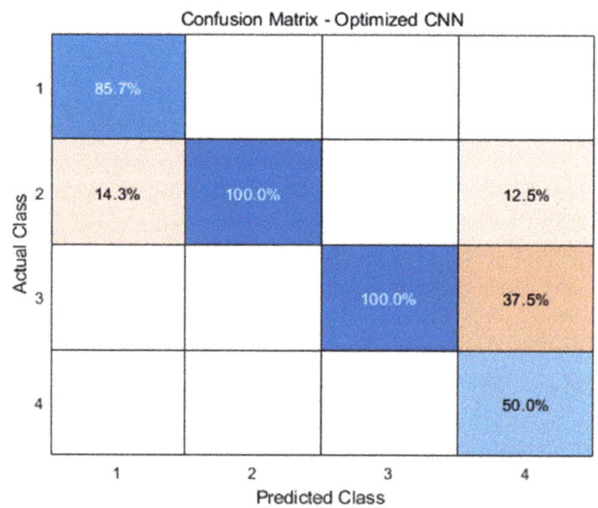

Figure 11. Confusion matrix of O-CNN.

The accuracy of the model is observed by examining the true positive (TP) values along the diagonal of the confusion matrix, with an overall accuracy of 0.87. The precision values for each class are 1.0, 0.85, 1.0, and 0.6, respectively. Similarly, the recall values for the classes are 0.83, 1.0, 0.75, and 1.0. Finally, the F1-scores for the classes are 0.9, 0.92, 0.86, and 0.75, in the same order. Both precision and recall values above 0.7 are generally considered satisfactory. Based on these results, the O-CNN metrics demonstrate satisfactory performance overall. However, for Class 4, the precision is significantly low at 0.6, indicating reduced reliability in classification. Additionally, the F1-score for Class 4 is also relatively low at 0.75, suggesting that O-CNN performs the least effectively for this specific class. The detailed metrics can be found in Table 4.

Table 4. Classification performance metrics of O-CNN.

Metric	Class 1	Class 2	Class 3	Class 4	Macro Average
Precision	1.0	0.85	1.0	0.6	0.86
Recall	0.83	1.0	0.75	1.0	0.9
F1-score	0.9	0.92	0.86	0.75	0.86
Accuracy					0.87

The results of the O-ResNet classification showed an accuracy of 0.95. The precision values were as follows: 1.0 for Class 1, 1.0 for Class 2, 1.0 for Class 3, and 0.8 for Class 4. The recall values were 1.0, 0.86, 1.0, and 1.0, respectively. Finally, the F1-scores were 1.0 for Class 1, 0.92 for Class 2, 1.0 for Class 3, and 0.89 for Class 4. These results indicate strong overall performance, with all metrics exceeding 0.7, which is generally considered a good threshold.

A detailed analysis revealed the following:

Class 1 achieved a perfect score (1.0) across all metrics, demonstrating that O-ResNet enables flawless classification for this category.

Class 2 showed a slightly lower recall compared to precision, indicating a minor shortfall in correctly predicting actual positive instances from the dataset.

Class 3 exhibited perfect scores across all metrics, confirming its reliable classification.

Class 4 had the lowest precision (0.8) among all metrics. As shown in the confusion matrix (Figure 12), while all other classes were classified with 100% accuracy, Class 4 was classified with an 80% accuracy rate. Nevertheless, since all metrics exceed 0.7, this is considered satisfactory. When precision and recall were harmonically averaged, Class 4 achieved an F1-score of 0.89, which is close to 1.

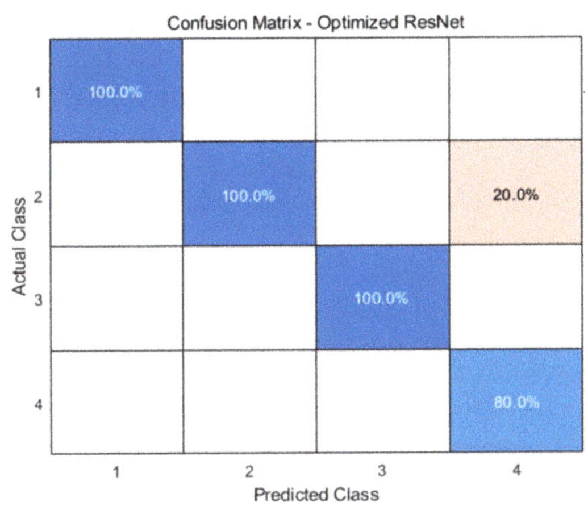

Figure 12. Confusion matrix of O-ResNet.

These results demonstrate that O-ResNet classification is well suited for this dataset, particularly as indicated by the F1-score, which accounts for both precision and recall. Furthermore, the results reflect the strong performance of the O-ResNet model, indicating appropriate parameter selection and effective use of the data. All evaluation metrics are summarized in Table 5.

To further assess the effectiveness of the O-ResNet model, its performance was compared to that of O-CNN using standard deep learning evaluation metrics. Accuracy, precision, recall, and the overall F1-score were used for comparison, and the results were analyzed for each breathing intensity. When comparing accuracy, O-CNN achieved 0.87, while O-ResNet achieved 0.96, both of which are considered satisfactory. A closer examination of the metrics revealed that the precision for Class 4 was 0.6 in O-CNN but significantly improved to 0.8 in O-ResNet. These findings indicate that O-ResNet enhances precision and demonstrates its suitability and reliability for the classification task in this study.

Table 5. Classification performance metrics of O-ResNet.

Metric	Class 1	Class 2	Class 3	Class 4	Macro Average
Precision	1.0	1.0	1.0	0.8	0.95
Recall	1.0	0.86	1.0	1.0	0.96
F1-score	1.0	0.92	1.0	0.89	0.95
Accuracy					0.96

In this study, a wearable respiratory sensor was employed to classify breathing patterns of varying intensities across cycles using deep learning techniques. The results demonstrate that wearable respiratory sensors can effectively capture and differentiate breathing patterns in both physical activities and daily life. The enhanced classification performance of O-ResNet highlights its potential for detecting abnormal breathing patterns, which could be valuable for health monitoring applications.

Despite these promising results, several challenges remain. First, the dataset used in this study was relatively small and was collected under controlled conditions, which may limit its generalizability. Future research will address this limitation by incorporating a more diverse dataset, considering various exercise conditions, postures, and sensor placements. Second, while O-ResNet demonstrated strong performance, it may encounter difficulties in processing more complex breathing patterns or real-time applications. To overcome these challenges, future work will explore alternative models, such as LSTM or GAN, which may enhance classification accuracy, particularly for long-term respiratory monitoring and data augmentation. Lastly, the real-world implementation of wearable respiratory sensors requires careful consideration of factors such as sensor placement variability, motion artifacts, and individual differences. Future studies will focus on refining sensor design and preprocessing techniques to improve the practicality of these models in real-world health monitoring systems.

4. Conclusions

This study investigated the classification of breathing patterns using a wearable respiratory sensor and deep learning models. The findings confirm that O-ResNet, with its skip connection architecture, outperformed O-CNN in accuracy and precision, particularly in detecting high-intensity breathing patterns. These results underscore the potential of wearable sensors and deep learning techniques for real-time health monitoring, offering a promising approach for capturing complex respiratory signals.

Despite these advancements, this study has limitations, including the relatively small dataset collected under controlled conditions. To address this, future research should incorporate larger and more diverse datasets while also exploring alternative models, such as LSTM or GAN, to improve classification performance. The insights gained from this study contribute to the advancement of wearable respiratory monitoring, paving the way for more accurate and reliable health applications.

Author Contributions: As a corresponding author, J.K. (Jooyong Kim) was responsible for the entire structure's construction, while J.K. (Jiseon Kim) was responsible for the experimental design and modeling. J.K. (Jiseon Kim) was responsible for data collection, data processing, and material selection. J.K. (Jooyong Kim) and J.K. (Jiseon Kim) drafted the manuscript. All authors have read and agreed to the published version of the manuscript.

Funding: This research was financially supported by Soongsil University, Seoul 156-743, Republic of Korea. This work was supported by the Technology Innovation Program (Industrial Strategic Technology Development Program: Materials/Parts Package Type, grant number 20016038; Development of textile–IT converged digital sensor modules for smart wear to monitor bio and activity signals in exercise and KS standard), the Ministry of Trade, Industry and Energy (MOTIE, Republic of Korea), and the Korea Institute for Advancement of Technology (KIAT, MOTIE), grant number P0012770 (The Competency Development Program for Industry Specialist).

Data Availability Statement: The original contributions presented in this study are included in the article; further inquiries can be directed to the corresponding authors.

Conflicts of Interest: The authors declare no conflicts of interest.

References

1. Nicolò, A.; Marcora, S.M.; Bazzucchi, I.; Sacchetti, M. Differential control of respiratory frequency and tidal volume during high-intensity interval training. *Exp. Physiol.* **2017**, *102*, 934–949. [CrossRef]
2. Nicolò, A.; Bazzucchi, I.; Haxhi, J.; Felici, F.; Sacchetti, M. Comparing continuous and intermittent exercise: An "isoeffort" and "isotime" approach. *PLoS ONE* **2014**, *9*, e94990. [CrossRef]
3. Nicolò, A.; Marcora, S.M.; Sacchetti, M. Respiratory frequency is strongly associated with perceived exertion during time trials of different duration. *J. Sports Sci.* **2016**, *34*, 1199–1206. [CrossRef]
4. Nicolò, A.; Massaroni, C.; Passfield, L. Respiratory frequency during exercise: The neglected physiological measure. *Front. Physiol.* **2017**, *8*, 922. [CrossRef]
5. Forster, H.V.; Haouzi, P.; Dempsey, J.A. Control of breathing during exercise. *Compr. Physiol.* **2011**, *2*, 743–777.
6. Duffin, J. The fast exercise drive to breathe. *J. Physiol.* **2014**, *592*, 445–451. [CrossRef] [PubMed]
7. Davies, R.C.; Rowlands, A.V.; Poole, D.C.; Jones, A.M.; Eston, R.G. Eccentric exercise-induced muscle damage dissociates the lactate and gas exchange thresholds. *J. Sports Sci.* **2011**, *29*, 181–189. [CrossRef]
8. Busse, M.; Maassen, N.; Konrad, H. Relation between plasma K+ and ventilation during incremental exercise after glycogen depletion and repletion in man. *J. Physiol.* **1991**, *443*, 469–476. [CrossRef]
9. Voduc, N.; Webb, K.A.; D'Arsigny, C.; McBride, I.; O'Donnell, D.E. McArdle's Disease Presenting as Unexplained Dyspnea in a Young Woman. *Can. Respir. J.* **2004**, *11*, 163–167. [CrossRef] [PubMed]
10. Bhaiyya, M.; Rewatkar, P.; Pimpalkar, A.; Jain, D.; Srivastava, S.K.; Zalke, J.; Kalambe, J.; Balpande, S.; Kale, P.; Kalantri, Y. Deep Learning-Assisted Smartphone-Based Electrochemiluminescence Visual Monitoring Biosensor: A Fully Integrated Portable Platform. *Micromachines* **2024**, *15*, 1059. [CrossRef] [PubMed]
11. Ohkubo, M.; Yamamura, M.; Uchiyama, H.; Nojima, T. Breathing clothes: Artworks using the hairlytop interface. In Proceedings of the 11th Conference on Advances in Computer Entertainment Technology, Madeira, Portugal, 11–14 November 2014; pp. 1–4.
12. Zięba, J.; Frydrysiak, M.; Błaszczyk, J. Textronic clothing with resistance textile sensor to monitoring frequency of human breathing. In Proceedings of the 2012 IEEE International Symposium on Medical Measurements and Applications Proceedings, Budapest, Hungary, 18–19 May 2012; pp. 1–6.
13. Chen, W.; McDuff, D. Deepphys: Video-based physiological measurement using convolutional attention networks. In Proceedings of the 15th European Conference on Computer Vision (ECCV), Munich, Germany, 8–14 September 2018; pp. 349–365.
14. Suriani, N.S.; Shahdan, N.S.; Sahar, N.M.; Taujuddin, N.S.A.M. Non-contact facial based vital sign estimation using convolutional neural network approach. *Int. J. Adv. Comput. Sci. Appl.* **2022**, *13*, 386–393. [CrossRef]
15. Goodfellow, I. *Deep Learning*; MIT Press: Cambridge, MA, USA, 2016.
16. Bengio, Y.; Simard, P.; Frasconi, P. Learning long-term dependencies with gradient descent is difficult. *IEEE Trans. Neural Netw.* **1994**, *5*, 157–166. [CrossRef]
17. Glorot, X.; Bengio, Y. Understanding the difficulty of training deep feedforward neural networks. In Proceedings of the Thirteenth International Conference on Artificial Intelligence and Statistics, Sardinia, Italy, 13–15 May 2010; pp. 249–256.
18. Srivastava, R.K.; Greff, K.; Schmidhuber, J. Highway networks. *arXiv* **2015**, arXiv:1505.00387.
19. He, K.; Zhang, X.; Ren, S.; Sun, J. Deep residual learning for image recognition. In Proceedings of the 2016 IEEE Conference on Computer Vision and Pattern Recognition (CVPR), Las Vegas, NV, USA, 27–30 June 2016; pp. 770–778.
20. Bello, I.; Fedus, W.; Du, X.; Cubuk, E.D.; Srinivas, A.; Lin, T.-Y.; Shlens, J.; Zoph, B. Revisiting resnets: Improved training and scaling strategies. *Adv. Neural Inf. Process. Syst.* **2021**, *34*, 22614–22627.
21. Mohsen, H.; El-Dahshan, E.-S.A.; El-Horbaty, E.-S.M.; Salem, A.-B.M. Classification using deep learning neural networks for brain tumors. *Future Comput. Inform. J.* **2018**, *3*, 68–71. [CrossRef]
22. Sadegh Ebrahimi, M.; Karkeh Abadi, H. Study of Residual Networks for Image Recognition. *arXiv* **2018**, arXiv:1805.00325.

23. Kim, J.-S.; Truong, T.; Kim, J. Development of embroidery-type sensor capable of detecting respiration using the capacitive method. *Polymers* **2023**, *15*, 503. [CrossRef]
24. Kim, J.; Kim, J. Classification of Breathing Signals According to Human Motions by Combining 1D Convolutional Neural Network and Embroidered Textile Sensor. *Sensors* **2023**, *23*, 5736. [CrossRef]
25. Kim, S.; Truong, T.; Jang, J.; Kim, J. The programmable design of large-area piezoresistive textile sensors using manufacturing by jacquard processing. *Polymers* **2022**, *15*, 78. [CrossRef]
26. *ISO 10993-5*; Biological evaluation of Medical Devices—Part 5: Tests for In Vitro Cytotoxicity. ISO: 2009. Available online: https://webdesk.jsa.or.jp/books/W11M0090/index/?bunsyo_id=ISO+10993-5:2009 (accessed on 21 January 2025).
27. Truong, T.; Kim, J.-S.; Kim, J. Design and optimization of embroidered antennas on textile using silver conductive thread for wearable applications. *Fibers Polym.* **2021**, *22*, 2900–2909. [CrossRef]
28. Pham, L.; Phan, H.; Palaniappan, R.; Mertins, A.; McLoughlin, I. CNN-MoE based framework for classification of respiratory anomalies and lung disease detection. *IEEE J. Biomed. Health Inform.* **2021**, *25*, 2938–2947. [CrossRef]
29. Seyfioğlu, M.S.; Özbayoğlu, A.M.; Gürbüz, S.Z. Deep convolutional autoencoder for radar-based classification of similar aided and unaided human activities. *IEEE Trans. Aerosp. Electron. Syst.* **2018**, *54*, 1709–1723. [CrossRef]
30. Raeis, H.; Kazemi, M.; Shirmohammadi, S. CAE-MAS: Convolutional Autoencoder Interference Cancellation for Multiperson Activity Sensing With FMCW Microwave Radar. *IEEE Trans. Instrum. Meas.* **2024**, *73*, 1–10. [CrossRef]
31. Brieva, J.; Ponce, H.; Moya-Albor, E. Non-contact breathing rate estimation using machine learning with an optimized architecture. *Mathematics* **2023**, *11*, 645. [CrossRef]
32. Chaichulee, S.; Villarroel, M.; Jorge, J.; Arteta, C.; McCormick, K.; Zisserman, A.; Tarassenko, L. Cardio-respiratory signal extraction from video camera data for continuous non-contact vital sign monitoring using deep learning. *Physiol. Meas.* **2019**, *40*, 115001. [CrossRef]
33. Hussain, F.; Abbas, S.G.; Husnain, M.; Fayyaz, U.U.; Shahzad, F.; Shah, G.A. IoT DoS and DDoS attack detection using ResNet. In Proceedings of the 2020 IEEE 23rd International Multitopic Conference (INMIC), Bahawalpur, Pakistan, 5–7 November 2020; pp. 1–6.

Disclaimer/Publisher's Note: The statements, opinions and data contained in all publications are solely those of the individual author(s) and contributor(s) and not of MDPI and/or the editor(s). MDPI and/or the editor(s) disclaim responsibility for any injury to people or property resulting from any ideas, methods, instructions or products referred to in the content.

processes

MDPI

Article

The Development and Optimization of a Textile Image Processing Algorithm (TIPA) for Defect Detection in Conductive Textiles

Sang-Un Kim [1] and Joo-Yong Kim [2,*]

1 Department of Smart Wearable Engineering, Soongsil University, Seoul 06978, Republic of Korea; tkddnsl0723@naver.com
2 Department of Materials Science and Engineering, Soongsil University, Seoul 06978, Republic of Korea
* Correspondence: jykim@ssu.ac.kr; Tel.: +82-10-8720-0631

Abstract: This study introduces a Textile Image Processing Algorithm (TIPA) designed to detect defects in conductive textiles, a crucial element of wearable technology. TIPAs employ image preprocessing, filtering, and classification to identify issues like uneven distribution of conductive particles. When applied to fabrics produced via dip-coating, our TIPA was optimized using a threshold ratio, achieving over 85% accuracy, with a maximum of 100% under ideal conditions. However, detection challenges were noted in fabrics with large, diffuse stains, particularly at extreme threshold ratios. This TIPA proves to be a valuable tool for improving quality control in smart textiles, with potential for further optimization.

Keywords: conductive textiles; textile image processing algorithm (TIPA); defect detection; dip-coating process; image filtering; algorithm optimization

Academic Editor: Jie Zhang

Received: 14 January 2025
Revised: 5 February 2025
Accepted: 9 February 2025
Published: 10 February 2025

Citation: Kim, S.-U.; Kim, J.-Y. The Development and Optimization of a Textile Image Processing Algorithm (TIPA) for Defect Detection in Conductive Textiles. *Processes* **2025**, *13*, 486. https://doi.org/10.3390/pr13020486

Copyright: © 2025 by the authors. Licensee MDPI, Basel, Switzerland. This article is an open access article distributed under the terms and conditions of the Creative Commons Attribution (CC BY) license (https://creativecommons.org/licenses/by/4.0/).

1. Introduction

Conductive textiles, which have gained attention as soft smart materials [1,2], are fusion materials used in various wearable products due to their ability to measure a range of biometric data [3,4], including heart rate [5], respiration [6,7], movement [6], voice [8], and pulse [9]. The mechanical properties of conductive textiles maintain the physical characteristics of traditional textiles, making them flexible [10], lightweight [11], and stretchable [12]. Conductive particles that are processed in textiles determine their electrical properties [10]. When they mechanically deform under external forces, their electrical properties are changed by changes in the distribution of conductive particles and the structure of the circuits. This property enables the measurement of parameters such as pressure and strain, facilitating the acquisition of diverse biometric information in the wearable technology field.

The manufacturing process of conductive textiles typically involves two main approaches. The first involves treating fibers with conductive particles at the fiber scale to produce the textile, or imparting conductivity by treating a textile with suitable mechanical properties using conductive particles [13]. The second approach, particularly the dip-coating method, which involves using a solution with dispersed conductive particles, is advantageous as it allows for the use of textiles with appropriate physical properties and can be applied in continuous processes such as roll-to-roll manufacturing [3,14]. However, textiles produced through this method may suffer from defects similar to those observed in dyeing processes, where uneven distribution of particles can result in blemishes or streaks (Figure 1).

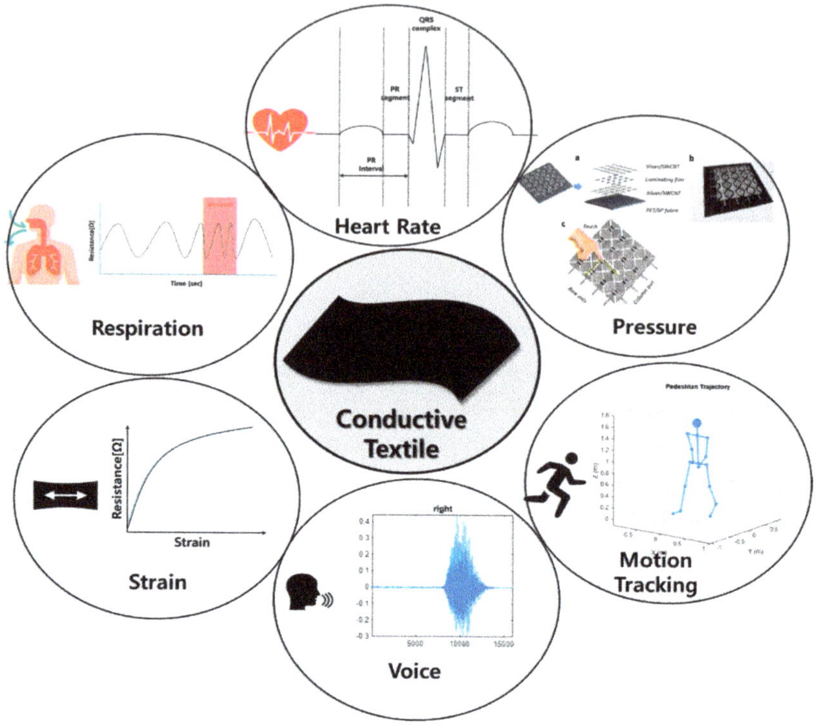

Figure 1. Conductive textile in wearable sensor.

The defects of dip-coated conductive textiles are particularly evident with common conductive particles such as carbon nanotubes (CNT) and Mxene, both of which are carbon-based and impart a black color to the textile [15–18]. One of the primary concerns of defects is the uneven distribution of conductive particles, which can lead to localized variations in electrical resistance. Such inconsistencies may result in signal distortions, reduced conductivity, and, in extreme cases, circuit discontinuities that impair the textile's sensing and actuation capabilities. Therefore, in the field of wearable technology, where precise signal transmission and control are essential, defects in conductive textiles directly lead to product malfunctions. To address this issue, this study aims to develop an image processing algorithm specifically tailored for detecting defects in conductive textiles, ensuring their reliability and functionality in wearable applications.

Due to recent advancements in computational capabilities, the development of machine learning (ML) [19], particularly in the area of image processing algorithms (IPA) using two-dimensional and three-dimensional images [20,21], has made it possible to replace humans in many fields. Specifically, in the quality inspection sector of textile manufacturing [22–25], cameras and various algorithms can now replace the human eye and brain to detect and classify defects.

However, textile materials, due to their diverse weaving methods, patterns, and colors, present significant challenges for feature extraction in images [15,26,27]. There are various types of defects in conductive textiles, including missing yarn, broken ends, needle lines, spots, stains, holes, press-off, mixed yarn, and gouts [28]. Of these, this study focuses specifically on stains defects. While other types of defects often occur during the textile manufacturing process or result from physical damage during use, changes in electrical properties that hinder resistance measurement typically manifest as stains or irregular patterns. Therefore, stain defects are particularly critical for assessing the functional integrity of conductive textiles [29,30].

Since the detection of stains through textile image processing algorithms (TIPAs) requires a precise definition of defects, threshold-based methods serve as a fundamental approach in defect detection [31,32]. Moreover, these thresholds act as a critical variable that directly influences the overall accuracy of TIPAs. Setting a standardized reference value in image data is inherently challenging, necessitating various preprocessing techniques and adaptive thresholding methods [33]. For instance, in the case of the textiles produced through the dip-coating process during this study, the degree of darkness varies depending on the penetration level of conductive particles. As a result, simply classifying areas as stains based solely on brightness or darkness could lead to misjudging the entire conductive fabric as defective (Figure 2).

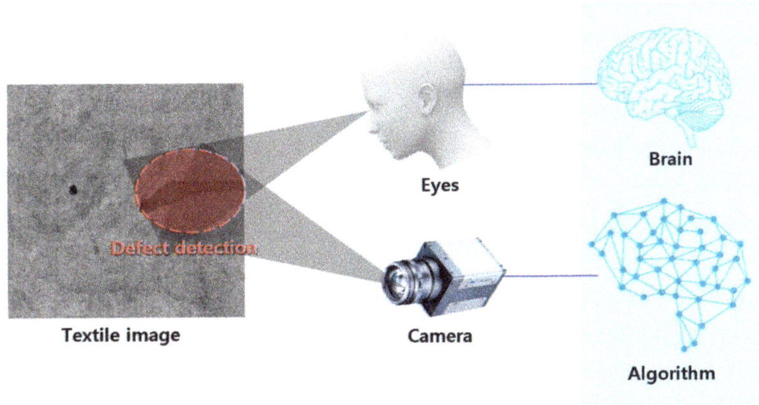

Figure 2. Defect detection through humans and through Textile Image Processing Algorithms.

One study proposed a TIPA that uses an encoder–decoder structure to detect defects in textiles, employing image preprocessing techniques such as dilation, median filtering, and edge detection to identify defects [34]. Another study utilized wavelet functions to improve detection performance by identifying coefficients from defect-free textiles and optimizing them through a subset selection using a genetic algorithm [35]. Additionally, another study demonstrated superior detection performance by addressing low performance in detecting defects of various shapes and sizes using histogram and threshold-setting techniques [36–38]. These examples underscore the importance of strategic structuring and optimization in TIPAs for effective image processing.

Overall, the objective of this study is to develop a Textile Image Processing Algorithm (TIPA) capable of detecting stain defects, which are caused by the agglomeration of conductive particles during the manufacturing process of conductive textiles, that lead to non-uniformity resistance. To achieve this, histogram equalization is employed as a preprocessing step to enhance contrast, enabling defect detection across various brightness conditions of conductive textiles. The algorithm's structure includes a multi-step detection process based on defect size, which is optimized by fine-tuning the threshold ratio. The first filtering step is designed to classify whether an image belongs to the normal or defective category by analyzing the spatial distribution of defect-suspect pixels. The second filtering step, on the other hand, specifically aims to detect and isolate abnormal regions within defective images, ensuring that only significant defect areas are identified while minimizing false positives from minor variations or noise. The final evaluation is conducted on 22 conductive textile samples fabricated through the dip-coating process.

2. Materials and Methods

2.1. Fabrication of Conductive Textile by Dip-Coating Process

The dip-coating process for manufacturing conductive textiles was conducted based on the fabrication procedures referenced in studies utilizing conductive fabrics. As illustrated in Figure 3, this process was implemented accordingly [15]. A 0.1 wt% aqueous dispersion of Single-Walled Carbon Nanotubes (SWCNT) (WART1200, KORBON Co., Ltd., Gangneung, Republic of Korea) was prepared by ultrasonically dispersing the solution at 1000 rpm for 1 h. The fabric used for dip-coating was a stretchable fabric consisting of 95% cotton and 5% spandex, with dimensions of 300 cm × 300 cm. This fabric was immersed in the 0.1 wt% SWCNT aqueous dispersion, followed by squeezing with a padding machine to ensure uniform coating. The coated fabric was then dried in an oven at 80 °C for 10 min to evaporate the water solvent, taking care to avoid boiling.

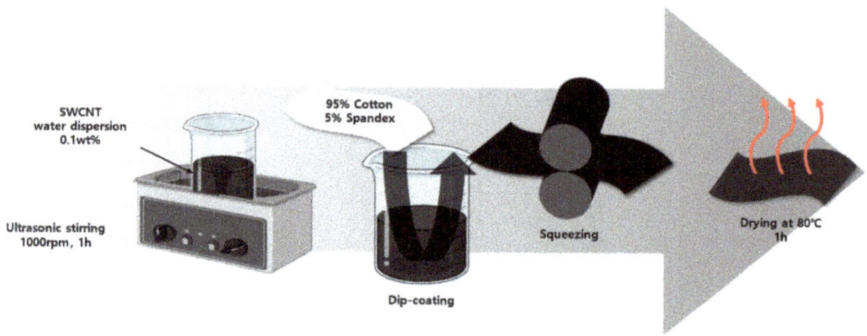

Figure 3. Dip-coating process of conductive textile fabrication.

The image files of conductive fabrics for the TIPA to be used on were produced using a consistent light source to prevent distortions caused by factors such as lighting, shadows, and lens effects. This was achieved by utilizing the scanner of a printer (LaserJet Pro MFP M428fdw, HP Korea Inc., Seoul, Republic of Korea.), which features an optical resolution of up to 1200 × 1200 dpi and a fixed focal distance optimized for contact image sensor (CIS) technology. The scanning process was performed with a maximum scan size of 216 × 297 mm (flatbed) and 216 × 356 mm (ADF), ensuring precise image acquisition for defect detection. To clearly distinguish between defect pixels and non-defect pixels, the images were saved in PNG format, which allows for minimal image loss due to its low compression ratio. The final image files were secured at a resolution of 1440 × 1440 pixels. A total of 22 conductive fabric image files were obtained, consisting of 15 normal images without defects and 7 images with defects. Figure 4 presents two representative normal images and two defect images from this collection.

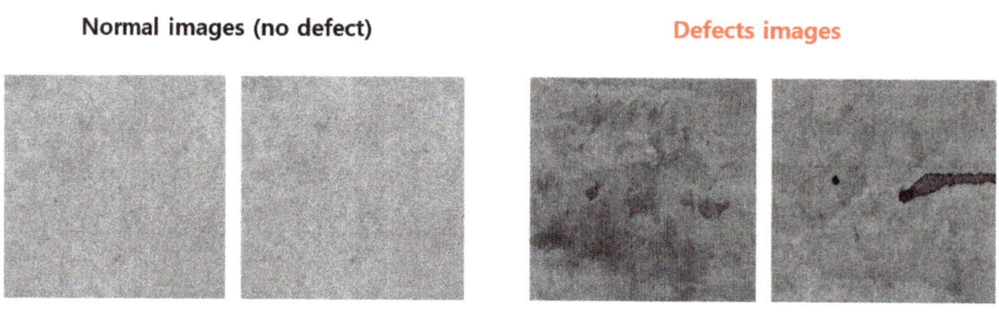

Figure 4. Images of normal conductive textile and textile with defects.

2.2. Textile Image Processing Algorithm (TIPA) for Defect Detection

To detect defects in the conductive fabric produced via the dip-coating process, a TIPA was designed, as illustrated in Figure 5. All steps of the TIPA were implemented using MATLAB 2023a to preprocess images and detect defects in each conductive textile image. The first step involved preprocessing the images, where histogram equalization was applied to smooth out the contrast. The histogram equalization involves calculating the cumulative distribution function (CDF) of the input image and using it to remap the pixel values to improve the contrast of the image. The mathematical equation of histogram equalization can be expressed as follows:

$$s = (L-1) \times \frac{\sum_{k=0}^{r} P_r(k)}{N} \tag{1}$$

where the r is original pixel value, L is the 256 of the 8-bit images in this study, $P_r(k)$ is the frequency of pixel value k, and N is the total number of pixels. In our TIPA, histogram equalization was performed using a MATLAB function. Histogram equalization was employed to enhance the contrast of the image by normalizing the histogram of pixel intensity values. This process involved measuring the occurrence frequency k of each pixel intensity, standardizing the histogram, and replacing each pixel value with its corresponding relative proportion. Subsequently, the cumulative distribution function (CDF) was utilized to compute and map the pixel values to new intensity levels, ensuring a more uniform distribution across the image.

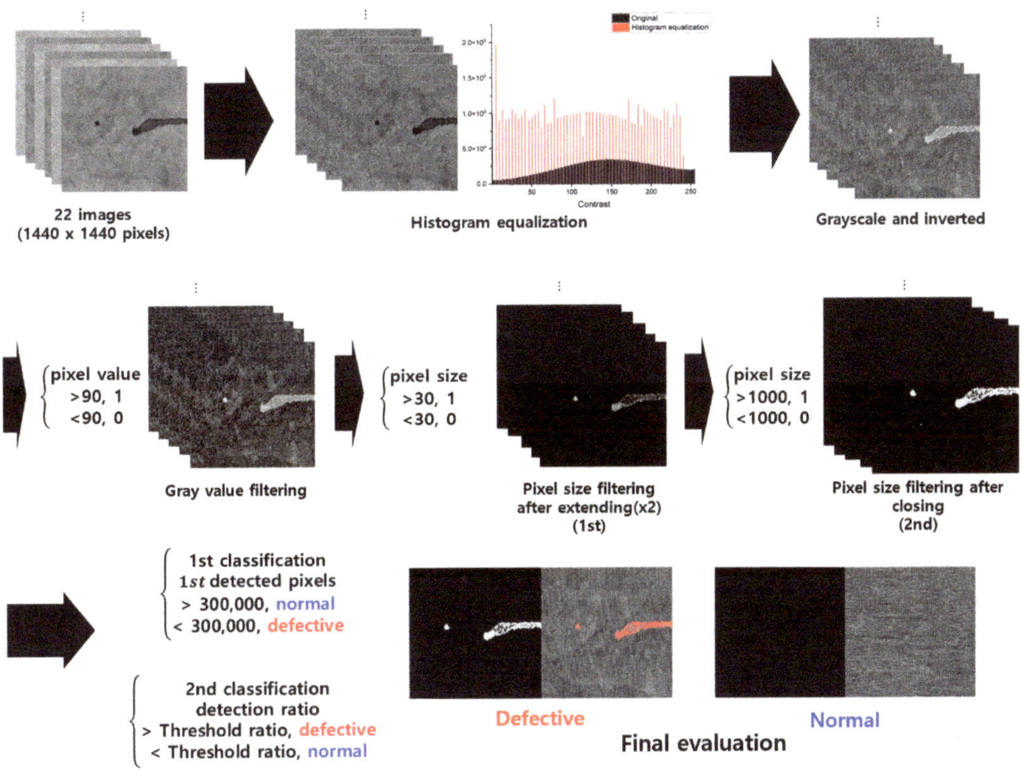

Figure 5. TIPA process using conductive textile images.

Grayscale conversion was performed using a MATLAB function, and the coefficients used for calculating the grayscale value were rounded to three decimal places. These values correspond to those specified in Rec. ITU-R BT.601-7 for computing luminance (gray value), which are as follows:

$$\text{Gray value} = 0.299\ \text{Red value} + 0.587\ \text{Green value} + 0.114\ \text{Blue value}. \quad (2)$$

These coefficients are widely adopted for grayscale conversion in image processing to ensure accurate luminance representation, particularly in video and image standardization processes.

The grayscale values, which range from 0 to 255, were binarized with a threshold set at 90. This threshold was chosen to represent approximately one-third of the full grayscale range, effectively excluding the most definitive background values while ensuring the detection of all potential defect areas. However, since the binarized values alone are not sufficient to accurately define defects, the pixels were expanded by doubling their width and height to enhance defect visibility and improve detection accuracy.

Primary size filtering was conducted with a threshold set at 30 after the pixel expansion step. This threshold was chosen based on the observation that, in defect images with high grayscale contrast, defect-suspect pixels tend to be spatially clustered, whereas in normal images, pixels are more evenly distributed. Upon expansion, defect pixels are more likely to connect, whereas isolated pixels remain sparse. Therefore, if the number of connected pixels is three or fewer, they are considered part of the background and filtered out. This step effectively reduces false positives by removing small, non-defect-related variations. Subsequently, gaps between defective regions were filled to enhance connectivity before secondary size filtering was performed with a threshold set at 1000 pixels. This threshold serves as a criterion to distinguish actual stains from random small spots, ensuring that only sufficiently large defect regions are classified as stains while filtering out minor noise or non-significant small defects.

The purpose of this TIPA, which is to detect defect areas and evaluate quality, was first approached by classifying images as normal or defective based on the number of pixels detected after primary size filtering. Specifically, the number of 1st detected pixels was compared to the threshold, which is 300,000 pixels, which comprised about 15% of the total pixel count of the image. This classification leverages the observation that more uniform fabrics tend to have a higher count of 1st detected pixels. Subsequently, the ratio of the number of 1st detected pixels to the number of 2nd detected pixels (obtained after secondary size filtering) was calculated as follows:

$$\text{detection ratio} = \text{The number of the 1st detected pixels} / \text{The number of the 2nd detected pixels}. \quad (3)$$

This ratio was then used to perform a second classification based on a threshold ratio. The threshold ratio in this TIPA was optimized to distinguish between defective and normal images, and the detected white defects were compared with actual defects in the images to verify accuracy, with discrepancies highlighted in red.

3. Results

3.1. Preprocessing Images in TIPA

The purpose of the preprocessing step is to maximize the continuous tonal values to improve the accuracy of defect detection. Figure 6a,b illustrate the preprocessing process of our TIPA before it was applied to a representative conductive fabric containing defects, showing a comparison between the histogram-equalized image and the original image. By examining the contrast values and the histogram after equalization, it is possible to observe a peak value in the dark contrast areas where the defects are detected. This indicates that the subsequent filtering step, which is based on pixel values and size, can be effectively performed.

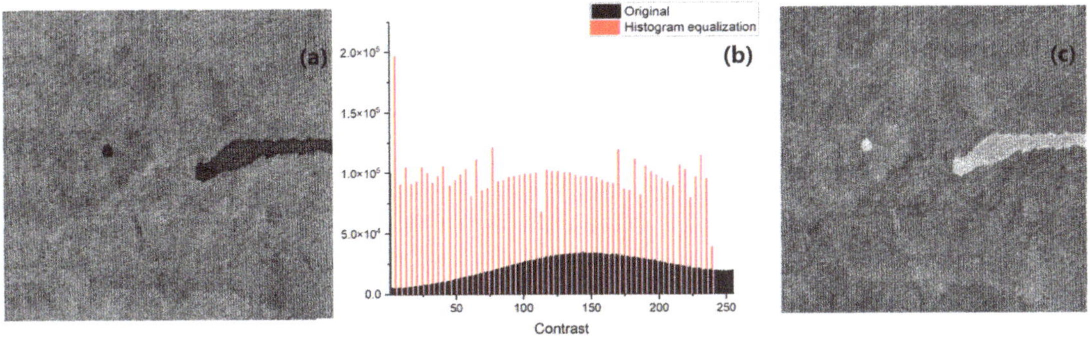

Figure 6. Images of preprocessing results for defective conductive textile: (**a**) histogram equalization image; (**b**) comparison of original and histogram equalization results; (**c**) grayscale image with inverted pixel values.

Based on this analysis, Figure 6c presents the grayscale conversion, where the RGB values have been removed, followed by an inversion of the image to facilitate easier defect detection.

3.2. Filtering Images in TIPA

Figure 7 illustrates the filtering results and the process of defect detection in our TIPA. Figure 7a shows the binarization filtering, where pixels are divided based on a threshold of 90 within the grayscale range of 0 to 255. The white areas represent pixels that are potential defects. While stains defined as defects in the conductive fabric are visible, many uniform fabric pixels also appear white, indicating that the binarization alone is insufficient for accurate defect detection. Figure 7b presents the results after expanding the filtered pixels by doubling their width and height, followed by size filtering based on a size threshold of 30. This process effectively filters out a significant number of pixels, leaving behind most of the stained areas. However, some of the stained pixels are still filtered out, indicating the need for additional processing. Figure 7c shows the results after a secondary size filtering process, where the edge pixels have been expanded into a circular shape with a radius of four pixels, followed by size-based filtering with a threshold of 1000 pixels. Ultimately, all non-defective pixels are removed, and the defect areas are largely filled, confirming the effectiveness of the filtering process.

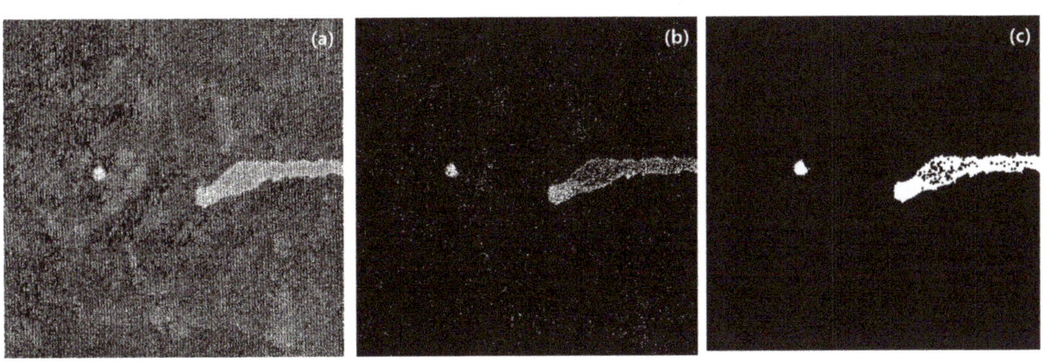

Figure 7. Images of filtering results for defective conductive textile: (**a**) pixel value filtering; (**b**) 1st pixel size filtering; (**c**) 2nd pixel size filtering.

Figure 8a shows the original image of a normal conductive textile with uniformly distributed conductive particles. Figure 8b presents the result after the first filtering stage of the TIPA. At this stage, numerous isolated pixels are detected, but they do not form any discernible pattern of stains and are uniformly distributed. After the second filtering stage, shown in Figure 8c, no pixels are detected, indicating that the fabric is defect-free and classified as normal.

Figure 8. Images of TIPA results for normal conductive textile: (**a**) original image; (**b**) 1st pixel size filtering; (**c**) 2nd pixel size filtering.

3.3. *Evaluation and Optimization of Conductive Textile Using TIPA*

The detection results and accuracy based on the threshold ratio, a key parameter in TIPAs, are summarized in Table 1. The results indicate that at lower threshold ratios, conductive fabrics that are normal were incorrectly detected as defective, while at higher threshold ratios, defective conductive fabrics were incorrectly classified as normal. The difficulty in detection is particularly evident in conductive fabrics like those shown in Figure 9, where the distribution of stains is large and diffuse, making it challenging to accurately classify the stains as defects. However, the overall accuracy remained above 85%, and through optimization, a maximum accuracy level of 100% was achieved. The analysis of our TIPA's precision, recall, and F1-score reveals that precision remained at 1 for threshold ratios above 15%, while for lower thresholds, it was measured at approximately 0.7 and 0.78. Conversely, the recall was highest for thresholds below 15% and decreased to a minimum of 0.57 at the highest threshold. The final F1-score ranged from a minimum of 0.73 to an average of approximately 0.86, with the 15% threshold achieving the highest F1-score of 1.0. Ultimately, while 15% yielded the best overall performance in this study, it was determined that threshold ratios in the range of 10% to 20% could be applicable for stain detection, depending on stain size and classification requirements.

Table 1. The results of all images processed by our TIPA, according to threshold ratio.

Threshold Ratio [%]	Defective [1]	Accuracy	Precision	Recall	F1-Score
5	10	0.86	0.7	1	0.83
10	9	0.90	0.78	1	0.88
15	7	1	1	1	1
20	6	0.95	1	0.86	0.92
25	5	0.90	1	0.71	0.83
30	4	0.86	1	0.57	0.73

[1] The number of defective conductive textiles was 7.

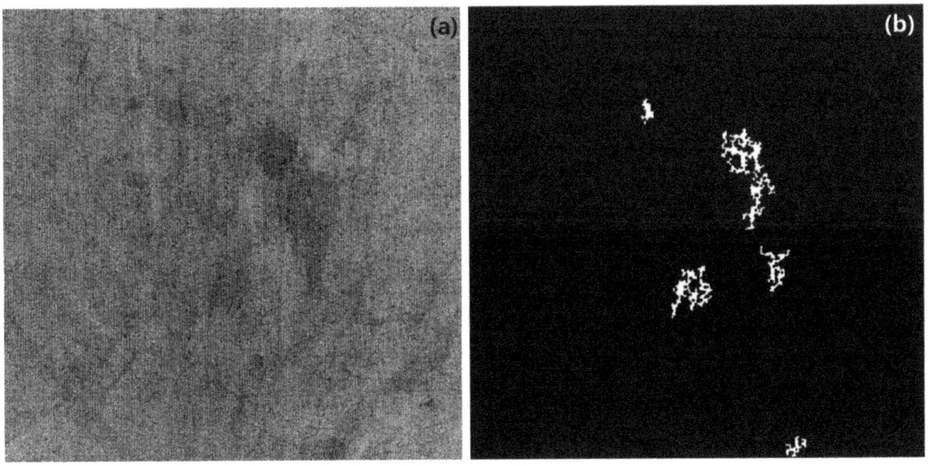

Figure 9. Blurred images of defective conductive textile: (**a**) original image; (**b**) TIPA result.

4. Discussion

This study presents a Textile Image Processing Algorithm (TIPA) specifically designed for the detection of defects in conductive textiles manufactured through the dip-coating process. The proposed algorithm demonstrates significant efficacy, achieving overall accuracy exceeding 85%, with a maximum performance of 100% under optimized conditions. These findings underscore the algorithm's potential to address critical quality control challenges in the domain of conductive textiles, a pivotal component in wearable technology applications. Table 2 presents a comparison between the results of this study and those obtained using various artificial intelligence models trained specifically for detecting stains or spot defects in actual textile fabrics. Except for the basic shallow CNN model, which achieved an accuracy level of 62.67%, other models demonstrated high accuracy, exceeding approximately 90%. The differences between the studies in terms of the image collection methods and training datasets used must be considered. However, this study optimized threshold criteria, achieving a remarkably high accuracy level of 100%, demonstrating the superiority of the proposed model.

Table 2. The comparative accuracy of various TIPAs.

Algorithm Model	Accuracy [%]
CNN [39]	62.67
VGG16 [39]	90.17
MobileNet [39]	97.64
InceptionV3 [39]	93.92
Xception [39]	96.89
DAGMC9 [34]	99.60
This study [2]	100

[2] Threshold ratio: 15%.

The primary contribution of this research lies in the multi-step processing framework encompassing preprocessing, filtering, and threshold optimization. By employing techniques such as histogram equalization and size-based filtering, our TIPA effectively identified and classified defect regions characterized by the uneven distribution of conductive particles. This approach builds upon existing studies, emphasizing the necessity of tailored image processing methodologies for detecting defects in textiles with intricate patterns and material heterogeneity. Moreover, this TIPA advances prior methodologies by optimizing threshold ratios to adapt to various defect types, thereby enhancing its robustness and applicability. A notable aspect of this study is the role of threshold ratios in controlling defect detection sensitivity. Adjusting the threshold ratio within the range of 5% to 30% allowed the algorithm to effectively adapt to diverse defect characteristics. This adaptability underscores our TIPA's flexibility and utility for a wide range of textile applications. While certain challenges persist in detecting large and diffuse stains, these cases were minimal and did not significantly impact the overall performance of the algorithm.

The limitations of this study lie in the classification criteria for grayscale-converted pixel values, which are influenced by various parameters. The filtering thresholds determined by the primary and secondary pixel sizes, as well as the classification of defective and normal images based on the first detected pixels and threshold ratio, are dependent on several factors. These include the base fabric color, the type of conductive particles, and the overall image resolution. As these factors can vary, it is essential to establish appropriate classification and filtering criteria tailored to specific conditions. This study aimed to propose a fundamental framework for the preprocessing and detection principles of a new TIPA, which is designed specifically to detect stain defects caused by the aggregation of conductive particles—one of the critical defects that can occur in conductive textiles.

Future research will focus on several key areas to further enhance the applicability and effectiveness of this TIPA. Expanding the dataset to include a broader variety of textiles and defect types will ensure greater generalizability across diverse manufacturing scenarios. Incorporating advanced machine learning techniques, such as convolutional neural networks (CNNs), could complement the current framework by enhancing its ability to detect complex defect patterns. Additionally, the development of real-time implementation strategies for industrial environments will be a critical step in validating the algorithm's practical utility. Further exploration of adaptive thresholding methods that respond dynamically to varying defect characteristics will also refine the algorithm's precision and robustness.

5. Conclusions

The findings of this study underscore TIPAs as transformative tools in the domain of textile quality control, particularly for conductive textiles employed in wearable technology. Our algorithm's high accuracy and adaptability render it highly effective for detecting defects in textiles with varied patterns and compositions. Beyond mere defect detection, the successful application of TIPAs promises numerous benefits, including enhanced production efficiency, reduced material wastage, and improved reliability of wearable devices that depend on conductive textiles.

In practical applications, TIPAs prove invaluable in identifying defects that occur when conductive textiles function as electronic circuits. For instance, these algorithms can detect anomalies such as stains that compromise both aesthetic and functional properties, or areas with abnormally high or low surface resistance that impede electrical performance. These capabilities are critical in contexts where consistent electrical properties are essential, such as wearable biosensors, flexible displays, or energy-harvesting electronic fabrics. By detecting such defects during the early stages of production, TIPAs significantly enhance the reliability and performance of the final products. The optimization and industrial deployment of TIPAs have the potential to set new benchmarks for quality control in smart textile manufacturing. Addressing current challenges and expanding our algorithm's functionality will establish TIPAs as a cornerstone for the integration of intelligent technologies in the textile sector, paving the way for innovative and dependable smart textile solutions.

Author Contributions: As a corresponding author, J.-Y.K. was responsible for the whole structure's construction, while S.-U.K. was responsible for reviewing, supervision, experiment design, and modeling. S.-U.K. was responsible for data collection, processing, and material selection. S.-U.K. and J.-Y.K. drafted the manuscript. All authors have read and agreed to the published version of the manuscript.

Funding: This research was partly funded by the Technology Innovation Program (or Industrial Strategic Technology Development Program—Materials/Parts Package Type) (20016038, for the development of textile–IT converged digital sensor modules for smart wear to monitor bio and activity signals during exercise, and KS standard), funded by the Ministry of Trade, Industry & Energy (MOTIE, Korea) and the Korea Institute for Advancement of Technology (KIAT) grant, funded by the Korean Government (MOTIE) (P0012770).

Data Availability Statement: All data are contained within the article.

Conflicts of Interest: The authors declare no conflicts of interest.

References

1. Shah, M.A.; Pirzada, B.M.; Price, G.; Shibiru, A.L.; Qurashi, A. Applications of nanotechnology in smart textile industry: A critical review. *J. Adv. Res.* **2022**, *38*, 55–75. [CrossRef] [PubMed]
2. Libanori, A.; Chen, G.; Zhao, X.; Zhou, Y.; Chen, J. Smart textiles for personalized healthcare. *Nat. Electron.* **2022**, *5*, 142–156. [CrossRef]
3. Nawawi, M.M.M.; Sidek, K.A.; Dafhalla, A.K.; Azman, A.W. Review on data acquisition of electrocardiogram biometric recognition in wearable smart textile shirts. *J. Phys.: Conf. Ser.* **2021**, *1900*, 012019. [CrossRef]
4. Nawawi, M.M.M.; Sidek, K.A.; Azman, A.W. Biometric Analysis on Smart Textile Garment in Real Life Scenario. *J. Adv. Res. Appl. Sci. Eng. Technol.* **2024**, *34*, 372–386. [CrossRef]
5. Shen, S.; Xiao, X.; Chen, J. Wearable triboelectric nanogenerators for heart rate monitoring. *Chem. Commun.* **2021**, *57*, 5871–5879. [CrossRef]
6. Kim, J.-S.; Truong, T.; Kim, J. Development of embroidery-type sensor capable of detecting respiration using the capacitive method. *Polymers* **2023**, *15*, 503. [CrossRef] [PubMed]
7. Min, S.D.; Yun, Y.; Shin, H. Simplified structural textile respiration sensor based on capacitive pressure sensing method. *IEEE Sens. J.* **2014**, *14*, 3245–3251.
8. Tang, C.; Xu, M.; Yi, W.; Zhang, Z.; Occhipinti, E.; Dong, C.; Ravenscroft, D.; Jung, S.-M.; Lee, S.; Gao, S. Ultrasensitive textile strain sensors redefine wearable silent speech interfaces with high machine learning efficiency. *NPJ Flex. Electron.* **2024**, *8*, 27. [CrossRef]
9. Lou, M.; Abdalla, I.; Zhu, M.; Wei, X.; Yu, J.; Li, Z.; Ding, B. Highly wearable, breathable, and washable sensing textile for human motion and pulse monitoring. *ACS Appl. Mater. Interfaces* **2020**, *12*, 19965–19973. [CrossRef]
10. Zhang, J.-w.; Zhang, Y.; Li, Y.-y.; Wang, P. Textile-based flexible pressure sensors: A review. *Polym. Rev.* **2022**, *62*, 65–94. [CrossRef]
11. Takamatsu, S.; Yamashita, T.; Imai, T.; Itoh, T. Lightweight flexible keyboard with a conductive polymer-based touch sensor fabric. *Sens. Actuators A Phys.* **2014**, *220*, 153–158. [CrossRef]
12. Büscher, G.H.; Kõiva, R.; Schürmann, C.; Haschke, R.; Ritter, H.J. Flexible and stretchable fabric-based tactile sensor. *Robot. Auton. Syst.* **2015**, *63*, 244–252. [CrossRef]
13. Su, M.; Li, P.; Liu, X.; Wei, D.; Yang, J. Textile-based flexible capacitive pressure sensors: A review. *Nanomaterials* **2022**, *12*, 1495. [CrossRef] [PubMed]
14. Alhashmi Alamer, F.; Almalki, G.A. Fabrication of conductive fabrics based on SWCNTs, MWCNTs and graphene and their applications: A review. *Polymers* **2022**, *14*, 5376. [CrossRef] [PubMed]
15. Kim, E.; Kim, S.; Chun, S.; Kim, J. Measurement of conductive fabrics electrical resistance by combining of image processing and convolutional neural network methods. *J. Ind. Text.* **2024**, *54*, 15280837241252182. [CrossRef]
16. Ojstršek, A.; Jug, L.; Plohl, O. A review of electro conductive textiles utilizing the dip-coating technique: Their functionality, durability and sustainability. *Polymers* **2022**, *14*, 4713. [CrossRef] [PubMed]
17. Cui, J.; Zhou, S. Highly conductive and ultra-durable electronic textiles via covalent immobilization of carbon nanomaterials on cotton fabric. *J. Mater. Chem. C* **2018**, *6*, 12273–12282. [CrossRef]
18. Khair, N.; Islam, R.; Shahariar, H. Carbon-based electronic textiles: Materials, fabrication processes and applications. *J. Mater. Sci.* **2019**, *54*, 10079–10101. [CrossRef]
19. Kahraman, Y.; Durmuşoğlu, A. Deep learning-based fabric defect detection: A review. *Text. Res. J.* **2023**, *93*, 1485–1503. [CrossRef]
20. Mahesh, B. Machine learning algorithms-a review. *Int. J. Sci. Res. (IJSR). [Internet]* **2020**, *9*, 381–386. [CrossRef]

21. Monga, V.; Li, Y.; Eldar, Y.C. Algorithm unrolling: Interpretable, efficient deep learning for signal and image processing. *IEEE Signal Process. Mag.* **2021**, *38*, 18–44. [CrossRef]
22. Li, C.; Li, J.; Li, Y.; He, L.; Fu, X.; Chen, J. Fabric defect detection in textile manufacturing: A survey of the state of the art. *Secur. Commun. Netw.* **2021**, *2021*, 9948808. [CrossRef]
23. Sikka, M.P.; Sarkar, A.; Garg, S. Artificial intelligence (AI) in textile industry operational modernization. *Res. J. Text. Appar.* **2024**, *28*, 67–83. [CrossRef]
24. Behera, B. Image-processing in textiles. *Text. Prog.* **2004**, *35*, 1–193. [CrossRef]
25. Conci, A.; Proença, C. A comparison between image-processing approaches to textile inspection. *J. Text. Inst.* **2000**, *91*, 317–323. [CrossRef]
26. Mohanty, A.K.; Bag, A. Detection and classification of fabric defects in textile using image mining and association rule miner. *Int. J. Electr. Electron. Comput.* **2017**, *2*, 28–33.
27. Nazir, M.K.; Ghani, M.A.N.U.; Ashraf, A.; Alam, S.M.; Farooq, R.U.; Latif, Z. A textile image classification based on texture and shape features. In Proceedings of the 2021 International Conference on Innovative Computing (ICIC), Online, 15–16 September 2021; pp. 1–6.
28. Shahrabadi, S.; Castilla, Y.; Guevara, M.; Magalhães, L.G.; Gonzalez, D.; Adão, T. Defect detection in the textile industry using image-based machine learning methods: A brief review. *J. Phys. Conf. Ser.* **2022**, *2224*, 012010. [CrossRef]
29. Wang, H.; Luo, H.; Zhang, X.; Zhao, Z.; Wang, J.; Li, Y. Automatic defect detection of carbon fiber woven fabrics using machine vision. *Mech. Adv. Mater. Struct.* **2024**, *31*, 10921–10934. [CrossRef]
30. Chen, M.; Yu, L.; Zhi, C.; Sun, R.; Zhu, S.; Gao, Z.; Ke, Z.; Zhu, M.; Zhang, Y. Improved faster R-CNN for fabric defect detection based on Gabor filter with Genetic Algorithm optimization. *Comput. Ind.* **2022**, *134*, 103551. [CrossRef]
31. Suryarasmi, A.; Chang, C.-C.; Akhmalia, R.; Marshallia, M.; Wang, W.-J.; Liang, D. FN-Net: A lightweight CNN-based architecture for fabric defect detection with adaptive threshold-based class determination. *Displays* **2022**, *73*, 102241. [CrossRef]
32. Shang, H.; Li, P.; Zhang, X.; Peng, X. Overview of fabric defect detection techniques based on computer vision. In Proceedings of the Third International Conference on Computer Graphics, Image, and Virtualization (ICCGIV 2023), Nanjing, China, 16–18 June 2023; pp. 108–115.
33. Guan, S. Fabric defect detection using an integrated model of bottom-up and top-down visual attention. *J. Text. Inst.* **2016**, *107*, 215–224. [CrossRef]
34. Wang, D.; Yu, W.; Lian, P.; Zhang, M. Textile Defect Detection Algorithm Based on Unsupervised Learning. In Proceedings of the 2022 7th International Conference on Image, Vision and Computing (ICIVC), Xi'an, China, 26–28 July 2022; pp. 81–86.
35. Heidari, N.; Azmi, R.; Pishgoo, B. Fabric textile defect detection, by selecting a suitable subset of wavelet coefficients, through genetic algorithm. *Int. J. Image Process. (IJIP)* **2011**, *5*, 25.
36. Aminzadeh, M.; Kurfess, T. Automatic thresholding for defect detection by background histogram mode extents. *J. Manuf. Syst.* **2015**, *37*, 83–92. [CrossRef]
37. Zhou, J.; Wang, J. Fabric defect detection using adaptive dictionaries. *Text. Res. J.* **2013**, *83*, 1846–1859. [CrossRef]
38. Kang, X.; Zhang, E. A universal and adaptive fabric defect detection algorithm based on sparse dictionary learning. *IEEE Access* **2020**, *8*, 221808–221830. [CrossRef]
39. Karegowda, A.G.; Pooja, R.; Rani, A.L.; Devika, G. Detection of Stain Defects in Textile Industry using State-of-Art Transfer Learning Models. In Proceedings of the 2024 International Conference on Smart Systems for applications in Electrical Sciences (ICSSES), Tumakuru, India, 3–4 May 2024; pp. 1–6.

Disclaimer/Publisher's Note: The statements, opinions and data contained in all publications are solely those of the individual author(s) and contributor(s) and not of MDPI and/or the editor(s). MDPI and/or the editor(s) disclaim responsibility for any injury to people or property resulting from any ideas, methods, instructions or products referred to in the content.

processes

MDPI

Article

Improving Human Activity Recognition Through 1D-ResNet: A Wearable Wristband for 14 Workout Movements

Sang-Un Kim [1] and Joo-Yong Kim [2,*]

1 Department of Smart Wearable Engineering, Soongsil University, Seoul 06978, Republic of Korea; tkddnsl0723@naver.com
2 Department of Materials Science and Engineering, Soongsil University, Seoul 06978, Republic of Korea
* Correspondence: jykim@ssu.ac.kr; Tel.: +82-10-8720-0631

Abstract: This study presents a 1D Residual Network(ResNet)-based algorithm for human activity recognition (HAR) focused on classifying 14 different workouts, which represent key exercises commonly performed in fitness training, using wearable inertial measurement unit (IMU) sensors. Unlike traditional 1D Convolutional neural network (CNN) models, the proposed 1D ResNet incorporates residual blocks to prevent gradient vanishing and exploding problems, allowing for deeper networks with improved performance. The IMU sensor, placed on the wrist, provided Z-axis acceleration data, which were used to train the model. A total of 901 data samples were collected from five participants, with 600 used for training and 301 for testing. The model achieved a recognition accuracy of 97.09%, surpassing the 89.03% of a 1D CNN without residual blocks and the 92% of a cascaded 1D CNN from previous research. These results indicate that the 1D ResNet model is highly effective in recognizing a wide range of workouts. The findings suggest that wearable devices can autonomously classify human activities and provide personalized training recommendations, paving the way for AI-driven personal training systems.

Keywords: 1D ResNet; neural network; artificial intelligence; human activity recognition; wearable healthcare

Academic Editors: Zhanxiao Kang and Qing Chen

Received: 12 December 2024
Revised: 6 January 2025
Accepted: 10 January 2025
Published: 13 January 2025

Citation: Kim, S.-U.; Kim, J.-Y. Improving Human Activity Recognition Through 1D-ResNet: A Wearable Wristband for 14 Workout Movements. *Processes* **2025**, *13*, 207. https://doi.org/10.3390/pr13010207

Copyright: © 2025 by the authors. Licensee MDPI, Basel, Switzerland. This article is an open access article distributed under the terms and conditions of the Creative Commons Attribution (CC BY) license (https://creativecommons.org/licenses/by/4.0/).

1. Introduction

In recent years, human activity recognition (HAR) using wearable sensors has gained considerable attention in the wearable technology field. HAR, which involves classifying and recognizing human movements and behaviors, is being explored in a wide range of applications, including virtual reality (VR) [1–4], sports [5–7], healthcare [8], robotics [9–11], and entertainment [12,13]. In these areas, various types of sensors are utilized to track or monitor movements. Broadly, these can be categorized into vision-based sensors and wearable sensors. Vision-based sensors, such as Microsoft's Kinect [14,15], radar [16,17], and optical devices like cameras, track movements externally by employing image processing or signal analysis techniques. In contrast, wearable sensors measure actual movements through data such as joint angles, acceleration, and gyroscopic information, often utilizing sensor fusion methods to provide more accurate motion tracking [18–20].

Vision-based sensors are capable of tracking human movement without limitations on the number of joint points or body parts, enabling real-time analysis and making them suitable for highly precise measurements and detailed analysis. However, there are significant limitations to this method. Optical sensors are prone to inaccuracies due to environmental factors such as lighting conditions, reflections, or obstacles, as well as the resolution limitations of the equipment itself. Consequently, the use of expensive equipment and the necessity of a controlled environment often make this approach less feasible for widespread use. Additionally, as the measurement duration increases, the volume of image processing data grows exponentially, requiring substantial computational resources and leading to increased costs and time for both measurement and analysis.

On the other hand, wearable sensors offer significant convenience by being minimally attached to critical joints or positions while still effectively analyzing or recognizing movements [21–23]. The smaller amount of data they generate allows for easy storage in compact modules or devices and enables real-time wireless transmission through technologies like Bluetooth or Wi-Fi. Furthermore, wearable sensors are less affected by external environmental factors, reducing the margin of error, which makes them particularly well-suited for outdoor activities, training, and exercise. Their accessibility and ease of use are enhanced by their relatively low cost, allowing for widespread adoption. Moreover, these sensors are typically lightweight and flexible, enabling prolonged use without causing discomfort to the wearer. Also, wearable sensors, due to their advantage of being wearable, can be utilized not only in special scenarios such as dangerous situations involving soldiers, firefighters, or police officers, artificial joints that move in coordination with the limbs of individuals with amputated body parts, and sensing abnormal muscle tremors caused by irregular brain signals in Parkinson's disease, but also in everyday applications such as monitoring athletes, supporting dieting efforts, and measuring sleep patterns.

Notably, numerous studies have utilized wrist-mounted IMU sensors for human activity recognition (HAR), as the wrist exhibits significant movement during various physical activities, making it an ideal location for capturing representative HAR data. In a related study, a wearable device was developed to recognize gestures from wrist movements occurring during motion [24]. Another study focused on detecting abrupt gestures rather than repetitive ones [25]. Such studies highlight the extensive research being conducted to interpret user intentions and states from the complex data generated by the hand and wrist, which are commonly involved in various activities [26]. In this study, we considered a watch-type wearable product, which is a representative wearable sensor, and attached an IMU sensor to the wrist to obtain Z-axis acceleration data for recognizing 14 types of movements.

Another reason why research on human activity recognition (HAR) is receiving increased attention is the rapid advancement of recognition algorithms driven by artificial intelligence (AI) [27–29]. With the development of machine learning and deep learning techniques, AI can now efficiently classify and recognize the complex and nuanced features present in human movement data. This has greatly simplified the process of analyzing and interpreting intricate motion patterns, which were previously difficult to capture and understand. AI's ability to handle large datasets and uncover hidden patterns has significantly enhanced the accuracy and efficiency of HAR systems. In previous research, a workout classification algorithm using a 1D CNN was designed to classify fitness exercises using artificial intelligence. To improve accuracy, a cascade structure was employed, where the first step involved pre-processing by grouping exercises into categories before final classification. This approach improved the classification accuracy from 82% to 92% [30]. Although traditional artificial intelligence models have demonstrated remarkable achievements across various domains by utilizing deep learning techniques, they face limitations due to vanishing or exploding gradient problems in deep learning layers. These issues result in significantly lower training accuracy as the gradient, propagated backward during training, diminishes or amplifies excessively as the network depth increases. Such limitations become more pronounced when attempting to utilize deeper network architectures, posing a significant obstacle to effectively learning complex data patterns [31].

In this study, we employed ResNet, a cutting-edge artificial intelligence image classification model that utilizes residual blocks, to develop a deeper-layer human activity recognition (HAR) algorithm and enhance its accuracy. ResNet was introduced during the 2015 ImageNet Large Scale Visual Recognition Challenge (ILSVRC), where it demonstrated a breakthrough by extracting features through 152 layers, significantly deeper than the 20–30 layers used in previous algorithms. One of the key innovations of ResNet is the use of residual blocks, which help overcome the vanishing gradient problem that often occurs in very deep neural networks. As a result, ResNet achieved an error rate of 3.57%, which is notably lower than the human error rate of around 5%, establishing itself as one of the most advanced recognition algorithms available [31]. One of the most prominent features of ResNet is the residual block, which effectively addresses the vanishing or exploding gradient problem that occurs as neural networks become deeper. In traditional deep learning models, the gradient can diminish to zero or explode as it propagates back through many layers due to extensive calculations and weight adjustments. ResNet resolves this issue by adding the initial input (residual) to the output of the deeper layers, a method known as skip connection (or short connection). These skip connections ensure that gradients can flow more easily through the network, preventing their loss or divergence. By designing networks with multiple residual blocks, ResNet not only preserves gradient flow but also enables much deeper architectures, resulting in improved accuracy and more stable learning.

In summary, this study developed an algorithm to recognize 14 different workouts using the Z-axis accelerometer sensor of a wristband. To enhance accuracy, we utilized ResNet, which features a residual block structure designed to mitigate the vanishing and exploding gradient issues common in deep neural networks. A comparative analysis was conducted against existing studies, focusing on improvements in classification accuracy and model performance. The use of the Z-axis accelerometer was particularly advantageous for capturing vertical motion data, making it suitable for recognizing a wide range of workouts.

2. Materials and Methods

2.1. Z-Axis Acceleration Data for Workout Recognition

In this study, we used an Inertial Measurement Unit (IMU) sensor as a wearable device to measure the Z-axis acceleration of the wrist. The IMU sensor is composed of a 3-axis accelerometer and a 3-axis gyroscope, providing a 6-axis measurement system. Additionally, some IMU sensors include a 3-axis magnetometer, forming a 9-axis system, which allows for the measurement of orientation relative to the Earth's magnetic field. All of these sensors are designed specifically for motion tracking. In this research, we employed the EBMotion V5.2, an IMU sensor from E2BOX (Hanam, Gyeonggi, Republic of Korea), which includes a wireless sensor (EBIMU24GV52) and a wireless receiver (EBRCV24GV5). This setup was used to accurately capture the Z-axis acceleration data. When measuring motion related to physical activities using Z-axis acceleration data, the Z-axis is more significant than the X and Y axes because it is the axis most affected by gravity when we are standing. One of the challenges with accelerometers is that it is difficult to determine the initial position of the object. When a person is standing still, the accelerometer detects gravitational acceleration along the Z-axis, resulting in a value of approximately -g. This value is recorded even when there is no motion, because the sensor continuously measures the gravitational force. However, during movement or when assuming a preparatory posture for specific activities, the Z-axis value is influenced by both internal and external factors due to gravity. These variations enable the extraction of meaningful features for motion analysis. This concept is depicted in Figure 1, and the equation representing the change in the accelerometer's angle with respect to its position is provided below.

$$\vec{A_z} = -\vec{g} \times cos\theta + \vec{l} \tag{1}$$

where $\vec{A_z}$ is the acceleration vector along the Z-axis, $\vec{g}$ is the acceleration vector of gravity, $\vec{l}$ indicates the acceleration vector of the arm movement over time, and θ is the angle between the $\vec{A_z}$ and $\vec{g}$. At the initial position, the sensor's acceleration value is determined by calculating the angle difference between the gravitational force and the sensor's Z-axis. This angle can be obtained through trigonometric functions or rotational transformations, depending on the orientation of the sensor. As the workout progresses, motion-induced acceleration is added to the initial value over time, allowing for the comprehensive tracking of dynamic movements.

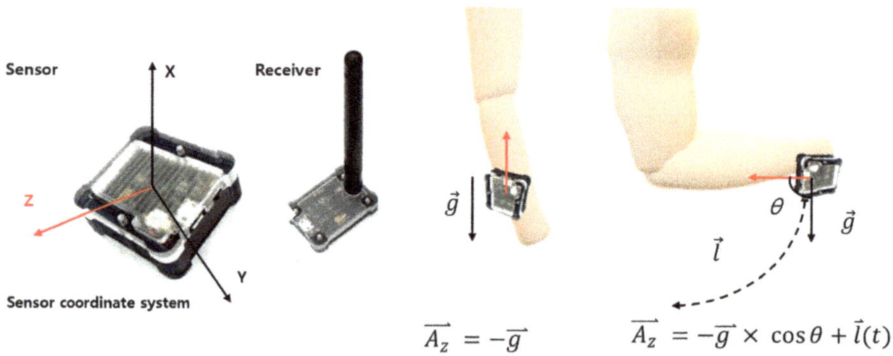

Figure 1. The sensing mechanism for workout recognition using the Z-axis acceleration of an IMU sensor using gravity.

2.2. 14 Workout Data

In this study, with reference to previous studies [30], a wristband equipped with an IMU sensor was designed, as shown in Figure 2, and worn on the right wrist to collect exercise data. The red arrow indicates the Z-axis of the sensor. A total of 14 different exercises, illustrated in Figure 3, were measured, with the initial movements based on the actions performed on the left side. To minimize the effects of body tremors and measurement noise during movements, a low-pass filter with a cutoff frequency of 40 Hz was applied during data processing. The filter was designed to remove high-frequency noise while preserving the relevant motion signals. The sampling frequency was set to 10 Hz, and a total of 18,020 data points were collected. Data from five male participants were used in this study, with an average height of 176 cm and an average weight of 77.4 kg. Detailed information about the participants, including variability in their physical characteristics, is provided in Table 1.

Table 1. The characteristics of 14 workout subjects.

Subject	Height (cm)	Weight (kg)
Subject 1	170	75
Subject 2	183	80
Subject 3	180	78
Subject 4	173	80
Subject 5	174	74

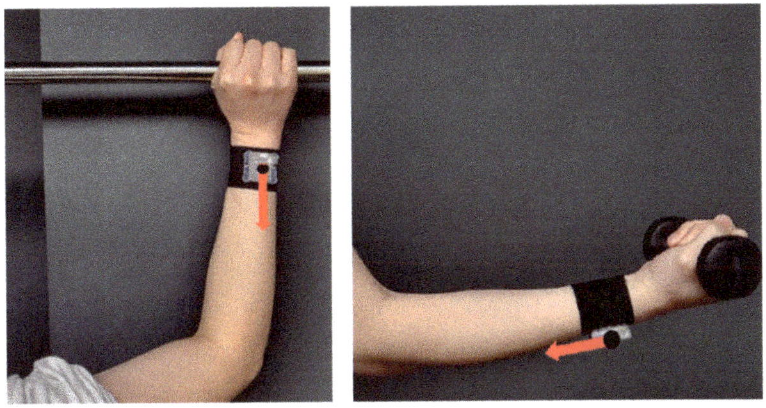

Figure 2. The wristband with IMU sensor and the red arrows are the Z-axis in the sensor coordinate system.

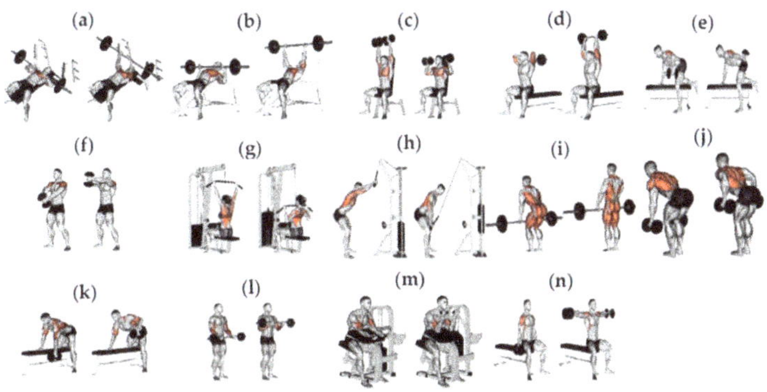

Figure 3. The 14 workouts were chosen for recognition. (**a**) Bench press, (**b**) Incline bench press, (**c**) Dumbbell shoulder press, (**d**) Dumbbell triceps extension, (**e**) Dumbbell kick, (**f**) Dumbbell front raise, (**g**) Lat pull down, (**h**) Straight arm lat pull down, (**i**) Deadlift, (**j**) Dumbbell bent row, (**k**) One-arm dumbbell row, (**l**) EZ-bar curls, (**m**) Machine preacher curl, (**n**) Seated dumbbell lateral raise [30].

2.3. 1D-ResNet

All algorithm structures and code in this study were designed and implemented using MATLAB R2023a. The original ResNet is designed for image classification and recognition, typically handling 2D and 3D image data. However, since the Z-axis data used in this study are one-dimensional, we modified the input data structure to accommodate 1D data by replacing the input layer with a sequence input layer. For the input data, 20 data points, corresponding to approximately 2 s of data with a sampling frequency of 10 Hz and one workout cycle, were grouped into sequences within single cells, resulting in a total of 901 sequences. The output data were labeled according to the corresponding workout type. Out of the 901 data samples, 600 were randomly allocated for training and 301 for testing. This 2:1 split ensures that the model has sufficient data to learn while maintaining enough data for evaluation. The data were randomly split to ensure an unbiased distribution between the training and test sets.

The key concept of ResNet, the residual block, can be divided into the main path and the skip connection. The main path consists of convolutional layers, batch normalization layers, and ReLu layers where the input data undergo processing. The skip connection, however, allows the initial input to bypass any processing and pass directly to the output. At the end of the process, the values from both the main path and the skip connection are combined through residual mapping, which helps mitigate the vanishing gradient problem and allows deeper networks to perform better. The equation for this process is as follows:

$$X + F(X) \tag{2}$$

where X is the input value and $F(X)$ is the value that passes through the main path. Additionally, the residual block where residual mapping occurs can be illustrated as shown in Figure 4.

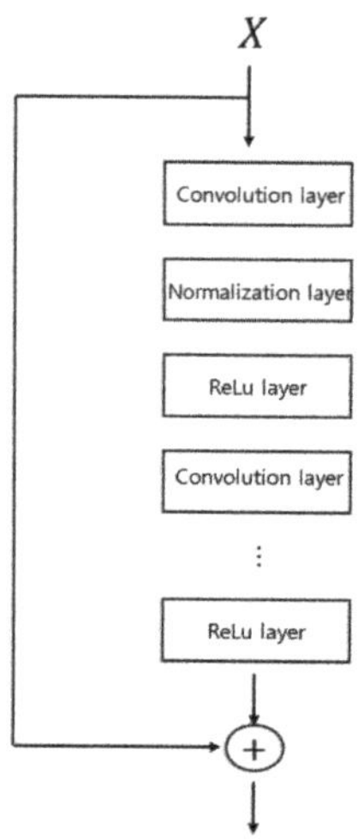

Figure 4. The main path and skip connection of residual block in ResNet.

The 1D-ResNet model in this study, as depicted in Figure 5, consists of three residual blocks. Each convolutional layer has a kernel size of 3 × 3, with 64 filters in the first block, 128 filters in the second block, and 256 filters in the third block. Following the residual blocks, a global average pooling layer is used to reduce the dimensionality while retaining important features. This is followed by two fully connected layers, a dropout layer with a dropout rate of 0.5 to prevent overfitting, and a softmax layer for calculating class probabilities. The final classification layer then recognizes the workout based on these features. The algorithm was trained using Root Mean Square Propagation (RMSprop) as the solver, chosen for its ability to adapt the learning rate during training, which helps in improving convergence. The model was trained for 200 epochs (800 iterations) with a batch size of 128, an initial learning rate of 0.001 and decrease factor of 0.5 per 25 epochs, allowing for a balance between learning speed and stability. Furthermore, we conducted the same training and validation processes with 3D input data comprising X-, Y-, and Z-axis dimensions to investigate the performance differences compared to the 1D ResNet model.

To validate the performance of the trained 1D ResNet algorithm, we utilized 301 test data samples as input to generate predictions. These predictions were then used to calculate not only the overall accuracy but also key classification metrics such as precision, recall, and *F1-score*. These metrics were calculated as follows:

$$Precision = \frac{TP}{TP + FP} \tag{3}$$

$$Recall = \frac{TP}{TP + FN} \quad (4)$$

$$F1\ score = 2\frac{Precision \times Recall}{Precision + Recall} \quad (5)$$

where the true positive (TP) refers to the number of positive samples correctly classified as positive. False positive (FP) is the number of negative samples incorrectly classified as positive, and false negative (FN) refers to the number of positive samples incorrectly classified as negative.

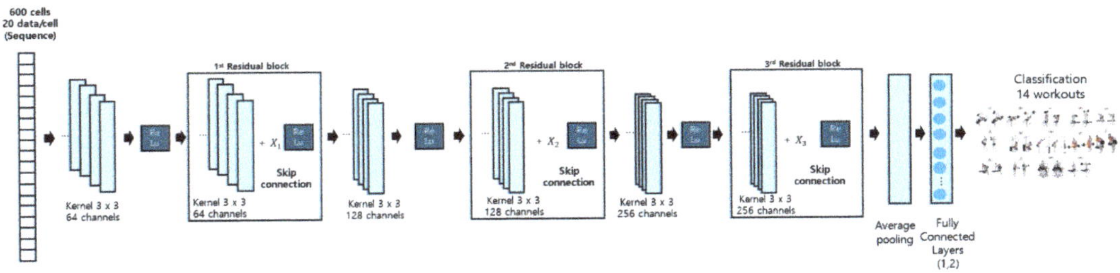

Figure 5. The architecture of 1D ResNet.

Precision measures the proportion of true positive predictions (TP) out of all predictions made for the positive class (TP + FP). It evaluates the model's ability to avoid false positives. Recall, also known as sensitivity, measures the proportion of actual positive instances correctly identified (TP) out of the total actual positives (TP + FN). It evaluates the model's ability to detect true positives. The *F1-Score* is the harmonic mean of precision and recall, balancing their trade-offs. It is particularly useful when dealing with imbalanced datasets.

Finally, to verify the reliability and precise applicability of the algorithm modeling method, we excluded 173 data samples from one participant (segment 2) who performed the 14 workout movements. The remaining 728 data samples from the other four participants were used for training, and the fresh data from segment 2 were utilized for testing.

3. Results and Discussion

3.1. Evaluation of 1D ResNet

Figure 6 presents the Z-axis acceleration data from a single sequence for the bench press exercise, recorded from subject 1 and used as input for the ResNet model in this study. With a sampling frequency of 10 Hz and an approximate workout duration of 2 s, each sequence is composed of 20 data points. For the bench press, the upward and downward movements of the bar, driven by the hands and arms, are clearly captured in the Z-axis data. This highlights the suitability of the Z-axis acceleration for accurately reflecting the motion dynamics of this exercise.

As illustrated in Figure 7, the training accuracy of both the 1D ResNet and 3D ResNet models was evaluated using the 600 training data samples across the training epochs. To monitor overfitting, the validation accuracy was also measured and compared using the 301 test data samples. While the 3D ResNet model achieved 100% training accuracy by epoch 200 and maintained it throughout, its validation accuracy fluctuated between 80% and 90%, ultimately stabilizing at 87.02%. This discrepancy can be attributed to the increased information complexity in the 3D data, which allowed the model to create more nuanced classifications for training data but led to overfitting, as evident from the lower validation accuracy compared to training accuracy. In contrast, the 1D ResNet model developed in this study also achieved a final training accuracy of 100% while maintaining a validation accuracy of 97.02%, closely matching its training performance. These results indicate that the 1D ResNet model achieved higher accuracy while effectively minimizing overfitting. This suggests that for workout data, which inherently include noise and human error, a simpler yet efficient model like 1D ResNet, which focuses on essential information rather than excessive data, is more suitable for classification tasks.

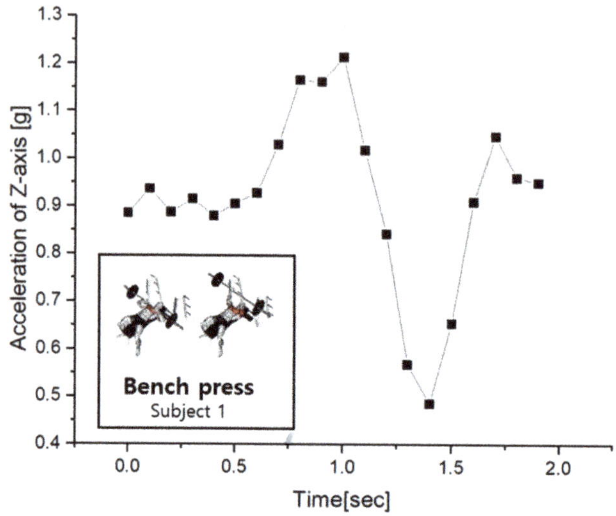

Figure 6. The Z-axis acceleration data of bench press for subject 1.

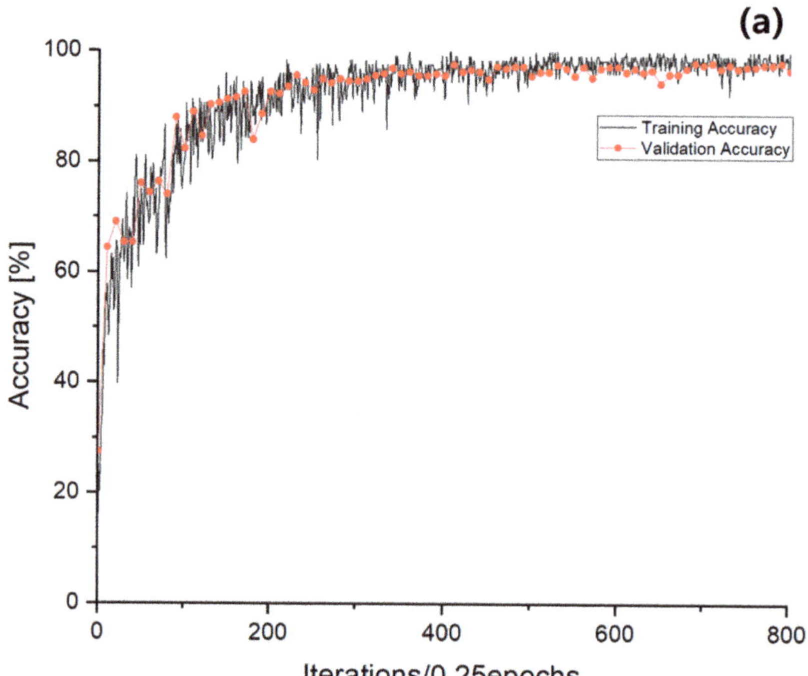

Figure 7. *Cont.*

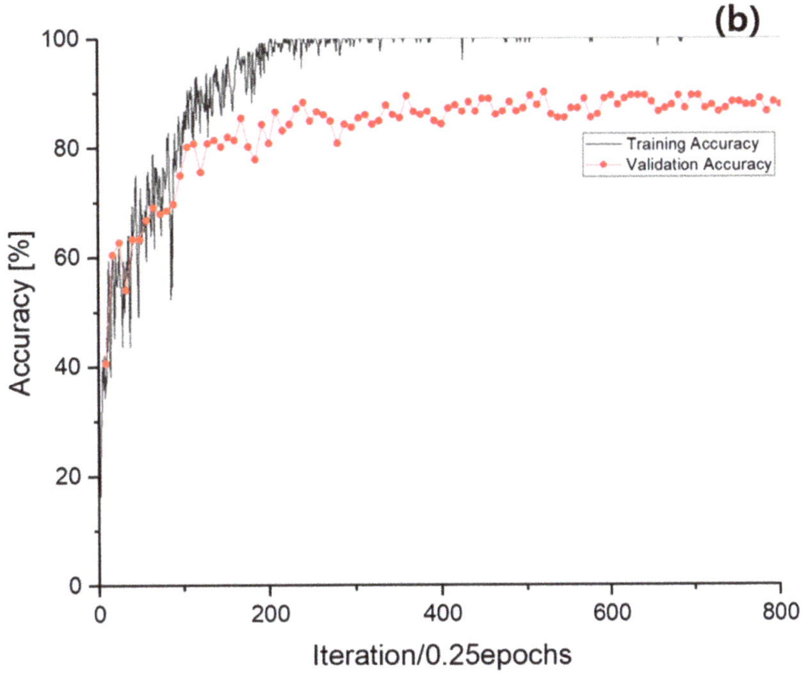

Figure 7. The accuracy of training and validation of (**a**) 1D ResNet and (**b**) 3D ResNet.

Figure 8 presents the confusion matrix, which compares the predicted workout results from the 1D ResNet model trained on the final 301 test data samples with the actual workout labels. The confusion matrix helps visualize how well the model predicts each workout category. First, misclassifications were observed in workouts (a), (b), and (c)—Bench press, Incline bench press, and Dumbbell shoulder press. Although these exercises involve different degrees of upper body inclination, the arm movements are highly similar, which likely contributed to the higher misclassification rate. Nevertheless, with a minimum accuracy of 88.2%, the model's performance is still sufficient for reliable workout recognition. Secondly, workouts (j) and (k)—Dumbbell kickback and One-arm dumbbell row—also had some misclassifications. The primary differences between these exercises involve whether both arms are engaged and whether the foot is elevated during the movement. Despite the similarity in arm motion, the misclassification rate remained low, with a maximum error rate of only 6.9%, suggesting that the algorithm is adequately robust for workout recognition.

Table 2 presents the classification metrics, including precision, recall, and F1-score, for each of the 14 workout categories evaluated in this study using the 1D ResNet model. The results demonstrate the model's high effectiveness in recognizing diverse workout types. The average precision, recall, and F1-score across all categories were 0.98, 0.97, and 0.97, respectively, indicating consistent performance across various exercises. Notably, exercises such as (d) Dumbbell triceps extension, (j) Dumbbell kickback, and (k) One-arm dumbbell row achieved perfect scores of 1.0 across all metrics, showcasing the model's robustness in identifying these movements.

Figure 9 and Table 3 present the results obtained using the same algorithm modeling method, excluding the data from one participant (segment 2) out of the total of five participants. Compared to training with the complete dataset, the accuracy decreased to 90.17%. The confusion matrix, which compares actual and predicted classifications, revealed that while the confusion among exercises (a), (b), and (c) remained the same, additional confusion was observed between exercises (d) and (m). This confusion is presumed to be influenced by the similar degree of arm rotation between the two exercises, despite their differing directions, and may have been further impacted by human error. As shown in Table 3, precision, recall, and F1-score decreased overall but remained approximately 90%. Considering the application to new data, this level of performance is deemed sufficiently high.

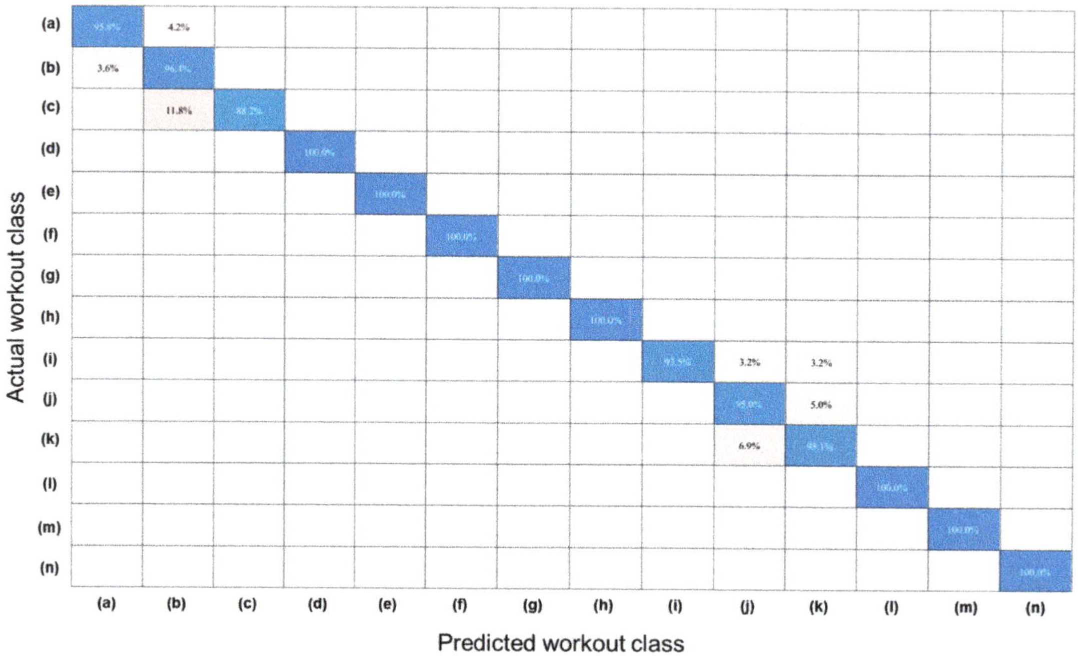

Figure 8. The confusion matrix of workout recognition 1D ResNet.

Table 2. The metrics of 1D Resnet model.

Workout	Precision	Recall	F1-Score
(a)	1	0.94	0.97
(b)	0.93	0.93	0.93
(c)	0.90	0.96	0.93
(d)	1	1	1
(e)	1	1	1
(f)	1	1	1
(g)	1	0.88	0.94
(h)	0.86	0.95	0.90
(i)	0.96	0.96	0.96
(j)	1	1	1
(k)	1	1	1
(l)	1	1	1
(m)	1	1	1
(n)	1	1	1
	0.98 *	0.97 *	0.97 *

* The average value of each metric.

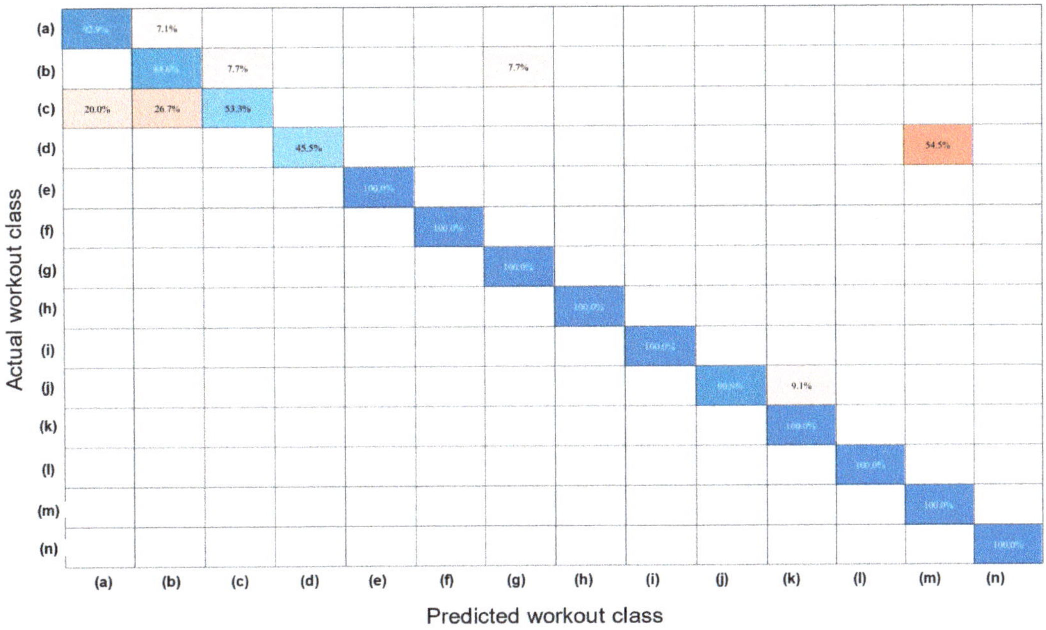

Figure 9. The confusion matrix of workout recognition 1D ResNet fresh test.

Table 3. The metrics of 1D Resnet fresh test model.

Workout	Precision	Recall	F1-Score
(a)	0.81	0.93	0.87
(b)	0.69	0.85	0.76
(c)	0.89	0.53	0.67
(d)	1	0.45	0.62
(e)	1	1	1
(f)	1	1	1
(g)	0.75	1	0.86
(h)	1	1	1
(i)	1	1	1
(j)	1	0.91	0.95
(k)	0.93	1	0.97
(l)	1	1	1
(m)	0.68	1	0.81
(n)	1	1	1
	0.91 *	0.91 *	0.89 *

* The average value of each metric.

3.2. Comparison with Previous Research

Finally, the accuracy of the 1D ResNet model for workout recognition in this study was compared with previous studies and a 1D CNN model without the residual block structure to assess the impact of the residual block on performance. This comparison highlights how the addition of residual blocks improves the model's ability to recognize workouts accurately. Although the 1D CNN without residual mapping had the same number of learning layers, its recognition accuracy was 89.03%. The cascaded 1D CNN structure from previous research achieved a higher accuracy of 92% [30]. However, the 1D ResNet model developed in this study achieved the highest recognition accuracy of

97.09%, demonstrating that the inclusion of residual mapping significantly improves model performance compared to both previous approaches.

The input values passed through residual mapping help retain crucial information from the initial IMU data related to the starting posture, minimizing the loss of valuable input features. Moreover, by preventing gradient vanishing and exploding issues in deep learning layers, the 1D ResNet model demonstrated higher accuracy in recognizing 14 different exercises compared to previous research that utilized a traditional 1D CNN algorithm. This shows that the use of residual blocks improves both information retention and overall model performance.

4. Conclusions

In summary, this study developed an algorithm for classifying 14 different workouts within the field of human activity recognition (HAR) using artificial intelligence in the wearable technology sector. By leveraging ResNet, which has demonstrated low error rates in recent studies, this research achieved substantial improvements in accuracy compared to previous algorithms. These findings suggest the potential for AI-powered personal training systems, where wearable devices can autonomously recognize user activities, gather data, and provide tailored workout techniques and plans to users. This system would enable more efficient and personalized training without requiring direct user input, paving the way for more advanced AI-driven personal training solutions.

The future research will focus on developing algorithms capable of classifying a wider variety of exercises and enhancing the applicability of the model by incorporating data from a larger and more diverse population. These efforts aim to improve the robustness and generalizability of the system, paving the way for broader adoption and effectiveness in real-world applications.

Author Contributions: As a corresponding author, J.-Y.K. was responsible for the whole structure construction, while S.-U.K. was responsible for reviewing, supervising, experiment design, and modeling. S.-U.K. was responsible for the data collection, processing, and material selection. S.-U.K. and J.-Y.K. drafted the manuscript. All authors have read and agreed to the published version of the manuscript.

Funding: This research was partly funded by the Technology Innovation Program (or Industrial Strategic Technology Development Program-Materials/Parts Package Type) (20016038, development of the textile-IT converged digital sensor modules for smart wear to monitor bio and activity signal in exercise, and KS standard) funded by the Ministry of Trade, Industry and Energy (MOTIE, Korea) and the Korea Institute for Advancement of Technology (KIAT) grant funded by the Korean Government (MOTIE) (P0012770).

Institutional Review Board Statement: Not applicable.

Data Availability Statement: Data are contained within the article.

Conflicts of Interest: The authors declare no conflicts of interest.

References

1. Fangbemi, A.S.; Liu, B.; Yu, N.H.; Zhang, Y. Efficient human action recognition interface for augmented and virtual reality applications based on binary descriptor. In *Augmented Reality, Virtual Reality, and Computer Graphics, Proceedings of the 5th International Conference, AVR 2018, Otranto, Italy, 24–27 June 2018, Proceedings, Part I 5*; Springer International Publishing: Cham, Switzerland, 2018; pp. 252–260.
2. Xia, C.; Sugiura, Y. Optimizing sensor position with virtual sensors in human activity recognition system design. *Sensors* **2021**, *21*, 6893. [CrossRef] [PubMed]

3. Xiao, F.; Pei, L.; Chu, L.; Zou, D.; Yu, W.; Zhu, Y.; Li, T. A deep learning method for complex human activity recognition using virtual wearable sensors. In *Spatial Data and Intelligence, Proceedings of the First International Conference, SpatialDI 2020, Virtual Event, 8–9 May 2020, Proceedings 1*; Springer International Publishing: Cham, Switzerland, 2021; pp. 261–270.
4. Jeyakumar, J.V.; Lai, L.; Suda, N.; Srivastava, M. SenseHAR: A robust virtual activity sensor for smartphones and wearables. In Proceedings of the 17th Conference on Embedded Networked Sensor Systems, New York, NY, USA, 10–13 November 2019; pp. 15–28.
5. Schuldhaus, D. *Human Activity Recognition in Daily Life and Sports Using Inertial Sensors*; FAU University Press: Boca Raton, FL, USA, 2019.
6. Host, K.; Ivašić-Kos, M. An overview of Human Action Recognition in sports based on Computer Vision. *Heliyon* **2022**, *8*, e09633. [CrossRef] [PubMed]
7. Pajak, G.; Krutz, P.; Patalas-Maliszewska, J.; Rehm, M.; Pajak, I.; Dix, M. An approach to sport activities recognition based on an inertial sensor and deep learning. *Sens. Actuators A Phys.* **2022**, *345*, 113773. [CrossRef]
8. Bibbò, L.; Vellasco, M.M. Human activity recognition (HAR) in healthcare. *Appl. Sci.* **2023**, *13*, 13009. [CrossRef]
9. Frank, A.E.; Kubota, A.; Riek, L.D. Wearable activity recognition for robust human-robot teaming in safety-critical environments via hybrid neural networks. In Proceedings of the 2019 IEEE/RSJ International Conference on Intelligent Robots and Systems (IROS), Macau, China, 3–8 November 2019; pp. 449–454.
10. Jaramillo, I.E.; Jeong, J.G.; Lopez, P.R.; Lee, C.-H.; Kang, D.-Y.; Ha, T.-J.; Oh, J.-H.; Jung, H.; Lee, J.H.; Lee, W.H. Real-time human activity recognition with IMU and encoder sensors in wearable exoskeleton robot via deep learning networks. *Sensors* **2022**, *22*, 9690. [CrossRef]
11. Martínez-Villaseñor, L.; Ponce, H. A concise review on sensor signal acquisition and transformation applied to human activity recognition and human–robot interaction. *Int. J. Distrib. Sens. Netw.* **2019**, *15*, 1550147719853987. [CrossRef]
12. Hoelzemann, A.; Romero, J.L.; Bock, M.; Laerhoven, K.V.; Lv, Q. Hang-time HAR: A benchmark dataset for basketball activity recognition using wrist-worn inertial sensors. *Sensors* **2023**, *23*, 5879. [CrossRef]
13. Wang, Z.; Wu, D.; Chen, J.; Ghoneim, A.; Hossain, M.A. A triaxial accelerometer-based human activity recognition via EEMD-based features and game-theory-based feature selection. *IEEE Sens. J.* **2016**, *16*, 3198–3207. [CrossRef]
14. Zhang, Z. Microsoft kinect sensor and its effect. *IEEE Multimed.* **2012**, *19*, 4–10. [CrossRef]
15. Han, J.; Shao, L.; Xu, D.; Shotton, J. Enhanced computer vision with microsoft kinect sensor: A review. *IEEE Trans. Cybern.* **2013**, *43*, 1318–1334.
16. Li, X.; He, Y.; Jing, X. A survey of deep learning-based human activity recognition in radar. *Remote Sens.* **2019**, *11*, 1068. [CrossRef]
17. Zhu, S.; Guendel, R.G.; Yarovoy, A.; Fioranelli, F. Continuous human activity recognition with distributed radar sensor networks and CNN–RNN architectures. *IEEE Trans. Geosci. Remote Sens.* **2022**, *60*, 5115215. [CrossRef]
18. Mohammadzadeh, F.F.; Liu, S.; Bond, K.A.; Nam, C.S. Feasibility of a wearable, sensor-based motion tracking system. *Procedia Manuf.* **2015**, *3*, 192–199. [CrossRef]
19. Longo, U.G.; De Salvatore, S.; Sassi, M.; Carnevale, A.; De Luca, G.; Denaro, V. Motion tracking algorithms based on wearable inertial sensor: A focus on shoulder. *Electronics* **2022**, *11*, 1741. [CrossRef]
20. Rana, M.; Mittal, V. Wearable sensors for real-time kinematics analysis in sports: A review. *IEEE Sens. J.* **2020**, *21*, 1187–1207. [CrossRef]
21. Poitras, I.; Dupuis, F.; Bielmann, M.; Campeau-Lecours, A.; Mercier, C.; Bouyer, L.J.; Roy, J.-S. Validity and reliability of wearable sensors for joint angle estimation: A systematic review. *Sensors* **2019**, *19*, 1555. [CrossRef]
22. Bakhshi, S.; Mahoor, M.H. Development of a wearable sensor system for measuring body joint flexion. In Proceedings of the 2011 International Conference on Body Sensor Networks, Dallas, TX, USA, 23–25 May 2011; pp. 35–40.
23. Teague, C.N.; Heller, J.A.; Nevius, B.N.; Carek, A.M.; Mabrouk, S.; Garcia-Vicente, F.; Inan, O.T.; Etemadi, M. A wearable, multimodal sensing system to monitor knee joint health. *IEEE Sens. J.* **2020**, *20*, 10323–10334. [CrossRef]
24. Zhao, H.; Ma, Y.; Wang, S.; Watson, A.; Zhou, G. MobiGesture: Mobility-aware hand gesture recognition for healthcare. *Smart Health* **2018**, *9*, 129–143. [CrossRef]
25. Digo, E.; Polito, M.; Pastorelli, S.; Gastaldi, L. Detection of upper limb abrupt gestures for human–machine interaction using deep learning techniques. *J. Braz. Soc. Mech. Sci. Eng.* **2024**, *46*, 227. [CrossRef]
26. Rivera, P.; Valarezo, E.; Choi, M.-T.; Kim, T.-S. Recognition of human hand activities based on a single wrist imu using recurrent neural networks. *Int. J. Pharma Med. Biol. Sci* **2017**, *6*, 114–118. [CrossRef]
27. Ayvaz, U.; Elmoughni, H.; Atalay, A.; Atalay, Ö.; Ince, G. Real-time human activity recognition using textile-based sensors. In Proceedings of the EAI International Conference on Body Area Networks, Tallinn, Estonia, 25–26 December 2020; pp. 168–183.
28. Zhang, S.; Li, Y.; Zhang, S.; Shahabi, F.; Xia, S.; Deng, Y.; Alshurafa, N. Deep learning in human activity recognition with wearable sensors: A review on advances. *Sensors* **2022**, *22*, 1476. [CrossRef] [PubMed]
29. Mani, N.; Haridoss, P.; George, B. Evaluation of a Combined Conductive Fabric-Based Suspender System and Machine Learning Approach for Human Activity Recognition. *IEEE Open J. Instrum. Meas.* **2023**, *2*, 2500310. [CrossRef]

30. Koo, B.; Nguyen, N.T.; Kim, J. Identification and Classification of Human Body Exercises on Smart Textile Bands by Combining Decision Tree and Convolutional Neural Networks. *Sensors* **2023**, *23*, 6223. [CrossRef] [PubMed]
31. Shafiq, M.; Gu, Z. Deep residual learning for image recognition: A survey. *Appl. Sci.* **2022**, *12*, 8972. [CrossRef]

Disclaimer/Publisher's Note: The statements, opinions and data contained in all publications are solely those of the individual author(s) and contributor(s) and not of MDPI and/or the editor(s). MDPI and/or the editor(s) disclaim responsibility for any injury to people or property resulting from any ideas, methods, instructions or products referred to in the content.

processes

Article

Performance Comparison of High-Temperature Heat Pumps with Different Vapor Refrigerant Injection Techniques

Yuqiang Yang [1], Yu Wang [2,3], Zhaoyang Xu [2,3], Baojiang Xie [4], Yong Hu [4], Jiatao Yu [4], Yehong Chen [4], Ting Zhang [4], Zhenneng Lu [5,*] and Yulie Gong [5,*]

1 State Grid Zhejiang Electric Power Co., Ltd., Hangzhou 310007, China
2 State Grid Electric Power Research Institute Wuhan Efficiency Evaluation Company Limited, Wuhan 430074, China
3 Nari Group Corporation State Grid Electric Power Research Institute, Nanjing 210000, China
4 Shaoxing Power Supply Company, State Grid Zhejiang Electric Power Co., Ltd., Shaoxing 312000, China
5 Guangzhou Institute of Energy Conversion, Chinese Academy of Sciences, Guangzhou 510640, China
* Correspondence: luzn@ms.giec.ac.cn (Z.L.); gongyl@ms.giec.ac.cn (Y.G.); Tel.: +86-13294195165 (Z.L.); +86-020-87058438 (Y.G.)

Abstract: In order to develop a highly efficient and stable high-temperature heat pump to realize high-efficient electrification in the industrial sector, performance of high-temperature heat pumps with a flash tank vapor injection and sub-cooler vapor injection are compared under different evaporation temperatures, condensation temperatures, compressor suction superheat degrees, subcooling degrees and compressor isentropic efficiencies. The results show that the COP, injection mass flow ratio and VHC of the FTVC are higher than those of the SVIC-0, SVIC-5, SVIC-10 and SVIC-20 under the same working conditions, while the discharge temperature of the FTVC is approximately equal to that of the SVIC-0 and lower than those of the SVIC-5, SVIC-10 and SVIC-20. When the evaporation temperature, the condensation temperature and injection pressure are 55 °C, 125 °C and 921.4 kPa, respectively, the system COP of the FTVC is 4.49, which is approximately 6.7%, 7.3%, 7.8% and 8.9% higher than those of the SVIC-0, SVIC-5, SVIC-10, and SVIC-20, respectively.

Keywords: high-temperature heat pump; vapor refrigerant injection technique; sub-cooler; flash tank; industrial electrification

Citation: Yang, Y.; Wang, Y.; Xu, Z.; Xie, B.; Hu, Y.; Yu, J.; Chen, Y.; Zhang, T.; Lu, Z.; Gong, Y. Performance Comparison of High-Temperature Heat Pumps with Different Vapor Refrigerant Injection Techniques. *Processes* **2024**, *12*, 566. https://doi.org/10.3390/pr12030566

Academic Editors: Zhanxiao Kang and Qing Chen

Received: 6 February 2024
Revised: 6 March 2024
Accepted: 7 March 2024
Published: 13 March 2024

Copyright: © 2024 by the authors. Licensee MDPI, Basel, Switzerland. This article is an open access article distributed under the terms and conditions of the Creative Commons Attribution (CC BY) license (https://creativecommons.org/licenses/by/4.0/).

1. Introduction

To cope with global climate change and carbon emission reduction, electricity-based and low-carbon alternatives are a global development trend in the heating sector. Industrial process heating is an important direction for the development of electricity-based alternatives in the future. According to the report of the International Energy Agency, global carbon dioxide emissions were 33.267 billion tons in 2019, of which the industrial sector accounted for about 40% of carbon dioxide emissions in energy consumption [1]. In Europe, 20% of total greenhouse gas emissions are currently from industrial processes. The heat pump is an important piece of equipment used to realize efficient industrial electrification to reduce greenhouse gas emissions.

The heat pump is an energy-saving and environmentally friendly technology. It extracts low temperature heat from a heat source by refrigerant state change and then transfers the high-temperature heat to a heat sink. During the heat transfer process, high grade energy (electric power or high-temperature heat) must be input to the heat pump. Normally, the heat transferred to a heat sink is several times greater than the electric power required to input the heat pump. Therefore, compared to conventional heating technologies such as boilers or electric heaters, heat the pump is far more efficient. However, the low output temperature of the traditional heat pump is insufficient at meeting the temperature requirements for heating in industrial processes, which limits the application of the heat pump in the industrial sector [2,3].

Refrigerant injection is a technique that uses some condensed refrigerant from the condenser outlet to inject to the compressor suction line, the sealed compressor pocket, or the condenser inlet in a vapor compression heat pump system. Refrigerant injection techniques can be used to lower the temperature of an electric motor, the compressor discharge temperature, reduce the compressor power consumption, and improve the cooling/heating capacity. Liquid refrigerant injection and vapor refrigerant injection are the two main types of refrigerant injection. Vapor refrigerant injection techniques have been widely used in the refrigeration system to improve the refrigeration system COP. The earliest vapor refrigerant injection technique was applied in cold storage, freezing engineering. When the heat pump system was applied in a cold region, the high compression pressure ratio would lead to a considerably high discharge temperature and low COP. So, the vapor refrigerant injection technique is also widely used in heat pump systems applied in cold regions. Xue et al. experimentally analyzed the refrigerant vapor-injection air source heat pump system with R32 used as the working fluid. The results showed that the injection ratio, heating capacity and compressor power consumption of the system increased with the increase in intermediate pressure, while the discharge temperature decreased. The heating COP decreased with the increase in intermediate pressure when the ambient temperature was higher than −5 °C. But when the ambient temperature was below −5 °C, the COP first increased and then decreased with the increase in intermediate pressure [4]. Roh et al. found that the advantage of the refrigerant vapor-injection technique under the condition of a small pressure ratio was not significant [5,6]. Zhang et al. established the refrigerant vapor-injection air source heat pump system in Lanzhou, China. The experimental results showed that the energy efficiency of the refrigerant vapor-injection air source heat pump system increased by 4~6% compared with that of the air source heat pump system without refrigerant vapor-injection [7]. Heo et al. investigated the influence of the refrigerant vapor-injection on the performance of the two-stage heat pump system. The results showed that under the ambient temperature of −25 °C, the COP and heat capacity increased by 10% and 25% compared with the two-stage heat pump system without refrigerant vapor-injection [8,9]. Mathison et al. studied the influence of the refrigerant vapor-injection technique with multiple injection ports on the performance of heat pumps. The results showed that the performance of heat pumps improved significantly for air conditioning and refrigeration applications when an economized refrigerant was injected at an infinite number of ports to maintain saturated vapor in the compressor [10].

Different types of refrigerant injection show different effects on the heating performance of heat pumps. Wang et al. found that two-phase suction, liquid injection and two-phase injection could effectively decrease the discharge temperature of an R32 scroll compressor used in a heat pump, and the two-phase injection showed better performance than the other two methods in cooling capacity and COP [11]. Qi et al. developed a hybrid vapor injection air-source heat pump cycle with sub-cooler and flash tank using R290 as a refrigerant. The simulation results indicated that the coefficient of performance and the volumetric heating capacity of the hybrid vapor injection cycle could be 2.8–3.3% and 6.4–8.8% higher than those of the conventional sub-cooler vapor injection cycle system and 1.1–2.0% and 3.2–6.0% higher than those of the flash tank vapor injection cycle system, respectively [12]. Kim et al. established a numerical model to compare and analyze the influence of liquid, vapor and two-phase injection techniques on the performance of heat pumps with a scroll compressor using R410A. Results showed that there was an optimum injection quality that made the two-phase injection heat pump exhibit the highest COP among all injection types and the two-phase injection heat pump with optimum injection quality more effective at improving the COP and reducing the discharge temperature when the outdoor temperature decreased [13].

The high-temperature heat pump needs to solve the problems of low efficiency and high discharge temperatures brought about by the large temperature rise, and the refrigerant vapor-injection technique is an important technical means to solve these problems. Previous studies mainly focus on the effect of the refrigerant vapor-injection technique on a conventional heat pump system. There are still only a few research studies which focus on analyzing refrigerant vapor-injection techniques on performance of the HTHP system, especially on the HTHP system with output temperatures higher than 100 °C. In this paper, the performance of high-temperature heat pumps with flash tank vapor injection (FTVI) and sub-cooler vapor injection (SCVI) are compared under different evaporation temperatures, condensation temperatures, compressor suction superheat degrees, sub-cooling degrees and compressor isentropic efficiencies.

2. System Description

There are two types of vapor injection cycles: the flash tank vapor injection (FTVI) cycle and the sub-cooler vapor injection (SCVI) cycle. In the FTVI cycle, the injected vapor refrigerant is provided by phase separation in the flash tank. Firstly, the vapor-liquid two-phase refrigerant absorbs heat from the heat source and the liquid refrigerant evaporates and becomes vapor refrigerant in the evaporator. Then the vapor refrigerant enters the compressor suction and is compressed to the intermediate pressure. The compressed vapor refrigerant from the compressor suction mixes with vapor refrigerant from the flash tank at the intermediate pressure location in the compressor. The mixed vapor refrigerant continues to be compressed to the condensing pressure and is discharged to the condenser. The compressed high-temperature vapor refrigerant releases heat to the heat sink and condensed to liquid refrigerant in the condenser. The condensed liquid refrigerant flows through expansion valve 1 and expands to be vapor-liquid two-phase refrigerant at the intermediate pressure. Then the vapor-liquid two-phase refrigerant is separated into liquid-phase refrigerant and vapor-phase refrigerant in the flash tank. The liquid refrigerant enters expansion valve 2 and then circulates through the evaporator. The vapor-phase refrigerant is injected into the compressor pocket at the intermediate pressure, as shown in Figure 1. In the SCVI cycle, the injected vapor refrigerant is generated from the sub-cooler. The vapor-liquid two-phase refrigerant absorbs heat from the heat source and the liquid refrigerant evaporates to vapor refrigerant just like evaporation part of the FTVI cycle. Then the vapor refrigerant flows into the compressor suction and is compressed to the intermediate pressure. The compressed vapor refrigerant from the compressor suction mixes with vapor refrigerant from the sub-cooler at the intermediate pressure location in the compressor. The mixed vapor refrigerant continues to be compressed to condensing pressure and then enters the condenser. The high-pressure and high-temperature vapor refrigerant condenses to liquid refrigerant by releasing heat to the heat sink. The liquid refrigerant from the condenser is separated into two paths. One path injects the refrigerant to the compressor pocket in the intermediate pressure. In this path, the liquid refrigerant from the condenser passes through expansion valve 1 and becomes two-phase refrigerant in the intermediate pressure. The two-phase refrigerant absorbs heat from refrigerant in another path in the sub-cooler and becomes vapor which is then injected into the compressor pocket in the intermediate pressure. In another path, the liquid refrigerant from the condenser is subcooled by two-phase refrigerant injected to the compressor pocket in the intermediate pressure. Then the sub-cooled liquid enters expansion valve 2 and flows to the evaporator, as shown in Figure 2.

Refrigerant has a decisive influence on the design of HTHPs, so the selection of the refrigerant is key in the design of HTHPs. A suitable refrigerant for HTHPs should have excellent thermodynamic properties at the operating conditions of high-temperature and pressure, it should also be highly energy efficient, commercially available, have low global warming potential (GWP), a zero or negligible Ozone Depletion Potential (ODP) and other factors. The critical temperature of refrigerant is an important property when selecting working fluids for HTHPs. The upper temperature limit of subcritical heat pump cycles is determined by the critical temperature of the refrigerant. A temperature gap of about 10 to 15 K from the desired condensation temperature has to be maintained to ensure subcritical heat pump operation [14]. Additionally, the pressure level of the heat pump system is also an important property for selecting working fluid. The higher pressure should be kept below a practical value of about 25 bar, and the pressure ratio should be as low as possible [15]. HFC-245fa belongs to hydrofluorocarbon chemicals and offers zero ODP and relatively high critical temperatures of 154.0 °C at moderate pressures of 36.5 bar, which is suitable for HTHPs, so R245fa is selected as the working fluid of the high-temperature heat pump. The parameters of HFC-245fa are shown in Table 1.

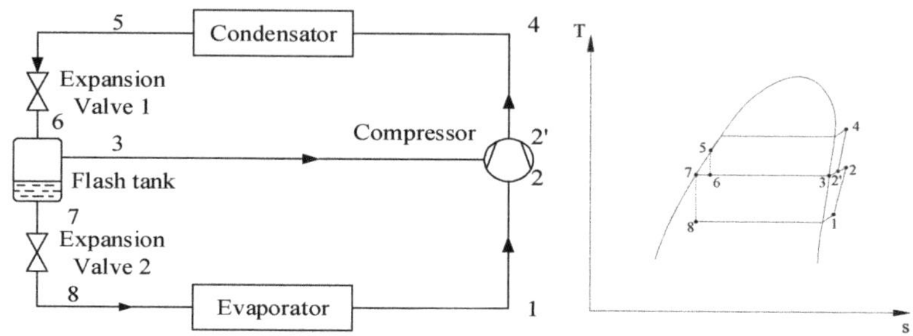

Figure 1. Principle diagram of the vapor injection heat pump with flash tank.

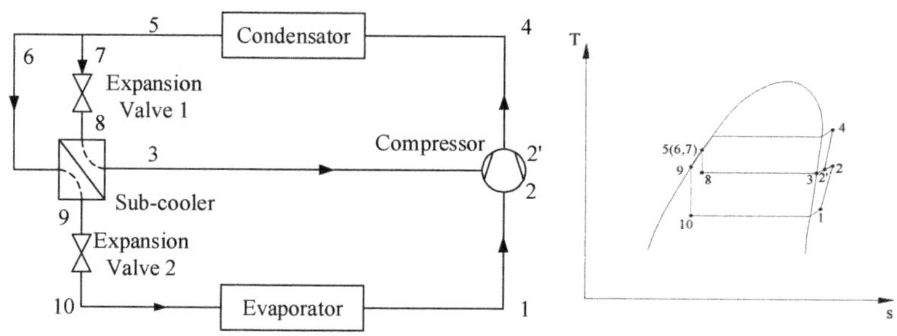

Figure 2. Principle diagram of the vapor injection heat pump with sub-cooler.

Table 1. Parameters of HFC-245.

Working Fluid	Chemical Formula	Group	M [g/mol]	t_{cr} [°C]	P_{cr} [bar]	NBP [°C]	GWP	ODP	SG
R245fa	$CHF_2CH_2CF_3$	HFC	134.05	154	36.5	15.3	858	0	B1

3. Model Establishment

3.1. Assumptions and Boundary Conditions

To analyze the performance of high-temperature heat pumps with different vapor injection techniques, mathematical models are developed based on assumptions as follows:

(1) The working fluid in the heat pump system is steady state;
(2) The heat loss and pressure loss in heat exchangers and pipelines are ignored;
(3) The compressor efficiencies in heat pumps are constant;
(4) The expansion valve process is considered isenthalpic.

The reference values and the boundary conditions are shown in Table 2.

Table 2. The reference values and the boundary conditions.

Parameters	Reference Values	Boundary Conditions
Refrigerant mass flow rate $\dot{m}_{ref}/(kg/s)$	1	-
Evaporation temperature $t_e/°C$	55	40~80
Superheat degree $t_{sup}/°C$	10	0~15
Condensation temperature $t_c/°C$	125	100~130
Subcooling degree $t_{sub}/°C$	5	0~15
Injection mass flow ratio A	0.3	-
Relative injection pressure RIP	1	-
Injection superheat degree $t_{sup,inj}/°C$	5	0~20
Compressor isentropic efficiency η_{is}	0.72	0~0.9

3.2. Model Establishment

Thermodynamics models of high-temperature heat pumps with flash tank vapor injection and sub-cooler vapor injection are established based on the energy conservation equation and the mass conservation equation.

In the condenser, heat transfers from the working fluid to the heat sink and the enthalpy of the working fluid decreases while the pressure loss is slight. So, the condensation process can be considered as heat release process with constant pressure. Condensation heat can be calculated as Equation (1) with the product of working fluid mass flow rate and the inlet and outlet condenser enthalpy difference.

$$Q_{cond} = (1+A) \times \dot{m}_{ref} \times (h_{cond,in} - h_{cond,out}) \tag{1}$$

The expansion valve throttling process is considered as the isenthalpic process, so

$$h_{th,in} = h_{th,out} \tag{2}$$

For the SCVI heat pump system, heat transfer takes place in the sub-cooler, so

$$\dot{m}_{ref} \times (h_6 - h_9) = \dot{m}_{ref} \times A \times (h_3 - h_8) \tag{3}$$

For the flash tank in the FTVI heat pump system

$$\dot{m}_{ref} \times (1+A) \times h_6 = \dot{m}_{ref} \times h_7 + \dot{m}_{ref} \times A \times h_3 \tag{4}$$

In the evaporators of the two heat pump systems, evaporation heat can be calculated as Equation (5) with the product of working fluid mass flow rate and the inlet and outlet evaporator enthalpy difference:

$$Q_{evap} = \dot{m}_{ref} \times (h_{evap,out} - h_{evap,in}) \tag{5}$$

In compression processes of the two cycles, processes 1–2 and 2′–4 are the adiabatic compression process, and process 3–2 can be treated as the mixing process with a constant pressure, so

$$h_2 = (h_{2s} - h_1)/\eta_{is} + h_1 \tag{6}$$

$$h_4 = (h_{4s} - h_{2'})/\eta_{is} + h_{2'} \tag{7}$$

$$\dot{m}_{ref} \times (1+A) \times h_{2'} = \dot{m}_{ref} \times h_2 + \dot{m}_{ref} \times A \times h_3 \tag{8}$$

$$W_{c1} = \dot{m}_{ref}(h_2 - h_1) \tag{9}$$

$$W_{c2} = \dot{m}_{ref} \times (1+A) \times (h_4 - h_{2'}) \tag{10}$$

$$W = W_{c1} + W_{c2} \tag{11}$$

Injection mass flow ratio A and relative injection pressure RIP are defined as follows:

$$A = \frac{\dot{m}_{\text{inj}}}{\dot{m}_{\text{ref}}} \tag{12}$$

$$RIP = \frac{P_{\text{inj}}}{\sqrt{P_{\text{cond}} P_{\text{evap}}}} \tag{13}$$

The compressor discharge temperature can be calculated from the condensation pressure and refrigerant enthalpy at the compressor outlet.

$$t_{\text{disch}} = f(P_{\text{cond}}, h_{\text{disch}}) \tag{14}$$

Most commonly, the coefficient of performance (COP) is used to evaluate the efficiency of a heat pump system. The COP is obtained from the heating capacity and the compressor power consumption, as shown in Equation (15). To compare the influence of the heating capacity and the volumetric flow rate at the suction line, the volumetric heating capacity (VHC) is calculated using Equation (16):

$$COP = Q_{\text{cond}} / (W_{c1} + W_{c2}) \tag{15}$$

$$VHC = Q_{\text{cond}} / \dot{m}_{\text{ref}} v_1 \tag{16}$$

4. Results and Discussion

Evaporation temperature, condensation temperature, compressor suction superheat degree, sub-cooling degree and compressor isentropic efficiency are the main factors that influence the performance of a high-temperature heat pump. So, the performance of a high-temperature heat pump with flash tank vapor injection and that with sub-cooler vapor injection are compared under these different main factors.

4.1. Performance Comparison under Different Evaporation Temperatures

Figure 3 shows the performance comparison at different evaporation temperatures of the high-temperature heat pump cycles with flash tank vapor injection and sub-cooler vapor injection. One of the important differences between the two cycles is that the refrigerant vapor injected into the compressor of the FTVC is in a saturated state, while that of the SVIC can be in either a saturated state or superheated state. Therefore, the SVIC with different injection superheat degrees are also investigated. SVIC-0, SVIC-5, SVIC-10 and SVIC-20 represent the SVIC with the injection superheat degrees of 0 °C, 5 °C, 10 °C and 20 °C, respectively. As can be seen from Figure 3a, the COP of the FTVC system is higher than the SVIC at the same evaporation temperature. At the same time, it can be seen from Figure 3a that the COP of the SVIC heat pump system decreases as the injection superheat degree increases. When the evaporation temperature, the condensation temperature and injection pressure are 55 °C, 125 °C and 921.4 kPa, the system COPs corresponding to the FTVC, SVIC-0, SVIC-5, SVIC-10 and SVIC-20 are 4.49, 4.19, 4.16, 4.14 and 4.09, respectively, and the COPs of the FTVC are 6.7%, 7.3%, 7.8% and 8.9% higher than those of the systems of the SVIC-0, SVIC-5, SVIC-10 and SVIC-20, respectively. Figure 3b shows that the injection mass flow ratio decreases with the increase in evaporation temperature, which is explained by the fact that the increase in evaporation temperature leads to the decrease in pressure difference between the condensation pressure and injection pressure. It can also be seen from Figure 3b that the injection mass flow ratio of the FTVC is higher than those of the SVIC under different injection superheat degrees, which causes the higher COP of the FTVC than that of the SVIC under the same injection pressure. At the evaporation temperature of 55 °C, condensation temperature of 125 °C and injection pressure of 921.4 kPa, the injection mass flow ratio of the FTVC is 18.6%, 23.6%, 28.0% and 35.2% higher than the SVIC-0, SVIC-5, SVIC-10 and SVIC-20, respectively. Figure 3c shows that the volumetric heating capacity of the FTVC is also higher than those of the SVIC at different injection superheat degrees. When the evaporation temperature, the condensation temperature and injection pressure are 55 °C, 125 °C and 921.4 kPa, the VHCs of the FTVC, SVIC-0, SVIC-5, SVIC-10 and SVIC-20 are 3976, 3714, 3710, 3706 and 3699 kJ/m^3, respectively. Figure 3d shows the discharge temperature of different systems at different evaporation temperatures. According to Figure 3d, the FTVC and SVIC-0 have equivalent discharge temperatures at different evaporation temperatures. When the evaporation temperature and the injection superheat degree increase, the discharge temperature increases.

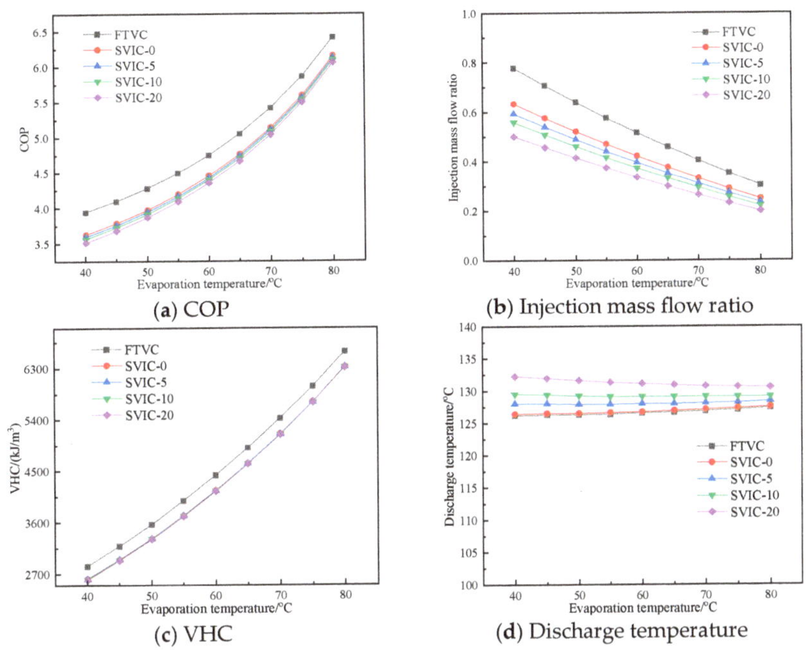

Figure 3. Performance comparison under different evaporation temperatures.

4.2. Performance Comparison under Different Condensation Temperatures

Figure 4 shows the performance comparison of the FTVC and SVIC at different condensation temperatures. From Figure 4a, it can be seen that at the same injection pressure and condensation temperature, the COP of the FTVC system is always higher than those of the SVIC with different injection superheat degrees. The higher condensation temperature leads to the larger COP difference between the FTVC system and the SVIC system. At the evaporation temperature of 55 °C, condensation temperature of 100 °C, compressor suction superheat degree and condensation subcooling degree of 5 °C, COPs corresponding to the FTVC, SVIC-0, SVIC-5, SVIC-10, and SVIC-20 are 6.15, 5.93, 5.92, 5.90 and 5.88, respectively. COP of the FTVC is 3.7%, 3.9%, 4.2% and 4.6% higher than those of the SVIC-0, SVIC-5, SVIC-10, and SVIC-20, respectively. When the condensation temperature is 130 °C and the other conditions are kept the same, COP of the FTVC is 7.9%, 8.9%, 9.8% and 11.3% higher than those of the SVIC-0, SVIC-5, SVIC-10 and SVIC-20, respectively. According to Figure 4b, the injection mass flow ratio increases with the increase in condensation temperature. The main reason is that as the condensation temperature increases, the pressure difference between the condensation pressure and the injection pressure increases, causing the injection vapor mass flow rate to increase. At the evaporation temperature of 55 °C, condensation temperature of 100 °C, compressor suction superheat degree and condensation subcooling degree of 5 °C, the injection vapor mass flow ratio of the FTVC is 24.2%, 29.9%, 35.5% and 46.6% higher than those of the SVIC-0, SVIC-5, SVIC-10 and SVIC-20. When the condensation temperature is 130 °C and the other conditions are kept the same, the injection vapor mass flow ratio of the FTVC is 22.5%, 31.4%, 40% and 57% higher than those of SVIC-0, SVIC-5, SVIC-10, and SVIC-20, respectively. Figure 4c shows that the VHC of the FTVC increases with the increase in condensation temperature and the SVIC-0, SVIC-5, SVIC-10 and SVIC-20 slightly decrease with the increase in condensation temperature. At the evaporation temperature of 55 °C, condensation temperature of 130 °C, compressor suction superheat degree and condensation subcooling degree of 5 °C, the VHCs of the FTVC, SVIC-0, SVIC-5, SVIC-10 and SVIC-20 are 4015, 3723, 3717, 3712 and 3702 kJ/m^3, respectively. Compared with the condensation temperature of 100 °C, the VHC of the FTVC increases by 3.5%, while the VHCs of SVIC-0, SVIC-5, SVIC-10 and SVIC-20 decrease by 0.5%, 0.7%, 0.8% and 1.1%, respectively. Figure 4d shows the discharge temperature at different condensation temperatures. When the condensation temperature and the injection superheat degree increase, the discharge temperature increases.

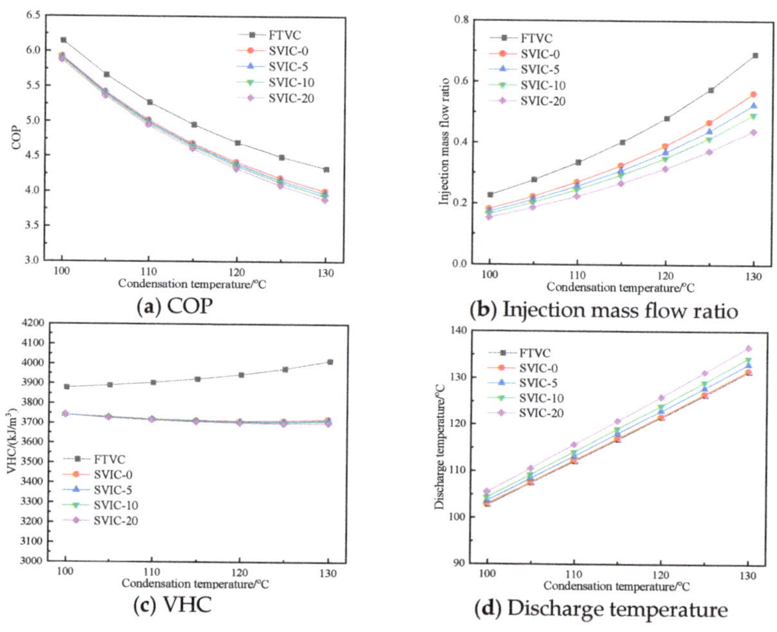

Figure 4. Performance comparison under different condensation temperatures.

4.3. *Performance Comparison under Different Suction Superheat Degrees*

Figure 5 shows the performance comparison of the FTVC and SVIC under different compressor suction superheat degrees. According to Figure 5a, the COPs of FTVC, SVIC-0, SVIC-5, SVIC-10 and SVIC-20 all increase with the increase in compressor suction superheat degree and the COP of the FTVC is higher than those of the SVIC-0, SVIC-5, SVIC-10 and SVIC-20 at the same conditions. At the evaporation temperature of 55 °C, condensation temperature of 125 °C, subcooling degree of 5 °C and compressor suction superheat degree of 15 °C, the COPs of the FTVC, SVIC-0, SVIC-5, SVIC-10 and SVIC-20 are 4.58, 4.29, 4.26, 4.24 and 4.19, which are 2.9%, 3.4%, 3.5%, 3.5% and 3.6% higher than those of the compressor suction superheat degree of 0 °C, respectively. From Figure 5b, the injection mass flow ratio does not change with the increase in the suction superheat degree. Figure 5c shows that the VHCs of the FTVC, SVC-0, SVIC-5, SVIC-10 and SVIC-20 increase with the increase in compressor suction superheat degrees. The VHC is decided by the total heat capacity, working fluid volume flow rate and suction specific volume. When the compressor suction superheat degree increases, the working fluid volume flow rate remains the same and the suction specific volume and total heat capacity increase. Meanwhile, the increase in total heat capacity is more than the increase in suction specific volume, thus causing the increase in VHC. Figure 5d shows the comparison of the discharge temperature of the FTVC and SVIC under different compressor suction superheat degrees. As can be seen from the figure, wet compression would take place for both the FTVC and SVIC-0 when the compressor suction superheat degree is low.

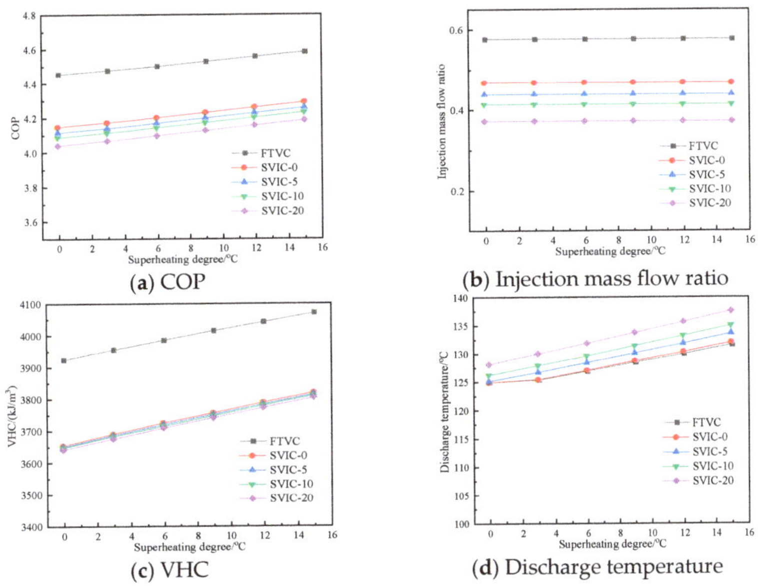

Figure 5. Performance comparison under different suction superheat degrees.

4.4. Performance Comparison under Different Subcooling Degrees

Figure 6 shows the performance comparison of the FTVC and SVIC under different condensation subcooling degrees. As shown in Figure 6a, the COPs of FTVC, SVIC-0 and SVIC-5 decrease with the increase in condensation subcooling degree; however, the COPs of SVIC-10 and SVIC-20 decrease first and then increase with the increase in condensation subcooling degree. The explanation for this phenomenon is that while the condensation subcooling degree increases, the initial temperature of refrigerant flowing into the sub-cooler or flash tank decreases, which causes the decrease in injection mass flow ratio and COP at a constant injection vapor pressure. On the other hand, the heat capacity increases with the increase in condensation subcooling degree, which leads to the increase in the COP. The joint action of two factors causes the trend in the COP. From Figure 6b, the injection mass flow ratio decreases sharply as the condensation subcooling degree increases. When the condensation subcooling degree changes from 0 to 15 °C, the injection mass flow ratios of the FTVC, SVIC-0, SVIC-5, SVIC-10 and SVIC-20 change from 0.74, 0.61, 0.57, 0.53 and 0.47 to 0.33, 0.27, 0.26, 0.24 and 0.20, decreasing by 55.4%, 55.7%, 54.4%, 54.7% and 53.2%, respectively. Figure 6c shows that the VHCs of the FTVC, SVC-0, SVC-5, SVIC-10 and SVIC-20 decrease with increasing condensation subcooling degrees. The VHC of FTVC changes more sharply than those of SVIC-0, SVIC-5, SVIC-10 and SVIC-20. Figure 6d shows the discharge temperature under different condensation subcooling degrees. According to Figure 6d, the discharge temperature of FTVC and SVIC-0 remain unchangeable with the increase of condensation subcooling degree, while those of SVIC-5, SVIC-10 and SVIC-20 decrease slightly.

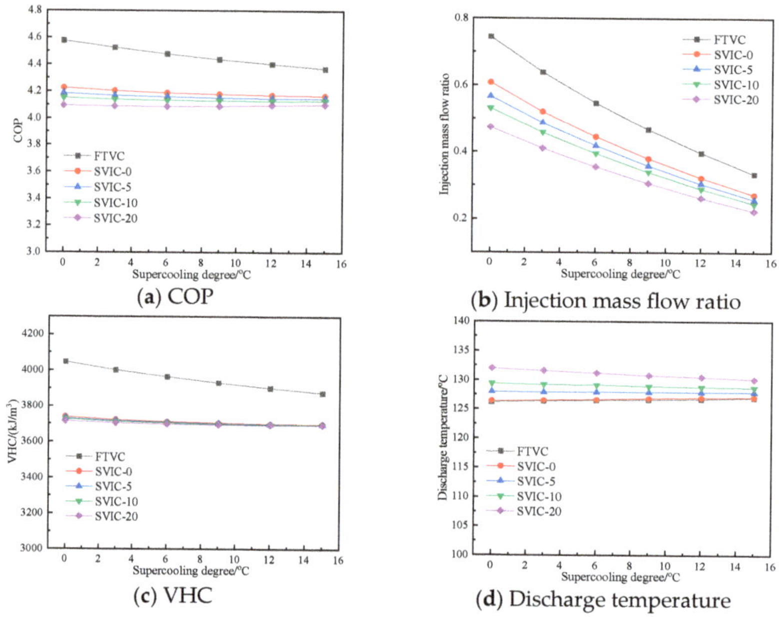

(**a**) COP (**b**) Injection mass flow ratio (**c**) VHC (**d**) Discharge temperature

Figure 6. Performance comparison under different subcooling degrees.

4.5. Performance Comparison under Different Compressor Isentropic Efficiencies

Isentropic efficiency is widely used in evaluating the performance of compressors, and the performance of compressors have great influence on the performance of heat pumps. Figure 7 shows the performance comparison of the FTVC and SVIC under different compressor isentropic efficiencies. According to Figure 7a, the COPs of the FTVC, SVIC-0, SVIC-5, SVIC-10 and SVIC-20 all increase with isentropic efficiency. Under the same isentropic efficiency, the COP of the FTVC is higher than those of the SVIC-0, SVIC-5, SVIC-10 and SVIC-20. At the evaporation temperature of 55 °C, condensation temperature of 125 °C, subcooling degree of 5 °C, compressor suction superheat degree of 15 °C and compressor isentropic efficiency of 0.75, the COPs of the FTVC, SVIC-0, SVIC-5, SVIC-10 and SVIC-20 are 4.63, 4.32, 4.29, 4.26 and 4.22, respectively. From Figure 7b, the injection mass flow ratio does not change with the increase in compressor isentropic efficiency. At the same compressor isentropic efficiency, the FTVC has the largest injection mass flow ratio. Figure 7c shows that the VHCs of the FTVC, SVC-0, SVIC-5, SVIC-10 and SVIC-20 decrease with the compressor isentropic efficiency. Figure 7d shows the comparison of the discharge temperature of the FTVC and SVIC under different isentropic efficiencies. From the figure, although the high isentropic efficiency will help to improve the performance of the system, the state of working fluid in the compressor is a two-phase state when the compressor suction superheat degree is small, which will form wet compression especially for the FTVC. So, the compressor suction superheat degree of the FTVC must remain large enough to prevent wet compression when the compressor isentropic efficiency is high. At the evaporation temperature of 55 °C, condensation temperature of 125 °C, subcooling degree of 5 °C, compressor suction superheat degree of 15 °C and compressor isentropic efficiency of 0.75, the discharge temperatures of the FTVC, SVIC-0, SVIC-5, SVIC-10, and SVIC-20 are 125.6, 125.8, 127.1, 128.2 and 130.3 °C, respectively.

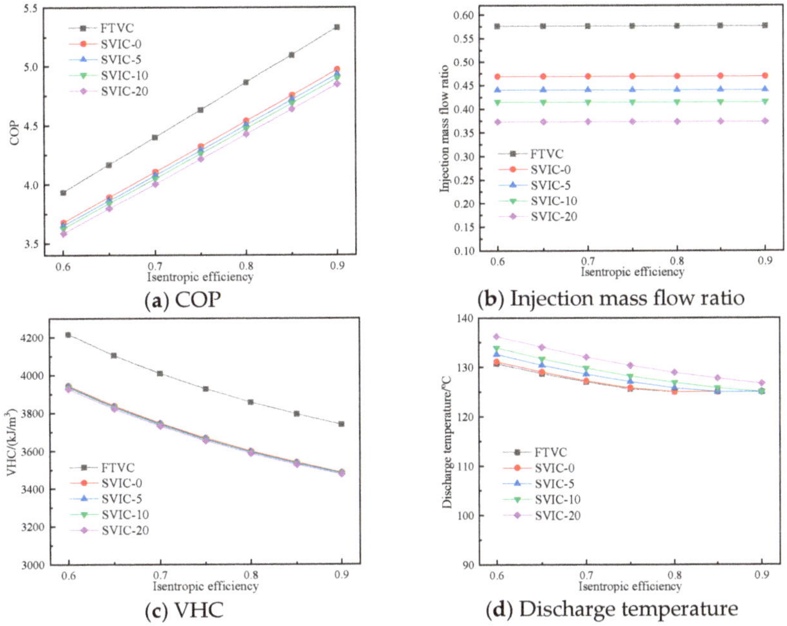

Figure 7. Performance comparison under different isentropic efficiencies.

5. Conclusions

The COP, injection mass flow ratio and VHC of the FTVC are higher than those of the SVIC-0, SVIC-5, SVIC-10 and SVIC-20 under the same working conditions, while the discharge temperature of the FTVC is approximately equal to the SVIC-0 but lower than the SVIC-5, SVIC-10 and SVIC-20. When the compressor suction superheat degree is low and the compressor isentropic efficiency is high, wet compression may happen for the FTVC system and the SVIC-0 system, which would lower the performance and even be harmful to the running safety of the HTHP system. Therefore, enough compressor suction superheat degree should be designed to avoid wet compression in the compressor. The COP of the SVIC heat pump system decreases as the injection superheat degree increases. The isentropic efficiency of the compressor, the evaporation and condensation temperatures have greater effects on the COP than the suction superheat degree and subcooling degree. Therefore, to achieve a higher COP, the isentropic efficiency of the compressor and the evaporation temperature should be increased, and the condensation temperature should be decreased. When the evaporation temperature, the condensation temperature and injection pressure are 55 °C, 125 °C and 921.4 kPa, the system COPs corresponding to the FTVC, SVIC-0, SVIC-5, SVIC-10, and SVIC-20 are 4.49, 4.19, 4.16, 4.14 and 4.09, respectively.

Author Contributions: Conceptualization, Y.Y.; methodology, Y.W.; investigation, Z.X.; resources, Z.L.; project administration, Y.G.; validation, B.X.; writing—original draft preparation, Y.H. and Y.C.; writing—review and editing, J.Y. and T.Z. All authors have read and agreed to the published version of the manuscript.

Funding: This research was funded by Science and Technology Projects of the State Grid Corporation of China, funding number 5400-202140401A-0-0-00.

Data Availability Statement: Data are contained within the article.

Conflicts of Interest: Author Yuqiang Yang was employed by the company State grid Zhejiang Electric Power Co., Ltd. Authors Yu Wang and Zhaoyang Xu were employed by the company State Grid Electric Power Research Institute Wuhan Efficiency Evaluation Company Limited and

Nari Group Corporation State Grid Electric Power Research Institute. Authors Baojiang Xie, Yong Hu, Jiatao Yu, Yehong Chen, Ting Zhang were employed by the company Shaoxing Power Supply Company, State Grid Zhejiang Electric Power Co., Ltd. The remaining authors declare that the research was conducted in the absence of any commercial or financial relationships that could be construed as a potential conflict of interest.

Nomenclatures

1,2, . . ., 10	thermodynamic state points
A	injection mass flow ratio
h	enthalpy, kJ/kg
$\dot{m}$	mass flow rate, kg/s
M	molecular mass, g/mol
P	pressure, kPa
Q	heat capacity, kW
RIP	relative injection pressure
t	temperature, °C
VHC	volumetric heating capacity, kJ/m^3
$\dot{W}$	mechanical power, kW
Abbreviations	
COP	coefficient of performance
FTVC	flash tank vapor injection cycle
GWP	global warming potential
HTHP	high-temperature heat pump
HTCHP	high-temperature cascade heat pump
LMTD	logarithmic mean temperature difference
NBP	normal boiling point
ODP	ozone depression potential
SG	safety grade
SVIC	sub-cooler vapor injection cycle
Subscripts	
c	compressor
cond	condensator
disch	discharge temperature
evap	evaporator
in	input
inj	injection
is	isentropic
out	output
sub	subcooling
sup	superheat
ref	refrigerant
th	throttle
Greek Symbols	
η	efficiency
v	specific volume, m^3/kg

References

1. IEA. *Net Zero by 2050-a Roadmap for the Global Energy Sector*; IEA: Paris, France, 2021; Available online: https://www.iea.org/reports/net-zero-by-2050 (accessed on 12 December 2023).
2. Yan, H.Z.; Zhang, C.; Shao, Z.; Kraft, M.; Wang, R.Z. The underestimated role of the heat pump in achieving china's goal of carbon neutrality by 2060. *Engineering* **2023**, *23*, 13–18. [CrossRef]
3. Jiang, J.T.; Hu, B.; Wang, R.Z.; Deng, N.; Cao, F.; Wang, C.C. A review and perspective on industry high-temperature heat pumps. *Renew. Sust. Energ. Rev.* **2022**, *161*, 112106. [CrossRef]
4. Xue, J.; Guo, X.M.; Xue, L.P. Experimental study on performance of air-source heat pump system using refrigerant R32 with flash-tank vapor injection. *Cryog. Supercond.* **2018**, *4*, 88–91+96. (In Chinese) [CrossRef]

5. Roh, C.W.; Kim, M.S. Effects of intermediate pressure on the heating performance of a heat pump system using R410A vapor-injection technique. *Int. J. Refrig.* **2011**, *34*, 1911–1921. [CrossRef]
6. Roh, C.W.; Kim, M.S. Effect of vapor-injection technique on the performance of a cascade heat pump water heater. *Int. J. Refrig.* **2014**, *38*, 168–177. [CrossRef]
7. Zhang, D.; Li, J.P.; Nan, J.H.; Wang, L.H. Thermal performance prediction and analysis on the economized vapor injection air-source heat pump in cold climate region of China. *Sustain. Energy Techn.* **2016**, *18*, 127–133. [CrossRef]
8. Heo, J.Y.; Jeong, M.W.; Kim, Y.C. Effects of flash tank vapor injection on the heating performance of an inverter-driven heat pump for cold regions. *Int. J. Refrig.* **2010**, *33*, 848–855. [CrossRef]
9. Heo, J.Y.; Jeong, M.W.; Baek, C.H.; Kim, Y.C. Comparison of the heating performance of air-source heat pumps using various types of refrigerant injection. *Int. J. Refrig.* **2011**, *34*, 444–453. [CrossRef]
10. Mathison, M.M.; Braun, J.E.; Groll, E.A. Performance limit for economized cycles with continuous refrigerant injection. *Int. J. Refrig.* **2011**, *34*, 234–242. [CrossRef]
11. Yang, M.H.; Wang, B.L.; Li, X.T.; Shi, W.X.; Zhang, L.P. Evaluation of two-phase suction, liquid injection and two-phase injection for decreasing the discharge temperature of the R32 scroll compressor. *Int. J. Refrig.* **2015**, *59*, 269–280. [CrossRef]
12. Qi, H.J.; Liu, F.Y.; Yu, J.L. Performance analysis of a novel hybrid vapor injection cycle with subcooler and flash tank for air-source heat pumps. *Int. J. Refrig.* **2017**, *74*, 540–549. [CrossRef]
13. Kim, D.W.; Jeon, Y.S.; Jang, D.S.; Kim, Y.C. Performance comparison among two-phase, liquid, and vapor injection heat pumps with a scroll compressor using R410A. *Appl. Therm. Eng.* **2018**, *137*, 193–202. [CrossRef]
14. Arpagaus, C.; Bless, F.; Uhlmann, M.; Schiffmann, J.; Bertsch, S.S. High temperature heat pumps: Market overview, state of the art, research status, refrigerants, and application potentials. *Energy* **2018**, *152*, 985–1010. [CrossRef]
15. Kontomaris, K. Zero-ODP, low-GWP, non-flammable working fluids for high temperature heat pumps. In Proceedings of the ASHRAE Annual Conference, Seattle, WA, USA, 1 July 2014; pp. 1–40.

Disclaimer/Publisher's Note: The statements, opinions and data contained in all publications are solely those of the individual author(s) and contributor(s) and not of MDPI and/or the editor(s). MDPI and/or the editor(s) disclaim responsibility for any injury to people or property resulting from any ideas, methods, instructions or products referred to in the content.

MDPI AG
Grosspeteranlage 5
4052 Basel
Switzerland
Tel.: +41 61 683 77 34

Processes Editorial Office
E-mail: processes@mdpi.com
www.mdpi.com/journal/processes

Disclaimer/Publisher's Note: The title and front matter of this reprint are at the discretion of the Guest Editors. The publisher is not responsible for their content or any associated concerns. The statements, opinions and data contained in all individual articles are solely those of the individual Editors and contributors and not of MDPI. MDPI disclaims responsibility for any injury to people or property resulting from any ideas, methods, instructions or products referred to in the content.

www.ingramcontent.com/pod-product-compliance
Lightning Source LLC
Chambersburg PA
CBHW061513210926
51850CB00008B/1899